DATE DUE			
Feb 7 '81			

Mechanics and Energetics of Animal Locomotion

MECHANICS AND ENERGETICS OF ANIMAL LOCOMOTION

EDITED BY

R. McN. Alexander
Department of Pure and Applied Zoology
University of Leeds

AND

G. Goldspink
Department of Zoology
University of Hull

LONDON
CHAPMAN AND HALL

A HALSTED PRESS BOOK
John Wiley & Sons, New York

First published 1977
by Chapman and Hall Ltd
11 New Fetter Lane, London EC4P 4EE
© *1977 Chapman and Hall Ltd*
Typeset and printed in Britain by
William Clowes & Sons Ltd
London, Colchester and Beccles

ISBN 0 412 13630 9

Distributed in the U.S.A. by Halsted Press
a Division of John Wiley & Sons, Inc., New York

Library of Congress Cataloging in Publication Data

Main entry under title:

Mechanics and energetics of animal locomotion.

   Includes bibliographical references and index.
   1. Animal locomotion. 2. Animal mechanics.
3. Bioenergetics. I. Alexander, R. McNeill.
II. Goldspink, G.
QP301.M38   591.1'852   77-6737
ISBN 0-470-99185-2

Contents

Contents

Contents

Contents

Preface

This book is about all aspects of animal locomotion. It is not only about how animals move but also about how much energy they use, how their muscles work and how they co-ordinate their movements. The animals discussed range from the Protozoa to the mammals but we have not tried to write a full account of every group. We have thought it better to concentrate on the groups of animals about which most is known, or on which research has recently been focused.

Our knowledge of animal locomotion has advanced remarkably in the past two decades. Twenty years ago we knew in outline how most major groups of animals moved in locomotion and how their movements exerted forces in appropriate directions on the environment. Much of this knowledge had come from the work of the late Professor Sir James Gray and his associates. Since then a wide variety of advances have been made. Treadmills, wind tunnels and water tunnels have come into widespread use. Animals have been trained to use them so that it is now possible for instance to have a bird in free flight stationary in a wind tunnel. These techniques have been used to get more exact information about animal movement, and have also made it possible to measure the oxygen consumption of animals as they run, fly or swim. General mathematical theories of running, swimming and flight have been developed. The role of elastic materials in saving energy in running and flight has been appreciated. The remarkable elastic protein resilin, which is so important in flight, has been discovered. Our understanding of muscle in general and particularly of insect flight muscle has advanced greatly. Our knowledge of nervous co-ordination has advanced to the point where we seem to be on the brink of being able to work out detailed neuronal circuits in some invertebrate nervous systems. It seems a good time to produce a book.

We have been very fortunate in recruiting authors. In almost every case the first author we invited to write a chapter accepted our invitation. However, our list of authors was very sadly depleted by the death in 1975 of Professor Torkel Weis-Fogh. He had agreed to write Chapters 9 and 10, on swimming and flight. He had contributed more than anyone else to our knowledge of flight so his chapter on flight would have been immensely authoritative. Swimming would have been a new topic for him and his chapter on it would surely have been as original as all his other work, and a revelation to us all. In the event, one of us had to write the chapters in his place.

CONTRIBUTORS

R. McN. Alexander   Department of Pure and Applied Zoology, University of Leeds, Leeds, LS2 9JT, U.K.

F. Delcomyn   Department of Entomology, University of Illinois, Urbana, Illinois 61801, U.S.A.

G. Goldspink   Department of Zoology, University of Hull, Hull, HU6 7RX, U.K.

D. V. Holberton   Department of Zoology, University of Hull, Hull, HU6 7RX, U.K.

H. D. Jones   Department of Zoology, University of Manchester, Oxford Road, Manchester, M13 9PL, U.K.

D. G. Stuart   Department of Physiology, University of Arizona College of Medicine, Tucson, Arizona, U.S.A.

E. R. Trueman   Department of Zoology, University of Manchester, Oxford Road, Manchester, M13 9PL, U.K.

Mary C. Wetzel   Department of Psychology, University of Arizona, Tucson, Arizona 85721, U.S.A.

D. C. S. White   Department of Biology, University of York, Heslington, York, YO1 5DD, U.K.

1 Design of muscles in relation to locomotion

G. Goldspink

1.1 Introduction

In metazoan animals, the main motile force for locomotion is produced by muscle. Muscle tissue shows considerable diversity in its structure and physiology as might be expected from the different modes of locomotion found throughout the Animal Kingdom. It is particularly interesting for the comparative anatomist and physiologist to see just how the different requirements and constraints have been met in different ways in different animal groups.

From the point of view of the muscle these requirements include:

(1) Velocity of shortening of the muscle. This is obviously important in enabling the animal to move rapidly and hence to catch its prey or escape from predators.

(2) Force production per unit volume of muscle. There are situations when muscle strength is more important than other considerations.

(3) Force production over a range of different muscle lengths. The force produced by a muscle is to some extent determined by its initial length. It is important in some situations that the muscle should be able to produce a reasonable force over most of the range of shortening of its fibres.

(4) Ability to work at low environmental temperatures. This aspect is particularly important in 'cold-blooded' (poikilothermic/ectothermic) animals. Warm-blooded animals are of course almost independent of environmental temperature.

(5) Cost of producing isometric force. Isometric force is important in postural activities and in certain kinds of locomotion such as gliding and galloping. Economies in producing this sort of force have apparently been an important factor in the evolution of certain kinds of muscle.

(6) Cost of producing work. The efficiency of producing isotonic force is relevant to many types of locomotion. The cost of producing this force is obviously an important factor in determining the overall cost of locomotion.

Both (5) and (6) are main factors in determining the rate of fatigue of the muscle. These two latter aspects are discussed in more detail in Chapter 3.

1

1.2 Muscle fibre structure

Basic information about muscle fibre structure may be found in most recent textbooks on physiology or cell biology. Therefore the intention here is only to summarize briefly those aspects which are particularly relevant to locomotion. This is best done diagrammatically and hence the reader is referred to Fig. 1.1. Striated muscle is composed of muscle fibres which are multi-nucleated tubes

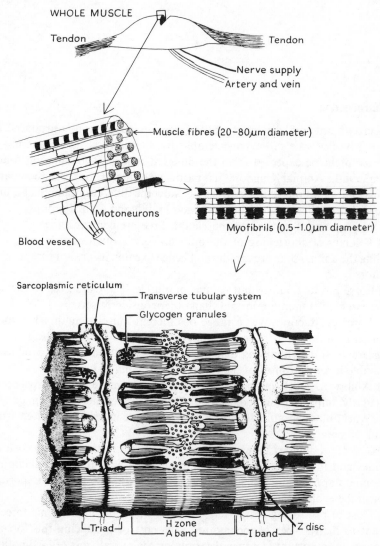

Fig. 1.1 Diagram of the structure of striated muscle showing the different levels of organization ranging from the whole muscles to the protein filaments. The lower part of the diagram is taken from Peachy (1965).

containing contractile proteins which in striated muscle are arranged in filaments. There are two types of filaments; thick filaments and thin filaments.
The thin filaments are held together by structures known as Z discs. The distance between one Z disc and the next is called a sarcomere. Each sarcomere
consists of two sets of thin filaments and one set of thick filaments. The thin
filaments interdigitate with the thick filaments. During contraction the thin
filaments are pulled between the thick filaments so that the sarcomeres shorten
and the degree of overlap between the filaments increases. The force required
to pull the thin filaments between the thick filaments is believed to be generated
by small projections on the thick filaments; these are the myosin cross bridges.
Each myosin cross bridge acts as an independent force generator which works
in a cyclical way. Each force generating cycle of the cross bridge requires
energy in the form of ATP, indeed the cross bridge has to be primed with
ATP before it can go through the cycle of force generation. Much more information on the force generating mechanism and the molecular control of
contraction is given in Chapter 2.

1.2.1 *Velocity of contraction and amount of force generated by muscle*

The total force developed by a muscle is proportional to the number of cross
bridges in parallel. In most muscles, the maximum force per square metre of
fibre cross-sectional area is of the order of 100–400 kN. In some animals the
force may be as high as 1400 kN m^{-2} as found in some molluscan muscles.
Of course, the force developed is only approximately proportional to the
fibre cross-sectional area because the amount of myofibril packing is variable.
Muscles which have a large fibre cross-sectional area and which are made up
of fibres with a high myofibril density are therefore well suited to producing
high forces. Methods of increasing the relative fibre cross-sectional area
using different fibre arrangements are mentioned below.

The rate at which force is developed by a muscle is proportional to the number of sarcomeres in series. The other important factor is the rate at which the
individual cross bridges work. This is called the intrinsic speed of shortening.
This parameter varies from muscle to muscle even within the same animal,
indeed many muscles contain two or three different populations of fibres
each with a different intrinsic speed of shortening. For rapid contractions the
muscle needs to have a high intrinsic rate of shortening and lots of sarcomeres
in series. In some reptilian and arthropod muscles the myosin filaments are
longer than the standard length (1.6 μm). This means that there are more
cross bridges per sarcomere and hence more force can be produced per unit
cross-sectional area. However, they have fewer sarcomeres in series hence the
rate of contractions will be slower. In these muscles, the force produced per
unit cross sectional area is higher although the power produced will remain
unchanged. At this point it should be mentioned that it is often not the force
or the velocity of contraction that is important, but a combination of both.

In other words what is important is the work output (power) of the muscle. Power is work done per unit time, and is given by the distance shortened multiplied by the load-lifted over time. Specific power is the power developed per unit of muscle mass, i.e. per gram, and this tends to be independent of the shape of that muscle mass. For a muscle to have a high power output it is necessary for it to contract both forcefully and rapidly. The maximum sort of power developed by muscles is in the region of 0.5–1.0 W gm^{-1} wet weight of muscle. However, the maximum sustainable power is usually considerably less than this, possibly in the order of 0.3 W gm^{-1} muscle.

1.2.2 Intrinsic rate of contraction

The force developed by different muscles does not differ very much from muscle to muscle. It ranges from about 1 to 4 kg cm^2 ($1–4 \times 10^5$ Nm^{-2}) and there are only a few exceptions that fall outside of these values. This range can usually be reduced even more if the force is expressed per mm^2 of myofibril cross-sectional area. In contrast the rate of shortening of muscles may differ as much as a thousand-fold, ranging from a very slow contracting sea anemone which may take several minutes to contract to the flight muscle of an insect or the limb muscle of a mouse, both of which will contract within a fraction of a second. Some examples of the rate of shortening of different muscles in different species are given in Table 1.1. The intrinsic rate of contraction may be measured either physiologically or biochemically. The maximum rate of contraction of a muscle depends on the rate at which its myosin cross bridges work the number of cross bridges per sarcomere. However, as already mentioned, it also depends on the number of sarcomeres in series. The intrinsic rate

Table 1.1 Intrinsic rate of shortening of different muscles in lengths, s^{-1}

Animal	Muscle	Rate	Temp. (°C)	Source
Mouse	Extensor digitorum longus	22	37	Close, R.I. (1972)
	Soleus	11	37	Close, R.I. (1972)
Rat	Extensor digitorum longus	17	37	Close, R.I. (1972)
	Soleus	7	37	Close, R.I. (1972)
Chicken	Posterior latissimus dorsi	18	35	Nwoye, L. (1975)
	Anterior latissimus dorsi (slow tonic muscle)	2	35	Nwoye, L. (1975)
Tortoise	Rectus femoris	1.5	20	Goldspink, G. (1976b)
Schistocerca	Wing muscle	9	35	Buchthal, F. et al. (1957)
Pecten	Striated adductor	3	20	Prosser, C.L. (1973)

has thus to be expressed as either the rate of shortening in muscle lengths per second, or better still, as the rate of shortening per sarcomere. Using either of these methods it is thus possible to compare the rate of shortening of muscles of different lengths. When the intrinsic rate of shortening is measured physiologically it is measured on zero loaded or very lightly loaded muscle in order to get the maximum rate of shortening and this is then divided by the muscle length or the number of sarcomeres in series. It is often difficult and inconvenient to measure the actual rate of shortening so sometimes isometric parameters are measured such as the half relaxation time of a single isometric twitch. These have been shown to be acceptable alternatives to the maximum rate of shortening (Close, 1965). Also Barany (1967) has shown that the maximum velocity of shortening as measured physiologically is related to the specific activity of the myosin ATPase of the muscle. Thus, the intrinsic rate of contraction may be estimated biochemically; however, it must be pointed out that the rate that isolated myosin hydrolyses ATP is considerably less than the maximum *in vivo* rate. Nevertheless, this biochemical approach is a very useful approach, particularly in cases where it is not possible to get a good physiological preparation. As well as carrying out measurements on isolated myosin, it is also possible and sometimes preferable to use isolated myofibrils. All of these measurements, whether they are physiological or biochemical, are really expressions of the rate at which the myosin cross bridges work.

1.2.3 Smooth muscle: helical and obliquely striated muscle

In many invertebrates, the locomotory muscles are smooth or obliquely striated muscle. In these muscles the filaments are arranged in an oblique or helical fashion. Smooth muscle has at least one major physiological advantage in that it has a very much broader length/tension peak than striated muscle (Fig. 1.3). The reason for this is that the actin filaments may, after interacting with one myosin filament, move to the next filament. In striated muscle, this is prevented by the presence of Z discs. In helical or double obliquely striated muscle, some of the length change is due to a change in the pitch of the helix. This is in addition to the change in the degree of overlap between the actin and myosin filaments so that these muscles are often capable of very considerable shortening and of developing a respectable tension over a wide range of shortening. These kinds of muscles are often found in soft-bodied invertebrates where locomotion usually involves very considerable changes in the shape of the animal.

1.3 Arrangements of muscle fibres

In some muscles the fibres run in a parallel manner and extend from tendon to tendon. This is referred to as a simple fusiform or parallel fibred arrangement. In many muscles, however, the fibres do not run in a parallel manner;

instead they have a pennate fibre arrangement (Fig. 1.2). This means that the fibres are shorter than they would be in a simple fusiform muscle. It also means that they have a greater total fibre cross-sectional area per unit volume and hence pennate muscles are able to produce more force per unit volume than fusiform muscles. As the direction of pull of the fibres in pennate muscle is not directly in line with the tendons, they do in fact exert slightly less force per mm² fibre cross-sectional area than fusiform muscles. Therefore, if the isometric force is to be related to the unit fibre cross-sectional area (stress), it is necessary to take into account the angle of pull (angle of pennation) of the

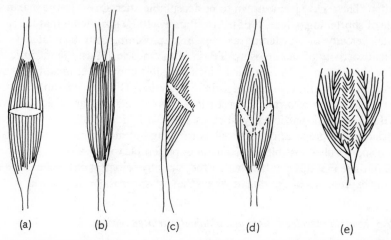

(a)　　　　(b)　　　　(c)　　　　(d)　　　　(e)

Fig. 1.2 Different fibre arrangements. (a) A simple fusiform muscle with fibres running in a more or less parallel fashion along the length of the muscle. (b) A fusiform muscle with two heads. (c) A unipennate muscle; note that this muscle has a larger fibre cross-sectional area than a fusiform muscle of comparable size. (d) A bipennate muscle; note that the total fibre cross-sectional area is increased even further than it would be in a unipennate or fusiform muscle of similar size. (e) A multipennate muscle; these are very powerful muscles, however the fibres are short and therefore they have a low overall rate of shortening.

fibres. The most convenient way of doing this is as described by Alexander (1974) viz.

$$P' = (V/2t)\,\sigma\sin 2\alpha$$

where P' is the force exerted by a muscle of volume V, t is the thickness of the fibre layer, σ is the force (stress) developed by the muscle fibres and α is the angle of pennation.

If the strain (fractional shortening) of the fibres is required, this can be calculated using a related expression, viz.

$$\varepsilon = (S'/2t)\sin 2\alpha$$

where ε is the strain, and S' the amount by which the whole muscle shortens; t and α are as above.

For further details the reader is referred to Alexander (1974) and Calow and Alexander (1973).

Given that the intrinsic speed of shortening is the same for a pennate muscle as for a parallel fibred muscle, the pennate muscle will, because of its shorter fibres, have a lower overall rate of shortening. For muscles that are required to contract isometrically or nearly isometrically this does not matter; indeed it is an advantage to have short muscle fibres because there will be fewer cross-bridges cycling in series in order to maintain a given isometric force. Hence a short fibred muscle will be expected to maintain isometric force more economically. Pennate muscles may have a length/tension curve of a different shape to that of a simple fusiform muscle. This is particularly true in the case of asymmetric pennate muscles as muscles with fibres of different length have a much broader length–tension peak than those with fibres all of approximately the same length (Fig. 1.3). This situation is also found in pennate muscles which have more than one head, however, it is generally associated with pennate muscles.

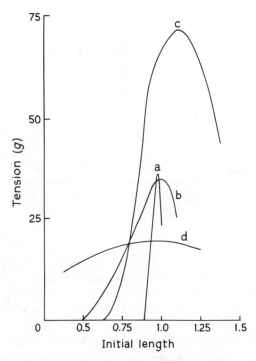

Fig. 1.3 Length/active tension curves for a range of different muscles. The tensions are not given as grams per unit cross-section therefore they cannot be directly compared. However, it is readily seen that the dependence on the initial length, as far as tension development is concerned, differs very considerably between muscles of different design. (a) Bumblebee (*Bombus*) flight muscle (Boettiger, 1957). (b) Locust (*Schistocerca*) flight muscle (Weis-Fogh, 1956). (c) Frog sartorius muscle 0°C (Wilkie, 1956). (d) Snail (*Helix*) pharynx retractor muscle 16°C (Abbott and Lowy, 1956). The resting tension data for these muscles is shown in the next chapter (Fig. 2.15).

1.3.1 Relationship of muscle fibres to the skeleton

In most animals the force developed by muscles during locomotion is used to move bones or parts of the exoskeleton. In some cases, however, muscles are attached to soft tissue and are used to develop hydrostatic pressure.

Where muscles are attached to the skeleton they are arranged so that they and the skeletal structures form a lever system. Lever systems may be classified as 1st, 2nd or 3rd order systems (Fig. 1.4). In the case of the 1st order system the pivot (joint) is in the centre with the muscle attachment at one end and the load at the other end, e.g. gastrocnemius muscle. In the 2nd order system, the

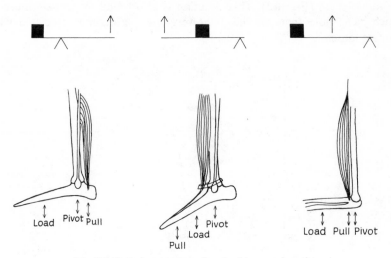

Fig. 1.4 Basic lever systems involved in muscle action.

load is between the pivot and the muscle attachment, e.g. tibialis anterior. In the 3rd order system, the muscle attachment is interposed between the pivot and the load, for example the mammalian biceps brachii.

Lever systems modify the forces produced by the muscles to a very considerable extent. The torque developed by the muscle working in conjunction with a lever system may be calculated using basic mechanics and the reader is referred to two excellent texts by Alexander (1968) and White (1974).

Although many muscles operate over only one joint, many others operate over two joints or more. For example, the soleus muscle of the mammalian lower limb only operates over the ankle joint, whereas the over-lying gastrocnemius has insertions on the femur as well as on the tibia; therefore it bends the knee as well as causing plantar flexion of the foot. Some muscles do in fact have a dual action and this is due often to their having more than one head with the separate heads inserting at different points. One head may operate over only one joint, whilst the other may operate over two joints. Examples of the dual action of a muscle include the human biceps brachii muscle which not only flexes the elbow but causes supination of the forearm.

A knowledge of the type of lever system, types of joints involved and the points of insertion of the muscle on the skeleton is of great importance when attempting to understand the forces involved and the mechanism of limb action during locomotion. More details of the actual mechanics are found in Chapter 7.

1.3.2 Means of attachment of muscle fibres to the skeleton

At first sight, it may seem that the means of attachment of the muscle fibres to the skeleton is of no particular importance in locomotion. However, in recent years it has become apparent that tendons (apodema of insect muscle) may not only transmit the forces generated by the muscle but may store some energy as elastic strain energy. This 'elastic storage' of energy is of particular importance in jumping and terrestrial locomotion in general and also in insect flight. In insects, the principal material used for storing energy is resilin. This is almost a perfect rubber (Anderson and Weis-Fogh, 1964) in that when it is stretched or deformed it will return extremely rapidly to its original state. Resilin is also an integral part of the jumping mechanism of several insects, e.g. fleas and click-beetles (Bennet-Clark and Lucey, 1967; Evans, 1972). The jump of the flea is by any standards a remarkable feat which requires a large amount of power. The power requirements are many times that which could be produced by using muscle alone. However, the muscles develop force over a relatively long period of time and this energy is stored in resilin. When the insect jumps the stored energy is released, just like releasing a catapult. The high power requirements are attained due to the fact that the rate of release of energy is so much more rapid than the rate at which it was originally produced.

Elastic storage of energy is important in insect flight because the frequency of wing beat in small insects may be as high as 600 Hz. If the flight muscles were to produce the power necessary to accelerate the wing to its full velocity at each wing beat, this would demand enormous power which would be lost at the end of each beat as the wings decelerate. However, much of the energy of the wing motion is stored in the resilin which is found in the hinges that hold the parts of the thoracic cuticle together.

It should also be mentioned that some elastic elements, principally ligaments, help in maintaining posture. Although the main function of ligaments is to hold bones together, some do it in such a way that they eliminate the need for an antagonistic muscle, and thus save energy. Examples of such ligaments are the ligamentum nuchae which runs along the neck of ungulates and which helps to maintain the head in an erect position. In birds there is a ligament (tensor patagii longus tendon and ligament) which runs from the shoulder to the second wing joint. When the bird is at rest this helps to maintain the wings in a folded position and during flight the tensioned ligament forms a sharp leading edge to the wing.

A similar sort of mechanism is found in the hinge of the scallop (*Pecten*) which is made out of a rubber-like material called abductin. The valves are closed by the retractor muscles and in this state the hinge material is compressed. When the muscles relax the abductin expands and causes the valves to gape. Thus, during swimming, the retractor muscles of the scallop are responsible for the power stroke and the hinge elasticity produces the recovery stroke. The physical properties of many of these biological elastic materials are extremely interesting as, like many polymers, their breaking stress increases with increasing strain rate. In other words, the more rapidly the force is developed the stronger the tendon or ligament becomes. A simple model for this system is a spring and a dashpot in series. At very high strain rates it seems that the breaking stress may decrease again as there is insufficient time for the applied force to be dissipated along the polymer molecules. These strain rates are usually outside the normal physiological range although they may represent upper limits as far as evolution is concerned (Goldspink, 1976).

Not only may energy be stored in the elastic elements of the skeleton whether they be arranged in series or parallel with the muscle, but it may also be stored in the activated muscle fibres. The myosin cross bridges are now known to be compliant structures which may be stretched to a considerable extent before detachment occurs. Some of this compliance is thought to be due to the rotation of the heads of the myosin cross bridges and some is due to the elongation of the arm (Huxley, 1974; Flitney, 1974). This latter part is believed to have a helical configuration and for this reason one would expect it to be extensible. More information about how the myosin molecules behave during contraction is given in the following chapter.

1.3.3 Arrangement of muscle fibres in fish

The muscle fibre arrangements in fish are of particular interest as in many cases the fibres are not attached directly to the skeleton but to the connective tissue of the myosepta. The other interesting aspect is the way the white myotomal fibres are arranged so that the rate of shortening of the innermost fibres is the same as that of the outermost fibres. This is very important because the maximum power produced by a muscle fibre is obtained when the fibre is shortening at about 0.3 of its maximum velocity. If all the white muscle fibres are to have the same intrinsic speed of shortening and hence the same optimum velocity, they have to be arranged in such a way that they shorten to a similar extent when the fish bends. This ensures that they will all be producing their maximum power when the white muscle is recruited. Alexander (1969) has shown that there are two basic patterns of fibre arrangement found in fish. One is found in the selachians and some of the primitive bony fishes, the other

10

arrangement is found in the teleosts in general. In the selachians (cartilaginous) fishes each myosepta has the shape of a series of cones with the apices of the cones pointing to the anterior and posterior of the fish in an alternating manner. In *Scyliorhinus*, each myoseptum has three anterior cones and two

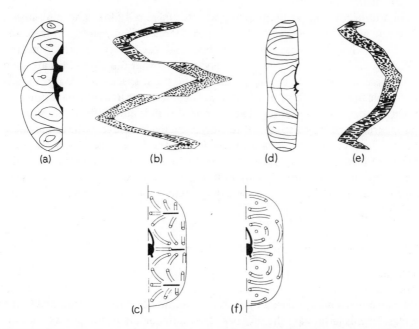

Fig. 1.5 The arrangements of the myotomal muscle fibres in cartilaginous fish are shown in (a), (b) and (c) and in bony fish in (d), (e) and (f). (a) A generalized transverse section through the fish showing the arrangement of the muscle fibre cones in cartilaginous and some primitive bony fish. In this case the cones are complete or almost complete whilst in bony fish (d) they tend to be only half cones. (b) shows the shape of a myomere viewed from the side and as will be seen, the fibre cones are relatively long and well pronounced. The cones in bony fish are much shorter and less pronounced (d). The trajectories of the fibres within the cones are shown in (c) and (f). From these diagrams it will be seen that the fibres within the cones in the cartilaginous fish have a simple arrangement (c) whereas in the bony fish they are arranged in a helical way (f). It is because of this helical arrangement that the cones in the bony fish can be much shorter and still ensure that all fibres are shortening to the same extent. Taken from Alexander (1969).

posterior cones as shown in Fig. 1.5. There is a horizontal septum which crosses the myoseptum between the main anterior cone and the hypaxial cones and each cone is attached to a tendon.

In bony fish, the arrangement of fibres is more complicated. The myomeres are still arranged in cones, although the cones are not as deep or pronounced as in cartilaginous fish. The fibres are arranged in the cones, in a helical way although the apices of the cones tend to lie nearer the medium plane of the fish and as a consequence of this the cones are not complete. This means that the fibres trajectories do not form a complete turn of the helix. Each muscle

11

fibre trajectory approximates to a segment of a helix which is rather more than half a turn, starting at the medium plane and returning to it (Fig. 1.5). The myomeres in bony fish are not usually attached to tendons except in the case of the caudal ones which have a similar configuration in bony fish as in cartilaginous fish.

In addition to the white muscle fibre, fish have red fibres and pink fibres. The red fibres are believed to be used to produce slow cruising movements whilst the white fibres are used only for bursts of speed. Recent work (Johnston, Davison and Goldspink, 1976) strongly suggests that the pink fibres are used for speeds intermediate between cruising and burst speeds. These three kinds of muscle fibres therefore constitute separate muscle systems and have different intrinsic speeds of shortening. The red and the pink muscle fibres are situated peripherally and they also constitute only a small percentage of the total muscle mass. Therefore no complex arrangement of the fibre is necessary. In most fish examined the red muscle is arranged in very simple myomeres with all the fibres running more or less parallel to the long axis of the body of the fish. Further details of the muscle fibre arrangements in fish are found in Alexander (1969).

1.3.4 Arrangement of flight muscles in insects

In some insects the wing movements are produced by wing muscles directly inserted into the base of the wing. However, in others, such as the Diptera, the movements are produced by muscles that are not directly connected to the wings. In this latter case the muscles cause a distortion of the thorax and this results in the wings moving up and down. This movement is aided by the fact that the wing hinge involves a click mechanism. That is to say the hinge has an unstable position and it moves very quickly from this unstable position to the stable positions of the wing either fully raised or fully lowered. There is a twisting movement superimposed on the up and down movement and there-fore this action produces thrust as well as lift. In some insects there is also a twisting movement on the up stroke and this produces some lift in addition to the lift from the down stroke.

Fig. 1.6 shows the arrangement of the wing muscles of insects with direct flight muscles and those with indirect flight muscles. The mechanism in insects with direct flight muscle is relatively straightforward. However, in the case of the indirect flight muscles these differ in that the down stroke is produced by the dorsal longitudinal muscles which pull on the anterior and posterior ends of the thorax. This causes the top of the thorax to rise which in turn depresses the wing as shown in Fig. 1.6. In both systems the elastic properties of the thorax are of great importance. The flight muscles cause the whole of the thorax and wings to vibrate at their characteristic frequency. As stated above, the elasticity of the thorax is due, in the main part, to the resilin pads found between the chitinous plates. In permitting the system to vibrate, the resilin enables some of the energy produced by the flight muscle in the up stroke to be stored and used during the down stroke.

As these insect flight systems are oscillating systems, the flight muscles are designed to shorten only by a small amount. They have sarcomeres with very short I bands and as they work over a very narrow range of shortening they tend to have very sharp length tension peaks as shown in Fig. 1.3. In insects with a high frequency of wing beat (asynchronous muscles), there is no direct relationship between the nervous stimulation and the cycles of contraction.

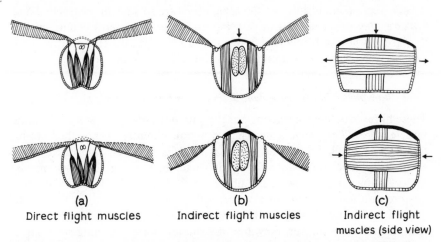

(a)   (b)   (c)
Direct flight muscles   Indirect flight muscles   Indirect flight
                                                          muscles (side view)

Fig. 1.6 Arrangement of direct and indirect flight muscles in insects. In insects with direct flight muscles (a), the muscles are connected to the wings. In those with indirect flight muscles (b), the downstroke is produced by the raising of the roof of the thorax which is brought about by the contraction of the dorsal longitudinal muscles as shown in (c). In this latter case, the wings are connected to the roof of the thorax with a double hinge system which involves a skeletal click mechanism. Modified from Pringle (1975a).

Instead the muscle fibres depend on stretch activation to initiate each cycle of contraction. For further details of the physiology of insect flight muscles the reader is referred to authoritative reviews by Pringle (1972, 1975b).

1.4 Muscle fibre composition of locomotory muscles

In the invertebrates, with the exception of the protozoa, locomotory movements are produced by striated muscle or obliquely striated muscle. In the vertebrates, however, only striated muscle is used. This shows considerable diversity in its structure, physiology and biochemistry. There are two main types of striated muscle fibres: tonic fibres and phasic fibres. The phasic fibres also show further diversity and unfortunately the multifarious methods of classification found in the literature are often very confusing. Here a very simple scheme of classifying muscle fibres has been adopted.

1.4.1 Tonic muscle fibres

These fibres are very slow contracting fibres which do not usually exhibit a propagated muscle action potential, hence when stimulated with a single stimulus they do not produce a significant response. They are multiply innervated (en grappe motor end plates) and thus have a graded response to stimulation of different frequencies. Tonic fibres usually have a very low intrinsic rate of contraction and therefore a very low specific myosin ATPase activity. These fibres are capable of developing and maintaining isometric tension very economically. As the tonic fibres are so slow in contracting and because they are inefficient in carrying out phasic (isometric) contractions, they are not normally involved in locomotory movements. However, in some animals, they do perform an important role in maintaining posture.

1.4.2 Slow phasic fibres (slow contracting, slow fatiguing fibres)

These fibres are slow contracting but they are not usually as slow as the tonic muscle fibres. They have a propagated action potential, hence they are referred to as twitch fibres. This type of muscle fibre is often innervated with a single 'en plaque' end plate although in some animals, e.g. fish, they may be multiply innervated. Slow phasic fibres are responsible for both maintaining posture and for carrying out slow repetitive movements and, as will be explained in Chapter 3, they are both economical and efficient in carrying out these functions. However, they are only suitable for performing slow movements. Slow phasic fibres use ATP at a slow rate and they usually contain a lot of mitochondria; hence they fatigue only very slowly.

1.4.3 Fast, phasic, glycolytic fibres (fast contracting, fast fatiguing fibres)

These fibres are adapted for a high power output and they have a reasonable thermodynamic efficiency for producing work. They have a high or very high intrinsic speed of shortening and hence a high myosin ATPase specific activity. Fast phasic fibres are usually recruited only when very rapid movement is required. They usually possess very few mitochondria because they cannot possibly supply ATP as fast as they use it. They therefore fatigue very rapidly and most of the replenishment of their energy supplies takes place after the exercise has ceased. During contraction the energy that is supplied comes mainly from glycolysis.

1.4.4 Fast, phasic, oxidative fibres (fast contracting fatigue resistant fibres)

This kind of fibre is essentially the same as the fast glycolytic fibre except that they contain reasonably large numbers of mitochondria. In some cases, but not in all, they may have a slightly lower intrinsic rate of shortening than the fast glycolytic fibres. These fast oxidative fibres are apparently adapted for

14

fast movements of a repetitive nature and are recruited next after the slow twitch fibres. Because they possess fairly large numbers of mitochondria they tend to be reasonably fatigue resistant and they are able to recover fairly quickly following a bout of exercise.

Out of necessity, this scheme of classifying muscle fibres is an over-simplification. If one looks at different animals, particularly those of different sizes, it is seen that there is a whole range of fibre types. For example, the fast muscle fibres of a horse are probably slower than the slow fibres of the mouse so one must exercise caution when describing fibres in different species. Nevertheless the types of fibres do have fairly discreet physiological roles and it is important to have some means of describing them. Several other schemes of classifying fibres are in common usage. Their relationship to the one given above is shown in Table 1.2. Some of the physiological properties of fast and slow twitch fibres are shown in Fig. 5.1.

Table 1.2 Classifications of fibre types in mammalian (phasic) muscle by different authors

Peter *et al.* (1972)	FG	FOG	SO
Burke *et al.* (1973)	FF	FR	S
Stein and Padykula (1962)	A	C	B
Romanul (1964)	I	II	III
Engel (1962)	II	II	I
Ashmore and Doerr (1971)	αW	αR	βR
Brooke and Kaiser (1974)	11B	11A (and 11C)	1
Henneman and Olson (1966)	Fast twitch white	Fast twitch red	Slow twitch intermediate
Main properties			
Mitochondrial content	low	high	intermediate
Z line	narrow	wide	intermediate
Neuromuscular junction	large and complex	small and simple	intermediate
Oxidative enzyme activities	low	high	high or intermediate
Glycolytic activities	high	low	intermediate
Myoglobin content	low	high	high
Glycogen content	intermediate	high	low
Myofibrillar ATPase at pH 9.4	high	high	low
pH sensitivity of myofibrillar ATPase	acid labile alkali stable	acid labile alkali stable	acid stable alkali labile
Formaldehyde sensitivity of myofibrillar ATPase	sensitive	stable	—
Rate of contraction	fast	fast or intermediate	slow
Recruitment	infrequent	fairly frequent	frequent
Fatigue time	very short	fairly long	long
Efficiency at option rate of shortening	reasonably high	?	high
Economy in maintaining isometric tension	low	?	high

1.4.5 Motor units

The group of muscle fibres that is innervated by the same motor neuron is called a motor unit. A motor axon originating from the spinal chord may innervate between one and 500 muscle fibres depending on the muscle. In most vertebrate muscles, the fibres act in an all or nothing way, therefore all the fibres within a given motor unit contract at the same time. Gradation of movement in these muscles is achieved by recruiting different numbers of motor units. Previously, it was thought that there was a rotation of motor units; that is to say, when the fibres of one motor unit began to fatigue, a different motor unit took over. However, it now seems that this is probably not the case and the recruitment is on a strictly hierarchical basis. It is likely that the slow muscle fibres are always recruited first and that the fast oxidative fibres are the next to become involved with the fast glycolytic fibres being recruited only when high speed or high force is required. The way motor units are recruited is discussed in more detail in Chapter 5.

All the fibres within a motor unit seem to have the same or similar characteristics (Burke *et al.*, 1973; also see Chapter 5). Indeed many of these characteristics are believed to be determined by the motor neuron (Buller *et al.*, 1960; Guth, 1968).

1.4.6 Methods of studying the motor unit/muscle fibre composition of different muscles

The different motor units (muscle fibres) within a muscle can be distinguished both electrophysiologically (Burke *et al.*, 1973) or histochemically (Plate 1.1). The electrophysiological methods involve very sophisticated techniques in which simple motor axons are stimulated in anaethetized animals by either teasing the axons apart and stimulating an individual axon or by impaling the cell body of the motor neuron in the spinal chord with a microelectrode. The contraction of the fibres in that motor unit is measured isometrically using a sensitive transducer. Further details of this approach are mentioned in Chapter 5.

Histochemical methods do not permit the distinguishing between different motor units *per se*, however, they do enable one to look at the relative percentages of the types of muscle fibre in a muscle. Histochemical studies are much more convenient than the electrophysiological methods and therefore much more information is available about the ratios of fibre types rather than motor units.

As stated above, the different types of muscle fibres have different intrinsic speeds of shortening, in other words they have different kinds of myosin molecules. One of the most useful histochemical tests is that for myosin ATPase. This test is based on the activity of the fibre with ATP or substrate after subjecting the muscle fibres to pre-incubation in either an acid or alkaline pH. Myosin of the fast type is known to be reasonably alkali stable and there-

16

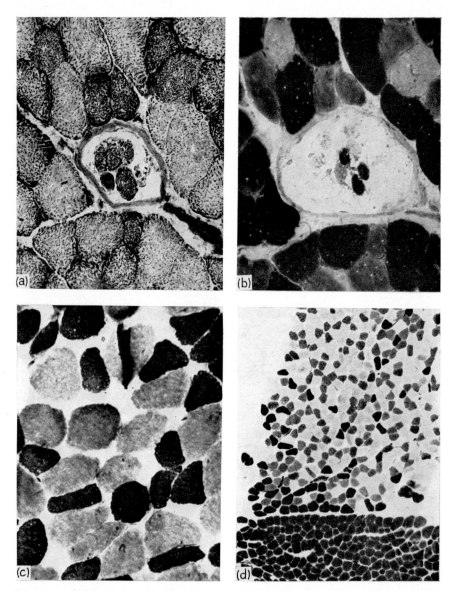

Plate 1.1 Different muscle fibre types as distinguished by different histochemical methods. (*a*) Oxidative enzymes—NADH-tetrazolium reductase. (*b*) ATPase after pre-incubation at alkaline pH (note that the large fibres stain darkly). (*c*) ATPase after pre-incubation at acid pH (note that the small fibres stain darkly). (*d*) ATPase staining of the mouse soleus muscle which has three types of fibres and the mouse bicep brachii muscle which has only one type of fibre. From Ward (1975).

fore after treatment at say pH 10.4 the fast muscle fibres still stain darkly, whereas the slow muscle fibres show no activity. If pre-incubation of the muscle fibre is carried out at pH 4.3 then the slow fibres are the ones that are still capable of reacting with the substrate. It is important, however, to realize that this histochemical method is not a measure of the specific activity of the myosin ATPase. Another cautionary note must be added to what has already been said; most of the histochemical methods available have been developed for mammalian muscle, therefore the methods must be suitably modified before they are applied to other kinds of muscle and the temperature at which the reactions are carried out should be appropriate to the environmental temperature of the animal. Nevertheless, this method is one of the most useful ways of distinguishing between different kinds of muscle fibre.

Another very useful group of histochemical tests is that for oxidizing enzymes such as succinic dehydrogenase. As stated above, the mitochondrial content of the different types of muscle fibres differs very considerably. Histochemical staining for mitochondrial enzymes, particularly if it is combined with the myosin ATP method, is a very useful approach to establishing the fibre type composition of a muscle. This is carried out by taking serial sections and staining alternate sections for myosin ATPase and for an oxidative enzyme as shown in Plate 1.1.

1.4.7 Muscle fibre types in locomotory muscles of different animals

Not only does the muscle fibre composition give some idea of the speed of contraction of the muscle, but it gives good information about the sort of action the muscle is primarily designed to perform. This will become more apparent after reading Chapter 3. There is also a good correlation between the mode of life of the animal and the fibre composition of its muscles. For example some mammals are able to sustain very high running speeds. Dogs, including wolves, are a good example of this and their muscles in general have a very high percentage of fast oxidative (FR) fibres. Many ungulates, for example the horse, also have the ability to sustain high running speeds and they too have a high percentage of fibres with a high oxidative capacity. Ungulates often have to sustain high running speeds in order to escape predators and dogs in the wild, e.g. African hunting dogs and wolves, hunt in such a way that they gradually pull down their prey often after a very extended chase. Dogs have been shown to have a very high VO_{2max} (Seeherman, Taylor and Maloiy, 1976) and this is presumably a direct consequence of the very high oxidative capacity of their muscle fibres and their very efficient blood system. Indeed, Seeherman *et al.* (1976) found that it was not possible to get dogs to exceed the power output equivalent to their VO_{2max}. In other words their muscles always work under aerobic conditions. On the other hand, large cats such as the lion are found to have anaerobic muscles containing fast fibres which are all of the glycolytic type (G. M. O. Maloiy, personal communication), in addition to some slow oxidative fibres which are presumably used in prowling. The lion and most cats for that matter are known to be lazy animals and the lion relies very much on stealth and strength to capture its prey.

Similarly, it is possible to see a strong correlation between the mode of life of different kinds of fish and their muscle fibre composition (Boddecke, Slipjer and Van der Stelt, 1959). Fast cruising fish such as the mackerel have a very high percentage of red (slow oxidative) myotomal muscle whilst the pike which approaches its prey very very slowly and then strikes, has very little red but a lot of white muscle. There are many other examples but space does not permit a description of all of these. However, it is fascinating to see how the muscle fibre composition has been adapted to the mode of locomotion and life of the animal.

1.5 Muscles designed to work at different environmental temperatures

Muscles in homeotherms (endothermic) work over a fairly restricted temperature range. However, this range is perhaps not quite so restricted as one may at first think. For instance, although the core body temperature of the mammal may be at 37°C, the temperature in some of the more peripherally situated muscles may be about 25°C or even less in very cold climates. Indeed the optimum working temperature for mammalian muscles seems to be somewhat lower than 37°C.

Muscles in poikilothermic (ectothermic) animals have, however, to be able to work at more extreme temperatures. For example, Antarctic fish experience environmental temperature of −1°C to 4°C whilst at the other end of the scale, fish in the hot springs of the Rift Valley in East Africa live at temperatures of 35°C to 38°C. How is it that the Antarctic fish can move about at these low temperatures? Recent work (Johnston, Walesby, Davison and Goldspink, 1975) has shown the contractile apparatus of Antarctic fish differs from that of warm water or even temperate water fish in that it is designed to have a high specific ATPase activity at the low temperature range (Fig. 1.7). The myosin of these cold water fish seems to have evolved a less rigid and more open tertiary structure and this is related in some way to its high activity at the low temperatures. However, because of the open structure, the myosin is very susceptible to heat denaturation. At higher environmental temperatures fish require a fairly heat stable contractile system, therefore these fish have evolved myosins with a more rigid type of molecular structure. This is reflected by the range of thermal denaturation rates for myofibrillar ATPase activity which varies over 350 times between fish from Antarctic and tropical environments (Johnston, Frearson and Goldspink, 1973; Johnston, Walesby, Davison and Goldspink, 1975).

The thermodynamic implications of designing a contractile system to work at particular temperatures has been studied (Johnston and Goldspink, 1975). It was found that there were small but significant lowerings of the Gibbs free-energy of activation for myofibrillar ATPase activity between Antarctic and tropical species. Furthermore, there was a striking, positive correlation between the enthalpies and entropies of activation of different muscle myofibrillar ATPases and environmental temperature (Johnston and Goldspink, 1975). It is considered that at lower habitat temperatures, where enthalpic activation is likely to be energetically more unfavourable, the enthalpic component is partially replaced by a larger entropic contribution to the free energy of activation. This confers an adaptive advantage on the

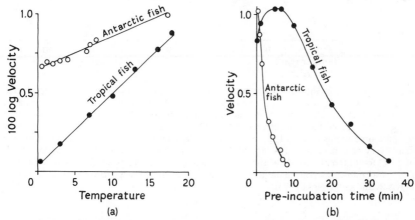

Fig. 1.7 Evolution of the contractile system for working at different environmental temperatures. (a) A plot of $\log_{10}$ ATPase activity (mol Pi mg^{-1} min^{-1}) of myofibrils prepared from an Antarctic fish *Notothenia* (o) and an Indian Ocean fish *Amphiprion* (●). Note that the myofibrils from the Antarctic fish have a much higher activity than the tropical fish particularly at the lower temperatures. Bearing in mind that the plot is a log plot the difference between the two species is very considerable. (b) A plot of the residual activity after incubating the myofibrils at 37° for different periods of time. Note that the myofibrils of the Antarctic fish are very susceptible to heat denaturation. In the case of the warm water fish there is an initial activation of the myofibrils. This is known to be true for myofibrils from mammalian muscle as well as those from warm water fish. From Johnston *et al.* (1975).

enzyme by partly compensating for the low heat content of the muscle fibres.

Many species of fish live within a very restricted temperature range, and thus they have evolved a myofibril system which is adapted for this range. However, a few species of fish are able to adapt to a wide range of temperatures; the carp family provide several examples of fish with this ability. For instance the common goldfish (*Carassius suratus* L.) can survive and move around reasonably well at temperatures of about +1°C in an ice-bound garden pool, whilst on the other hand it can live quite happily in an ornamental pool in the tropics. The Carp family can acclimate to these temperatures because apparently they have the ability to alter the characteristics of the contractile proteins or

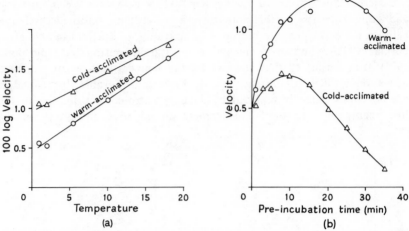

Fig. 1.8 Changes in the contractile system of fish acclimated to living at low and at high temperatures. (a) A plot of $\log_{10}$ ATPase activity (mol Pi mg^{-1} min^{-1}) of myofibrils prepared from the white myotomal muscle of goldfish acclimated to 1°C ($\triangle$) and 26°C (o). Note that the myofibrils from the cold acclimated fish have a higher activity particularly at the lower temperatures and as this is a log plot the difference between the cold- and warm-acclimated fish is very considerable. (b) A plot of the residual activity after incubating the myofibrils at 37°C for different periods of time. Note that the myofibrils from the warm acclimated fish (o) are more temperature stable than those of the cold acclimated fish ($\triangle$). From Johnston, Davison and Goldspink (1975).

they are able to synthesize different kinds of proteins (Johnston, Davison and Goldspink, 1975) (Fig. 1.8). If the latter case is true they must presumably have the genes to produce either a high temperature or a low temperature myofibril system. Most animals no doubt do possess the genes for producing fast myosin and slow myosin and perhaps other intermediate types of myosin, therefore it is feasible that some animals such as the goldfish also possess genes for producing either a high temperature or a low temperature myosin. Only temperature adaptation in the contractile system has been mentioned here; however, molecular adaptation to environmental temperature has also been demonstrated for a number of enzyme systems in addition to the contractile system (see Hazel and Prosser, 1974) and, of course, is also very important from the locomotory point of view.

References

Abbott, B.C. and Lowy, J. (1956) Resting tension in snail muscle. *Nature* **178**, 147–148.
Alexander, R.McN. (1968) *Animal Mechanics.* Sedgwick and Jackson, London.
Alexander, R.McN. (1969) The orientation of muscle fibres in the myomeres of fishes. *J. Mar. Biol. Ass.* **49**, 263–289.
Alexander, R.McN. (1974) The mechanics of jumping by a dog (*Canis familiaris*). *J. Zool., Lond.* **173**, 549–573.
Anderson, S.O. and Weis-Fogh, T. (1964) Resilin. A rubberlike protein in arthropod cuticle. *Adv. Insect Physiol.* **2**, 1–6.

Ashmore, C.R. and Doerr, L. (1971) Postnatal development of fibre types in normal and dystrophic skeletal muscle of the chick. *Exp. Neurol.* **30,** 431–439.

Barany, M. (1967) ATPase activity of myosin correlated with speed of muscle shortening. *J. gen. Physiol.* **56,** 197–218.

Bennet-Clark, H.C. and Lucey, E.C.A. (1967) The jump of the flea: A study of the energetics and a model of mechanism. *J. exp. Biol.* **47,** 59–76.

Boddecke, R., Slipjer, E.J. and Van der Stelt, A. (1959) Histological characteristics of the body musculature of fishes in connection with their mode of life. *Koninklijke Nederlands Akademic van Wetenschappen, Ser. C* **62,** 576–588.

Boettiger (1957) In: *Recent Advances in Invertebrate Physiology*, p. 117. B.T. Sheer, ed. University of Oregon Publishers.

Brooke, M.H. and Kaiser, K.K. (1974) The use and abuse of muscle histochemistry. In: The Trophic Functions of the Neuron. *Ann. N. Y. Acad. Sci.* **228,** 121–144.

Buchthal, F., Weis-Fogh, T. and Rosenfalck, P. (1957) Twitch contractions of isolated flight muscle of locusts. *Acta physiol. scand.,* **39,** 246–276.

Buller, A.J., Eccles, J.C. and Eccles, R.M. (1960) Interaction between motoneurons and muscles in respect to the characteristic speed of their responses. *J. Physiol., Lond.* **150,** 417–438.

Burke, R.E., Levine, D.N., Tsairis, P. and Zajac, F.E. (1973) Physiological types and histochemical profiles in motor units of the cat gastrocnemius. *J. Physiol., Lond.* **234,** 723–748.

Carlow, L.J. and Alexander, R.McN. (1973) A mechanical analysis of a hind leg of a frog. *J. Zool., Lond.* **171,** 293–321.

Close, R. (1965) The relationship between intrinsic speed of shortening and the duration of the active state of muscle. *J. Physiol., Lond.* **180,** 542–559.

Close, R.I. (1972) Dynamic properties of mammalian skeletal muscles. *Physiol. Rev.* **52,** 129–197.

Davison, W., Goldspink, G. and Johnston, I.A. (1976) Division of labour between fish myotomal muscles during swimming. *J. Physiol., Lond.* **263,** 185P.

Engle, W.K. (1962) The essentiality of histo- and cytochemical studies in the investigation of neuromuscular disease. *Neurology (Minn.)* **12,** 778.

Evans, M.E.G. (1972) The jump of the click beetle (Coleoptera, Elateridae) a preliminary study. *J. Zool. Lond.* **167,** 319–336.

Flitney, F.W. (1974) Light scattering associated with tension changes in the short-range elastic component of resting frog's muscle. *J. Physiol., Lond.* **244,** 1–14.

Goldspink, G. (1975) Biochemical energetics for fast and slow muscles. In: *Comparative Physiology: Functional Aspects of Structural Materials.* L. Bolis, H. P. Maddrell and K. Schmidt-Nielsen, eds. North Holland, Amsterdam.

Goldspink, G. (1976a) Mechanics and energetics of muscles in animals of different sizes In: *Scale Effects in Locomotion.* T. S. Pedley, ed. Academic Press, London.

Goldspink, G. (1976b) Unpublished findings.

Guth, L. (1968) Trophic influence of nerve on muscle. *Physiol. Rev.* **48,** 645–687.

Hazel, J.R. and Prosser, C.L. (1976) Molecular mechanisms of temperature compensation in poikilotherms. *Physiol. Rev.* **54,** (3) 620–677.

Henneman, E. and Olson, C.B. (1965) Relations between structure and function in the design of skeletal muscles. *J. Neurophysiol.* **28,** 581–598.

Huxley, A.F. (1974) Review Lecture, Muscular Contraction. *J. Physiol.* **243,** 1–44.

Johnston, I.A., Frearson, N. and Goldspink, G. (1973) The effects of environmental temperature on the properties of myofibrillar adenosine triphosphatases from various species of fish. *Biochem. J.* **133,** 735–738.

Johnston, I.A., Davison, W. and Goldspink, G. (1975) Adaptations in myofibrillar ATPase induced by temperature acclimation. *FEBS Letters* **50,** 293–295.

Johnston, I.A., Walesby, N.J., Davison, W. and Goldspink, G. (1975) Temperature adaptation in the myosin of an antarctic fish. *Nature* **254,** 74–75.

Johnston, I.A. and Goldspink, G. (1975) Evolutionary adaptations to temperature; thermodynamic activation parameters of fish myofibrillar ATPase enzyme. *Nature* **257,** 620–622.

Johnston, I.A., Davison, W. and Goldspink, G. (1976) Studies on the energy metabolism and the division of labour of the swimming muscles of the common carp. (in press).

21

Nwoye, L. (1975) Chemical energetics of mammalian and other vertebrate fast and slow muscles. Ph.D. Thesis, University of Hull, U.K.

Peachy, L.D. (1965) The sarcoplasmic reticulum and transverse tubules of the frog's sartorius. *J. Cell Biol.* **25**, 209–231.

Peter, J.B., Barnard, R.J., Edgerton, V.R., Gillespie, C.A. and Stempel, K.E. (1972) Metabolic profiles of three fibre types of skeletal muscle in guinea-pigs and rabbits. *Biochemistry* **11**, 2627–2635.

Pringle, J.W.S. (1972) Arthropod muscle. In: *The Structure and Function of Muscle*, Vol. 1, 2nd Ed. G. H. Bourne, ed. Academic Press, New York and London.

Pringle, J.W.S. (1975a) *Insect Flight*. Oxford Biology Reader 52, Oxford University Press.

Pringle, J.W.S. (1975b) Insect fibrillar muscle and the problem of contractility. In: *Comparative Physiology, Functional Aspects of Structural Materials*. L. Bolis, H. P. Maddrell and K. Schmidt-Nielsen, eds. North-Holland, Amsterdam.

Prossor, C.L. (1973) *Comparative Animal Physiology*. 3rd edition, Saunders, Philadelphia.

Romanal, F.C.A. (1964) Enzymes in muscle. I. Histochemical studies of enzymes in individual muscle fibres. *Arch. Neurol. (Chig.)* **11**, 355.

Seeherman, H.J., Taylor, C.R. and Maloiy, G. (1976) Maximum aerobic power and anaerobic glycolysis during running in lions, horses and dogs. *FASEB Abstracts* **32**, 797, Abs 32/62.

Stein, J.M. and Padykula H.A. (1962) Histochemical classification of individual skeletal muscle fibre of the rat. *Am. J. Anat.* **110**, 103–123.

Ward, P.S. (1975) Histochemical studies of developmental and disease processes in rodent skeletal muscle. Ph.D. Thesis, University of Hull, U.K.

Weis-Fogh, T. (1956) Tetanic force and shortening in locust flight muscle. *J. exp. Biol.* **33**, 668–684.

White, D.S.C. (1974) *Biological Physics*, Chapman and Hall, London.

Wilkie, D.R. (1956) The mechanical properties of muscle. *Br. med. Bull.* **12**, 177–182.

2 Muscle mechanics

D. C. S. White

2.1 Introduction

Muscle provides the motive power for locomotion, converting the chemical energy of the incoming fuel (carbohydrates, fats) into mechanical work. It serves two functions in locomotion (a) by shortening and thereby getting a limb moving, or changing the shape of the animal, and (b) acting as a means of stopping a moving limb by exerting a tension (or force) opposing the movement.

In the previous chapter, the structure of muscle is discussed. The purpose of this chapter is to describe how the muscle operates. Obviously a chapter of this length cannot include the detail to be found in longer or more specialized texts. The interested reader will find the following worth reading: Alexander (1968), Aidley (1971), Prosser (1973), Offer (1974), White and Thorson (1973), Simmons and Jewell (1974), Simmons (1977).

The tension exerted by the muscle as a whole is made up from the contributions from each of the individual fibres. The exact way in which the individual fibres contribute to the tension depends upon the arrangement of the fibres within the muscle, as discussed in Chapter 1. It is certainly not the case that each fibre contributes equally to the tension. For one thing, the individual fibres in a muscle are not identical, and, in fact, may be extremely dissimilar, and for another the degree of activation of the individual fibres at any instant may be extremely different, some fibres perhaps being maximally activated whilst others are not activated at all. Most of this chapter will be concerned with the properties of individual fibres. It is important to bear in mind that the important effects for locomotion are produced by whole muscles.

The initial command to a muscle telling it to contract is a nerve impulse arising in the central nervous system of that animal. Figure 2.1 illustrates the processes that are initiated by this nerve impulse. Before the process of 'calcium release' the flow is entirely in the forward direction, and there is no feedback from a later box in the diagram to an earlier one. These processes, giving the command to the fibre to contract are known as the *activation* processes or as

23

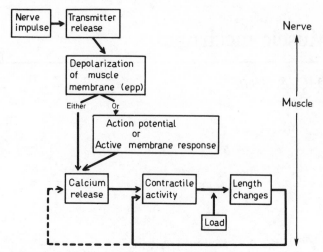

Fig. 2.1 Block diagram of the sequence of activation of muscle.

'*excitation-contraction coupling*'. Figure 2.2 shows the relative time scales of the different processes. As will be discussed later, different muscles can contract at widely differing rates.

Between the last boxes in the sequence shown in Fig. 2.1 there is feedback. The activity of the contractile proteins, as we shall see in Section 2.4, is dependent upon the length changes which take place in the muscle, and is therefore dependent upon the load on the muscle which determines the nature of these length changes. In other words, the amount of energy released by the muscle can be directly affected by the load on the muscle, and is not a constant for a single contraction. This is why there is an arrow going from 'length changes' to 'contractile activity'. It appears likely that the amount of calcium released is also dependent upon the length changes that take place in the muscle.

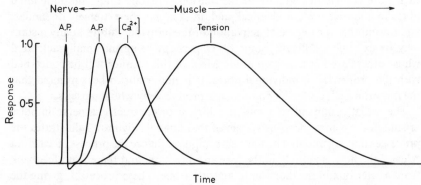

Fig. 2.2 Time course of the activation processes of muscle. Schematic. A.P. = action potential.

2.1.1 Isometric and isotonic contraction

The term muscle 'contraction' is used by different people in different ways. When a muscle is stimulated it is said to contract. However, the exact nature of this contraction is dependent upon the load on it. The essential feature of contraction is a *tendency* to shorten. Whether shortening occurs or not depends upon whether the force exerted by the muscle is sufficient to move the load. Obviously if the muscle is fixed between two massive supports, no overall shortening of the muscle can occur, and the contraction is said to be *isometric* (iso:equal, metric:length). We still talk about contraction, even though no length change has taken place. On the other hand, if the muscle is stimulated under conditions in which there is no external load, then the muscle shortens as fast as it can. By applying increasing loads (e.g. making the muscle lift increasing weights) the muscle will shorten, but with decreasing velocity. A contraction against a constant load is said to be *isotonic* (iso:equal, tonus: force).

Experiments to investigate the nature of the contraction have to be performed under controlled conditions. Most experimental contractions are performed under either isometric or isotonic conditions (or as closely as the experimenter can approach these). However, it is very rare for a real contraction in locomotion to be either one or the other. During normal operation the load on a muscle will vary greatly, and so the length changes that are obtained will also be more complex. Of course in real life an animal does not simply send commands to the muscles and hope that the force is correct for the desired response to occur. There are intricate feedback mechanisms operating which involve measurements of the length changes that actually occur, and subsequent automatic corrected signals to the muscles to change the degree of contraction to compensate for any errors in response. These mechanisms are dealt with in Chapters 4 and 5. In this chapter we are concerned solely with the response of a muscle to simple nerve inputs.

2.1.2 Twitch and tetanus

The mechanical response of a muscle to a single nerve impulse is known as a *twitch*. In many muscles the duration of a twitch is several tens of milliseconds, and may be hundreds of milliseconds or longer. Since the activation processes (the sequence up to and including 'calcium release' in Fig. 2.2) last a much shorter time than the mechanical processes, it is possible to stimulate a muscle several times during the time-course of the response to a single input, and thereby build up a greater response. At a low enough frequency the individual twitches are separate, but above a certain frequency this is no longer the case, the twitches fuse, and the response is known as a *tetanus* (see Fig. 2.3).

2.2 Activation processes

This section is concerned with the processes up to and including 'calcium release' in Fig. 2.1.

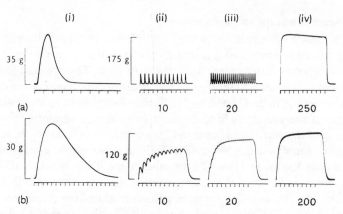

Fig. 2.3 Effect of stimulus rate upon (a) a fast (extensor digitorum longus) and (b) a slow (soleus) muscle from the rat. (i) A single twitch. Time pulses every 10 ms. (ii)–(iv) Effect of stimulation at the frequency indicated to show development of tetanus. Time pulses every 100 ms. From Close (1967).

2.2.1 Depolarization of the muscle membrane

A nerve impulse, reaching the neuromuscular junction or synapse between the nerve and the muscle fibre initiates the release of transmitter (Katz (1969) describes this process very clearly). If its effect is excitatory, the transmitter then acts on the membrane of the muscle to increase the permeability of the membrane to Na^+ and K^+, thereby causing the muscle membrane to be depolarized. The extent of the depolarization (known as an end-plate potential – e.p.p.) is dependent upon the amount of transmitter released. This depolarization is initially a local response in the region of the synapse. What happens next depends upon the kind of muscle being excited. In vertebrate fast fibres with only one, or a group of just a few, neuromuscular junctions at one point of the fibre, the synaptic depolarization then initiates a propagated impulse down the muscle membrane, similar in most respects to that propagated along a nerve fibre. With multiterminally innervated muscle the e.p.p.'s are normally smaller, and occur all along the length of the muscle fibre. Their small size arises because the nerve has released less transmitter than in the case of the vertebrate fast fibres. There is no propagated action potential with multiterminal innervation. Often this is because the membrane potential changes are too small. In this case, the muscle membrane in the region of the end plates is depolarized because of the synaptic phenomena, and the regions between the end plates are depolarized by passive electrotonic spread of this depolarization. The end plates are sufficiently close together that the decay of potential is small in these regions. However, even when the depolarizations are sufficiently large to exceed the threshold, the response is not propagated along the fibre in the normal sense because it is initiated at all points along the fibre. Since the propagation velocity in nerve is about ten times faster than in muscle, the

26

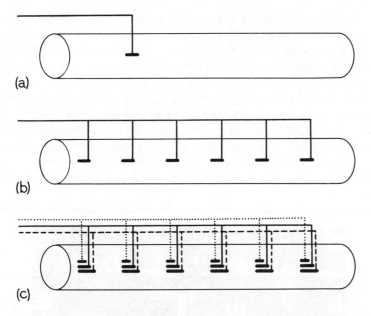

Fig. 2.4 Types of muscle innervation. (a) Single fibre, single end-plate, as found in vertebrate fast fibres. (b) Multiterminal innervation. (c) Polyneuronal and multiterminal innervation, as found in many invertebrates.

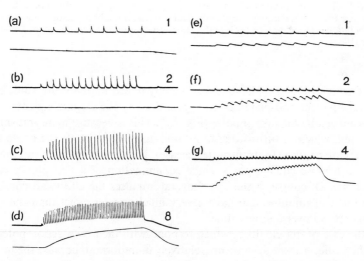

Fig. 2.5 Slow and fast innervation of arthropod muscle. Electrical (upper traces) and mechanical (lower) recordings from muscle fibres of king crab muscle. Stimulation by slow fibres on left (traces a–d) and by fast fibres on right (e–g) at the frequencies indicated. Note the facilitation of the electrical records in the responses to the slow axon, but not to the fast. From Lang, Sutterlin and Prosser (1970).

muscle potentials are initiated more-or-less simultaneously as far as the muscle is concerned. If the potential does exceed the threshold for active processes to occur in the muscle then the response is known as an *active membrane response*. The effect of having multiterminally innervated muscle is to allow fine control of the extent of depolarization.

Two phenomena help in effecting this fine control. Because end-plate potentials have no refractory period, they can be *summated* by making them occur closely enough together that the second starts before the depolarization of the first has decayed to the resting potential. It is thus possible to build up membrane depolarizations to any desired level by putting in the right frequency of nerve inputs to the muscle. The phenomenon of *facilitation* is also of importance

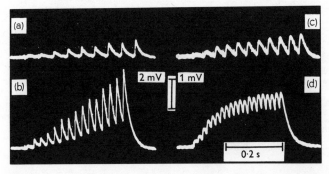

Fig. 2.6 Facilitation of junction potentials at a crayfish neuromuscular junction. Traces a and b are from an excitatory axon, and c and d from an inhibitory axon. Traces a and c show facilitation alone, and traces b and d show summation in addition. From Dudel and Kuffler (1961).

here. Facilitation is the phenomena whereby if two impulses are sufficiently close together the end-plate potential produced by the second is greater than that produced by the first impulse (Fig. 2.6). This is because more transmitter is released, which in turn is due to a build-up of Ca^{2+} inside the nerve fibre (the Ca^{2+} are required for transmitter release). Thus now, instead of summation of two otherwise equal e.p.p.'s, there is summation of a first plus a larger second one. Of course, if there are several impulses the effect can continue, and a burst of impulses can have a very much greater effect than the same number spread over a longer time.

In the case of muscle fibres which respond with their own action potential then these effects at the synapse are relatively unimportant, because the muscle action potential has a refractory period and so intensity is coded once again in terms of frequency of these action potentials. Muscle action potentials are easily distinguished from those in nerves due to their much longer duration (about 10 ms instead of 1–2 ms).

28

In the case of multiterminal innervation, however, in which the entire muscle membrane in effect takes up a potential change similar to that of the end-plates, the effects of summation and facilitation are of prime importance for grading the contraction, since the amount of calcium released into the muscle (see next section) is dependent upon the magnitude of the potential change.

Invertebrates, because they are small, have muscles with few fibres by comparison with most vertebrates. In order to effect fine control of muscular activity they cannot just change the number of fibres which are active; they also need to be able to change the activity of single fibres with precision. Multiterminal innervation, by allowing there to be small depolarizations of the

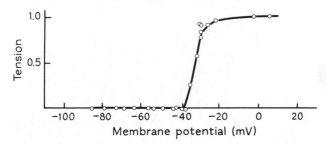

Fig. 2.7 Tension versus membrane potential. This curve was obtained from snake fibres bathed in tetrodotoxin (to prevent any active membrane response), voltage clamped. The potential was raised from −100 mV to the required potential with 90 ms pulses. Redrawn from Heistracher and Hunt (1969).

muscle membrane along its entire length, enables this to happen, the fine control then being effected by varying the nerve input frequency to that muscle. Vertebrates on the other hand, whose muscles may contain tens of thousands of separate fibres, can produce much more effective control by varying the number active at any one time. Although they still can, and do, produce variation in contraction in singly innervated fibres by varying the input frequency, this is a less fine control than in multiterminally innervated fibres (compare the traces of Figs. 2.3 and 2.5).

If a muscle fibre is depolarized it develops tension. A maintained depolarization can be effected experimentally by changing the external K^+ ion concentration, or, if the fibres are sufficiently short for there to be no passive decay of potential, by voltage clamp techniques. The relationship found between depolarization and tension is shown in Fig. 2.7 for snake fibres. There is a threshold below which there is no appreciable development of tension. The tension then increases sharply with increasing depolarization to its maximal value. Similarly shaped curves have been found for other muscles. The value of the threshold depolarization differs slightly from muscle to muscle, and with different ionic conditions. Figure 2.7 was obtained under steady-state conditions; i.e. the fibres were left for long enough at the depolarized level for the equilibrium tension to be obtained. It does not follow that these tensions will be obtained for short depolarizations, such as are obtained during an action potential (which in a muscle lasts about 10 ms). Figure 2.8 shows that this is indeed not the case; because the production of tension by the muscle is a slow process by comparison with the rapid changes in membrane depolarization which can be obtained, the depolarization must be maintained for a long time for the tension maximum to be produced.

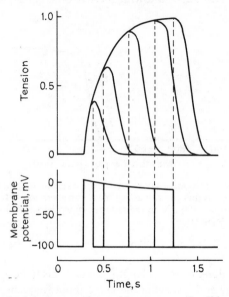

Fig. 2.8 Effect of pulse length upon the tension response. Conditions as for Fig. 2.7. The membrane potential was raised from -100 mV to 0 mV for the durations indicated. Redrawn from Heistracher and Hunt (1969).

2.2.2 Release of calcium from the sarcoplasmic reticulum

The previous section has shown that the effect of stimulating the nerve to a muscle is to depolarize the muscle membrane by an amount, and for a time, which are both variable.

The effect of this depolarization is to cause Ca^{2+} to be released into the main body of the muscle. (We shall see in Section 2.3.3 that it is the Ca^{2+} concentration which activates the contractile proteins.)

In nearly all muscle fibres the endoplasmic reticulum – known as the sarcoplasmic reticulum in muscle – which is in the form of closed sacs of various shapes in different muscles, acts as a store of Ca^{2+}. The membrane has a very

Plate 2.1 Longitudinal section of frog sartorius muscle illustrating the T-system (T), the sarcoplasmic reticulum (SR) and the scalloped region between them (S). (*a*) ×42 750. (*b*) ×95 000. From Franzini-Armstrong (1970).

high ability to transport Ca^{2+} into the sacs actively (and a correspondingly high ATPase activity). In unstimulated muscle, the sarcoplasmic reticulum (SR) acts to lower the Ca^{2+} in the sarcoplasm to very low levels (of the order of 10^{-9} M – the sarcoplasm is then said to have a pCa of 9 in a fashion analogous to that used for pH).

Depolarization of the outer membrane of the muscle (the sarcolemma) causes the SR to release its Ca^{2+} into the sarcoplasm. The link between the outer membrane and the SR, which is often well inside the body of the muscle within easy reach of the contractile proteins, is a specialized membrane system known as the T-system. T stands for transverse or tubular, even though the T-system is not always transverse (although it usually is) and is certainly not always tubular (although it often is). The T-system is a series of invaginations of the outer membrane running into the body of the fibre. In most vertebrate muscles, and many invertebrates also, there is either one or two series of invaginations at the level of each sarcomere. The invaginations at one level may link up to form a network, and the total area of membrane in the T-system may be many times that of the sarcolemma. The fluid inside the T-system is continuous with the fluid surrounding the fibre, and is presumably extracellular fluid. Certainly large molecules and comparably sized substances such as ferritin, which are unable to cross the sarcolemma, can readily penetrate and enter the T-system.

The T-system makes intimate contact with the SR at juxtapositions of the two membrane systems which have a peculiar and characteristic structure (Plate 2.1). Depolarization of the outer membrane system is propagated down the T-system (as an action potential in frog muscle which is the only muscle in which it has been studied, and presumably in a similar way in other muscles). It would seem probable, although there is as yet no experimental evidence to test this point, that the depolarization of the T-system is passed across the special junctions between the T-system and SR (due to presumed low resistance pathways between the two membrane systems in this region, maybe at the junctions shown in Plate 2.1), and that depolarization of the SR causes an increase in permeability to Ca^{2+}. The Ca^{2+} ions are thereby released into the sarcoplasm. When the electrical events of the membrane are finished the permeability presumably reverts to its normal low level, the Ca^{2+} pump starts pumping Ca^{2+} back into the SR and the concentration in the sarcoplasm falls once again, allowing the muscle to relax.

The time course of the release of calcium into the sarcoplasm has been measured in barnacle fibres (which have a large diameter and so can be cannulated) by injecting aequorin into the fibres. Aequorin is a protein which gives a light flash when it binds Ca^{2+}; thus by observing the intensity of light emitted from the fibres containing aequorin it is possible to follow the time course of Ca^{2+} changes. Figure 2.9 shows such a result. Notice that the peak of Ca^{2+} concentration occurs well before the tension has reached its peak, by which latter time the Ca^{2+} concentration has reverted almost to its resting level.

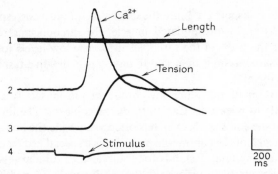

Fig. 2.9 Time course of calcium concentration changes and tension following a depolarizing stimulus in barnacle fibres. From Ridgway and Gordon (1975).

The steady-state relationship between Ca^{2+} and tension development (i.e. under conditions in which the Ca^{2+} concentration is maintained for long enough for the equilibrium tension at that concentration to be attained) has also been determined for a number of muscles. Figure 2.10 illustrates the relationship found for frog fibres by Hellam and Podolsky (1969). The pCa for 50% maximum tension to be developed is very dependent upon the Mg^{2+} concentration.

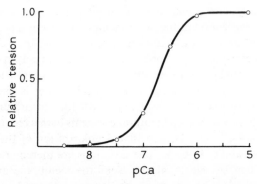

Fig. 2.10 Tension versus Ca^{2+}. Frog semitendinosus muscle skinned (i.e. sarcolemma removed so as to allow the experimenter to control the intracellular ionic concentrations). Redrawn from Hellam & Podolsky (1969).

2.3 The contractile machinery

Chapter 1 has briefly described the ultrastructure of muscle and has shown that a common feature of all muscle is the presence of two types of filament arranged longitudinally along the length of the fibre. In striated muscle there is a very regular packing of the two filament types, whereas in non-striated muscle the filaments are arranged in a much more haphazard fashion. There are a number of muscles which have intermediate degrees of order, such as the obliquely striated muscles found in annelids.

The basic mechanisms of contraction are common to all types of muscle yet investigated. In the main, both because there is much more information, and also because this book is concerned with locomotion, most of this section will concern itself with striated muscle. Most locomotory muscles are either striated or obliquely striated; the non-striated or smooth muscles are usually used for non-locomotory functions.

There are a number of contractile proteins (i.e. those which constitute the filaments and associated structures such as the Z-line, M-line and dense bodies, as opposed to the glycolytic and other enzymes and proteins of the cell). The molecular weights and locations of these are summarized in Table 2.1.

Table 2.1 The contractile proteins

Protein	Location	Molecular weight
Myosin	A filament	460 000
Light meromyosin 'backbone'		120 000
Heavy meromyosin 'cross bridge'		340 000
Actin	I filament	41 700
Tropomyosin	I filament	70 000
Troponin	I filament	80 000
C-protein	A filament	140 000
α-actinin	Z line	180 000

The function of C-protein, α-actinin and M-line protein is not properly understood, although it is thought that they are responsible for the integrity of the structure of striated muscle. Thus α-actinin is found in the Z-line, and may do little more than keep the I filaments correctly positioned; M-line protein may do the same for the A filaments; C-protein, which is found attached to the thick filaments, could perhaps be responsible for controlling the overall length of the A filament. We shall be concentrating on the properties of just four of the contractile proteins – actin, myosin, tropomyosin and troponin. It appears that actin and myosin are the only two proteins needed for contraction, and that troponin and tropomyosin are used to control the switch from relaxation to contraction of muscle.

2.3.1 The arrangement of the contractile proteins into filaments

Of the four main proteins we are considering, three – actin, tropomyosin and troponin – form the thin or I filament, and the other – myosin – aggregates to form the thick or A filament. Their approximate shapes and dimensions, and the way they pack together to form the two filaments are illustrated in

Fig. 2.11. Their approximate molecular weights are given in Table 2.1. Note that myosin, which is a very heavy and large molecule, can be subdivided in various ways to form subcomponents. Basically, it is composed of two equivalent components, the heavy chains, which are the same length as the myosin molecule and which aggregate side by side in the whole molecule, together with three light chains which combine with the heavy chains in the head of the molecule. The molecule has a long tail and a double globular head region. It can be subdivided in another way, using light proteolytic enzyme treatment

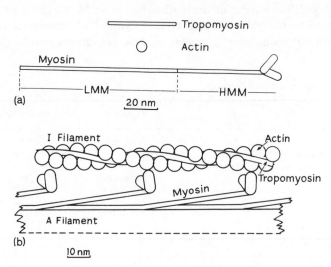

Fig. 2.11 (a) Relative sizes of the tropomyosin, actin and myosin molecules. (b) Short length of A and an I filament to show the arrangement of the molecules.

with trypsin and papain, into a tail region known as light meromyosin (LMM), and a head region with a very much shorter tail known as heavy meromyosin (HMM). The HMM can itself be subdivided in a similar way into two subfragments known as subfragment-1 (S1) and subfragment-2 (S2), S1 being the head region of HMM (there are two S1 per HMM – the two heads) and S2 being the short tail region.

The LMM forms the backbone of the A filament, and the HMM forms the projections or 'cross bridges' (XB) which can be seen on the A filament in electron microscope pictures of thick filaments. Myosin is the ATPase of the contractile proteins, and so has an ATP binding site. It is also able to attach to actin, and so has an actin binding site. These two sites are both located on the S1 head of the myosin molecule. S1 shows its ATPase activity and actin binding ability quite happily in the absence of the rest of the myosin molecule.

There are two classes of light chain known (1) as the alkali light chain – this is dissociated from myosin by the action of alkali – and (2) the DTNB light chain – which is dissociated from myosin by the action of DTNB. The DTNB light chain has a Ca^{2+} binding site, and in some muscles certainly, although maybe not in all, confers Ca^{2+} sensitivity to the muscle. The alkali light chain is required for the ATPase activity of the myosin.

The myosin molecules aggregate to form the thick filament. Halfway along

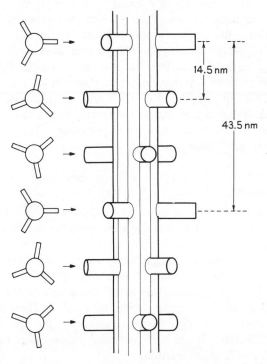

Fig. 2.12 Probable structure of an A filament from vertebrate striated muscle, to illustrate the relative positions of the cross bridges. Three cross bridges originate at each level of the A filament, and the angular position of the origination points changes by 40° between adjacent levels.

the length of the filament the molecules lie back to back, and there is a clear zone, free from side projections or cross-bridges. In either half of the filament the myosin molecules are all aligned in the same direction. The exact way in which they pack is still not generally agreed, and is almost certainly different in different types of muscle. In vertebrate striated muscle, the cross bridges occur every 14.5 nm, but it is not certain whether there are 3 or 4 cross bridges leaving the A filament at these repeat distances (Squire, 1973). The most probable structure of the A filament (Squire, 1974) is shown in Fig. 2.12.

In some muscles, the thick filaments, also contain another protein – paramyosin. This is a rod-shaped protein, very similar in many respects to LMM. In some molluscan muscles it forms the core of the thick filament. Some insect flight muscles also contain paramyosin. It is possible, though not yet certain, that here also the paramyosin forms an inner core of the thick filament.

The thin filament (Fig. 2.11) is formed from the proteins actin, tropomyosin and troponin. The globular actin monomers aggregate to form a filamentous structure of two strings of monomers loosely wound round one another to form the structure illustrated, with about 13 monomers in a complete turn of one string. The tropomyosin, which is a long rod-shaped molecule (a pure coiled-coil α helix), fits into the groove between the two actin chains. When coiled in this fashion the tropomyosin extends alongside exactly 7 actin monomers. Although we have so far described troponin as a single molecule, it is in fact composed of three separate molecules – known as troponin T (TNT), troponin C (TNC) and troponin I (TNI). These aggregate together to form a functional and structural unit and we shall continue to treat them as one entity – troponin. This is attached to the tropomyosin. There is one troponin per tropomyosin per 7 actin molecules.

2.3.2 Filament sliding

When muscle shortens, it does so without there being any detectable length changes in the filaments. Overall length changes in the muscle occur with the filaments moving relative to one another (see Fig. 2.17b); the extent of overlap of the A and I filaments thus changes, being increased as the muscle shortens. If the muscle is extended sufficiently, then a stage is reached when there is no overlap at all. This explanation of how length changes occur was first proposed by A. F. Huxley and Niedergerke (1954) and by H. E. Huxley and Hanson (1954), based upon the banding patterns observed in the light microscope at different muscle lengths. It was later confirmed with the electron microscope, and with the use of X-ray diffraction.

The contractile force is generated between the A and I filaments; the evidence that this is brought about by the action of the cross-bridges is discussed in Section 2.4.

2.3.3 Control of activity

We have seen in Section 2.2.2 that the effect of depolarizing the muscle membrane is to cause a release of Ca^{2+} into the sarcoplasm, changing the Ca^{2+} concentration from 10^{-9} M to about 10^{-5} M. This change in Ca^{2+} concentration is able to activate the muscle from its relaxed state, when there is negligible interaction between the filaments (most relaxed muscles are readily extensible although some muscles, such as insect flight muscle, show high resistance to stretch even in the relaxed state) and very low metabolic activity, to its maximally active state in which the muscle contracts and the metabolic activity of the cell increases by some 1000-fold. Two different ways of controlling this remark-able change in activity have been discovered – the one mediated via the A filament, and the other via the I filament – and different muscles have either one or the other or in some cases both types of control mechanism present. Figure 2.13 illustrates the presence of these control mechanisms in different animal groups.

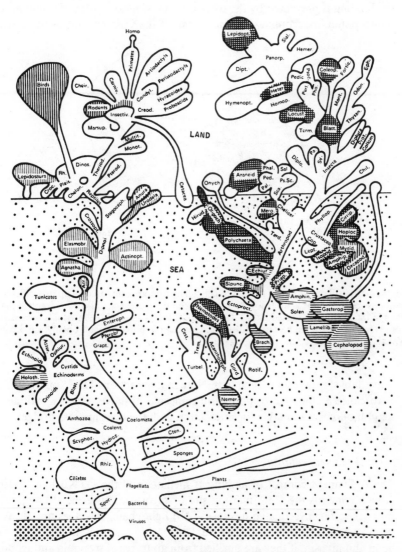

Fig. 2.13 Evolutionary tree to illustrate the presence of muscles with Ca^{2+} control systems on the A filament (horizontal shading), the I filament (vertical shading) or both filaments (cross-hatching). From Lehman and Szent-Gyorgyi (1975).

(i) *I filament control*

Contraction of muscle takes place when, in the presence of Mg-ATP, myosin is able to interact with actin. I filament control is mediated by the tropomyosin and troponin. The effect of Ca^{2+}, as indicated by X-ray diffraction studies, is to cause the tropomyosin to shift its position in the groove as illustrated in Fig. 2.14, the implication being that if the binding of myosin is located as indicated in the figure, then in the absence of Ca^{2+}, the binding is prevented by the tropomyosin, whereas with Ca^{2+}, bound to the troponin, the tropomyosin moves out of the way, and the binding site

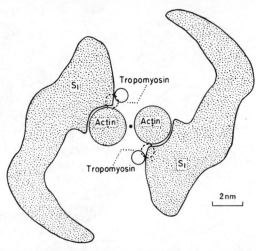

Fig. 2.14 Control of binding of myosin to actin. The tropomyosin can take up two different positions relative to the actin (shown by the full and the dashed circles); it is suggested that when in the position indicated by the dashed circle, the S1 head of the myosin is prevented from binding to the actin because of the steric restraint imposed by the tropomyosin. From Wakabayashi, Huxley, Amos and Klug (1975).

is available for myosin to bind to it. Note that the action of Ca^{2+} on troponin is to cause a complete tropomyosin molecule to move, thereby making 7 actin monomers available for interaction with the myosin cross bridges.

(ii) *A filament control*

The A filament control of activity is mediated by the DTNB light chain (which contains the Ca^{2+} binding site of myosin). However, although all muscles so far investigated have a DTNB light chain, this does not always seem to be effective in mediating Ca^{2+} control in that muscle. Thus, rabbit actin and myosin, if they are purified from all traces of troponin and tropomyosin, are not sensitive to changes in the Ca^{2+} concentration, even though the DTNB light chain is still present (under these conditions rabbit actomyosin is fully active irrespective of the Ca^{2+} concentration). Frog muscle also is one in which only the I filament control is effective in mediating contraction/relaxation. Despite this, there are structural changes in the A filament when the Ca^{2+} concentration is raised. In frog muscle which has been extended so that there is no overlap between the thick and thin filaments, X-ray diffraction studies show that upon stimulation of the muscle the cross bridges change their orientation (Hazlegrove, 1975).

2.4 The mechanism of contraction

We have already seen that a complete muscle is composed of fibres which may not be identical. Obviously, in order to investigate the fundamental mechanism of contraction it is essential to experiment with a preparation which does not have such inhomogeneities, and for this reason most experiments nowadays are performed on single fibres. In principle, it would be convenient to use a smaller unit than this, but there are a number of reasons why this has not been done with any great success. (1) Technical problems with apparatus. Although these problems will certainly be overcome the present range of tension transducers are not sensitive enough to work with appreciably smaller tensions than are produced by single fibres, while still maintaining the high frequency response required. (2) Technical problems with muscle. The smallest *complete* unit for contraction is the fibre, since the natural control mechanism for the release of Ca^{2+} is initiated by membrane potential changes in the sarcolemma or outer membrane of the cell. Thus, if it is desired to make use of this natural method of control, it is essential to work with a single whole fibre.

It is possible to disrupt the outer membrane system of the fibre, and to work either with a complete fibre whose membrane has been destroyed, or to subdivide the fibre and work with smaller units. There are various means of doing this, including removing it mechanically with fine needles, disrupting it with glycerol or making it permeable to most substances with EDTA. The advantage of doing this is that the experimenter can now control the concentration of chemical species surrounding the myofibrils, circumventing the cell's natural control mechanisms which tend to maintain constant concentrations of most substances. Although a few of the experiments described have used these preparations, most of the experiments of this section were done on intact fibres.

2.4.1 Passive muscle

Relaxed (i.e. unstimulated) muscle is characterized by being flaccid and in most cases, when at the normal length in the animal, exerting very little tension. It can be readily extended. In fact, recent work on frog muscle (Hill, 1968) has shown that there is a small tension associated with the relaxed state, but this tension is so small that it is irrelevant to the functioning of the muscle in the animal (although it is of considerable interest to the physiologist attempting to elucidate the mechanism of contraction in muscle).

If relaxed muscle is stretched, then it will develop passive tension. The stiffness of different types of relaxed muscle is very different. In particular, insect flight muscles are very stiff by comparison with other types of muscle; Fig. 2.15 gives some examples. With the exception of the insect flight muscles it is thought that the mechanical properties of the relaxed muscle are almost entirely due to the membrane systems of the muscle. Thus, stretching frog muscle to twice its body length stretches the membrane systems which act in a passive way. Of course, in intact vertebrate muscles, as opposed to single

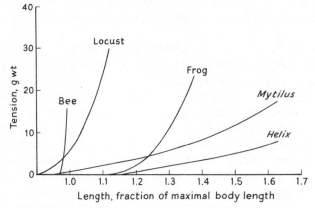

Fig. 2.15 Resting tension-length curves from a variety of muscles. From Hanson and Lowy (1960).

fibres dissected out from a whole muscle, there are also connective tissue sheaths around bundles of fibres, which increase the stiffness of the relaxed muscle.

The high resting elasticity of insect flight muscles, in particular the asynchronous or fibrillar muscles of certain orders of insect (Pringle, 1967), is thought to be due to connections between the A filaments and the Z line (Auber and Couteaux, 1963; White and Thorson, 1973; Trombitas and Tigyi-Sebes, 1974).

2.4.2 Contractile force – evidence for involvement of cross bridges

In 1957 A. F. Huxley published a paper which included a model for how contraction in muscle might be developed by cross-links between the two sets of filaments. This paper was written before such cross-links, which are now universally known as cross bridges (XB), had been observed by the electron microscope. Since that time several lines of evidence, of which the four most important are given below, have suggested that these cross bridges are responsible for the muscle force.

40

(i) *Tension-length diagrams of active muscle*

The force that a whole muscle can develop is strongly dependent upon its length. Such a curve for cat soleus muscle is shown in Fig. 2.16. The problem with whole muscles for investigating such relationships is that inhomogeneities in the fibres will mask out details of the curves, although of course from the animal's point of view, it is the whole muscles which count. Furthermore, even with single fibres it was found that the response of sarcomeres near the end of the fibres was different from that of sarcomeres in the centre. For this reason Gordon, A. F. Huxley and Julian (1966a) developed an apparatus known as the spot-follower, for investigating the mechanics of

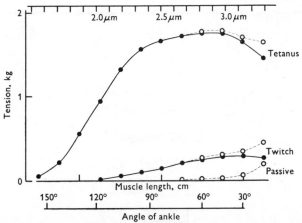

Fig. 2.16 Tension-length curves obtained from whole cat soleus muscle. The upper continuous line shows the tetanic tension, and the lower continuous line the peak tension in a twitch. The lowest dashed line is the passive curve. The upper dashed lines show the total tensions during a twitch or tetanus (i.e. they are the full lines to which the passive tension has been added). The upper scale indicates the sarcomere length. From Rack and Westbury (1969).

small regions in the centre of a single fibre. Using this apparatus they obtained the curve shown in Fig. 2.17a for the relationship between active tension and sarcomere length. Active tension is defined as the difference between the full tetanic tension obtained at that sarcomere length and the tension in the unstimulated muscle. Three particular features of this curve are of importance: (a) The finding that very little or no tension was developed in the fibres for sarcomere lengths greater than 3.7 μm. (b) The linear increase of tension with decreasing sarcomere length between 3.7 μm and 2.25 μm. (c) The plateau of tension between 2.25 and 2.0 μm. These three findings correlate very well with the hypothesis that tension is generated by cross bridges. Figure 2.17b shows the appearance of the sarcomeres at the sarcomere lengths indicated by numerals in Fig. 2.17a. The lengths of the filaments are such that when the sarcomere length is about 3.7 μm, there is just no overlap between the A and I filaments. The cross bridges cannot make contact with the I filament, and so no tension is generated. As the sarcomere length is decreased there is a linear increase in tension, corresponding to a linear increase in overlap. Maximum tension is reached at 2.25 μm, at which point there is maximal overlap of the cross bridges by

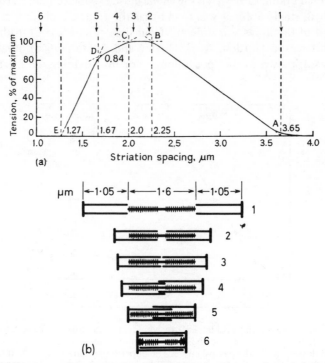

Fig. 2.17 (a) Summary of the isometric, active tension–length curve from a single frog semi-tendinosus muscle fibre using the spot follower apparatus as described in the text. From Gordon, Huxley and Julian (1966b). (b) Appearance of the sarcomere at the striation spacings corresponding to the numbered points of part a.

the filament, although not maximal filament overlap. Between 2.25 μm and 2.0 μm there is a plateau of tension. Between these sarcomere lengths the I filaments are moving across the clear zone in the centre of the A filament. The correspondence of this plateau of tension with the structural lack of the possibility of more cross bridge formation is perhaps the best single piece of evidence implicating the role of cross bridges in producing tension.

For sarcomere lengths shorter than 2.0 μm several phenomena occur to reduce the magnitude of the tension developed. First, the I filaments interfere with one another (as indicated in position 4 of Fig. 2.17b). Next the I filament from one side starts to interact with the cross bridges on the 'wrong' side of the A filament (position 5). Whether the cross bridges can actually

42

bind to the I filament from the opposite half-sarcomere, or whether the 'proper' I filament is prevented from binding is not known. Next the A filament abuts the Z-line, and at shorter sarcomere lengths starts to crumple (in some species it can pass through into the adjacent sarcomere), and finally, at very short sarcomere lengths the I filaments hit the Z line. All these occurrences reduce the ability of the muscle to develop tension. It has also been found that the activation processes (the release of Ca^{2+}) from the sarcoplasmic reticulum are reduced for sarcomere lengths less than about 1.6 μm, thereby also reducing the tension generated.

(ii) *Electron micrographs of muscle*

Micrographs indicate the presence of cross bridges, and furthermore, work using electron microscopy and X-ray diffraction together, indicates that these cross bridges can take up different conformations depending upon the particular nucleotide (or lack of) bound to the muscle. Thus, Reedy, Holmes and Tregear (1965) demonstrated that in the presence of ATP the cross bridges were detached from the I filament, whereas in the absence of ATP the cross bridges were bound, and angled at approximately 45° to the axis of the filaments. More recently, Rodger, Marston and Tregear (1976) have demonstrated a third conformation of the cross bridges – attached and perpendicular to the filament axis – in the presence of the ATP analogue, AMP.PNP. Biochemical kinetic studies have indicated that the predominant biochemical state to be expected in the presence of this analogue is intermediate in the states between ATP binding and ADP and inorganic phosphate release by the actomyosin system, and it is tempting therefore to suggest that the three conformational states observed structurally correspond to three states in a mechanical cycle of activity of the cross bridges, involved in tension generation (see Section 2.4.4).

(iii) *X-ray diffraction studies of actively contracting muscle*

These indicate a change from the ordered, relaxed, pattern found in unstimulated muscle, to a much more disordered pattern (Huxley, 1971). This is what would be expected if, in active muscle, the cross bridges are going through a cycle of shape changes, in an unsynchronized fashion. Similar types of results are obtained by intensity fluctuation spectroscopy (Carlson, 1975).

(iv) *Biochemical studies*

It is known from biochemical studies that the primary source of chemical energy for contraction is ATP (Davies, 1964). The cross bridges (i.e. subfragment 1 of the HMM) contain both the actin-binding and the ATPase sites of the muscle. It seems to be the case therefore that the cross bridge is involved in the conversion of chemical energy to mechanical energy in muscle, although it does not necessarily follow that the active force in muscle is generated by mechanical attachment of the cross bridge to the I filament.

2.4.3 Steady-state properties of vertebrate striated muscle

(i) The isometric tension versus sarcomere length relationship has been covered in the previous section.

(ii) *Force–velocity (P–V) curves*

When muscle contracts against a constant force (i.e. an isotonic contraction), the speed of contraction depends upon the magnitude of that force; the smaller the force, then the faster the contraction. The quantitative relationship between force and velocity has been fitted by a number of equations, the best known of which is A. V. Hill's characteristic equation for muscle shortening

$$V(P + a) = b(P_0 - P) \tag{2.1}$$

in which P is the force, P_0 the isometric tension developed by the muscle, V the velocity of shortening and a and b are constants whose values are different for different muscles. The constant a has the units of force, and for most vertebrate muscles its value is in the range $0.15P_0$–$0.25P_0$. The constant b has units of velocity. The maximal shortening velocity is obtained when $P = 0$; from Hill's equation $b = (a/P_0) V_{max}$. Hill's equation provides a good description of the relationship between force and velocity only for muscle shortening. Figure 2.18c gives an example of the dependence of force and velocity for both muscle shortening and extension (obtained when forces greater than the isometric tension are applied).

One point must be made about Hill's equation and force–velocity curves. This relationship is a steady-state relationship, and is only the situation that occurs once the muscle settles down to shortening at a constant velocity after applying whatever experimental procedure is being used to measure it. The steady-state takes about 30–50 ms to be attained in frog semitendinosus muscle at 0°C. If the conditions in the muscle are not steady-state conditions, for example if the load on the muscle is changing (as occurs in every loco-motory contraction of a muscle in an animal), then the force–velocity curve, as measured experimentally cannot strictly be considered to apply. Exactly

---

Fig. 2.18 (a) Force–velocity curve as predicted by Hill's equation with $a = 0.2P_0$. (b) Force–power curve derived from Hill's equation, and force–efficiency relationship as determined by Hill (1964), both for values of $a = 0.2P_0$. (c) Force–velocity curve in the region of P_0. Redrawn from Aubert (1956).

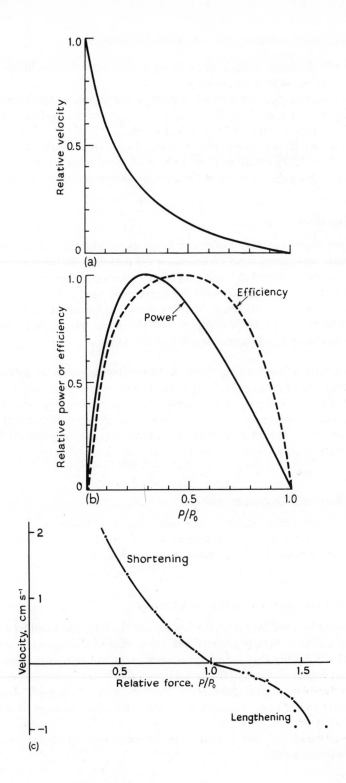

(a)

(b)

Efficiency

Power

P/P_0

(c)

Shortening

Lengthening

Velocity, cm s^{-1}

Relative force, P/P_0

how valid the relationship is depends upon how close the conditions are to having a constant load on the muscle.

The power output from a muscle is given by the product of the force and the velocity of shortening. Power output can be determined directly from the force–velocity curves, and for the force–velocity relationship of Fig. 2.18a the power versus force relationship is as shown in Fig. 2.18b. The curve has a maximum, which from Hill's equation is at a value of force of $(a^2 + aP_0)^{1/2} - a$; for $a = 0.2P_0$ this gives the value of force for maximum power output of $0.29P_0$.

(iii) *Rate of energy use*

A very important feature of muscle contraction is that the rate at which the muscle uses energy is dependent upon the nature of the contraction. Fenn (1924) demonstrated that the total energy released in the form of heat plus mechanical work is greater when the muscle is allowed to shorten, than when the muscle is held isometric. This phenomenon, known as the Fenn effect, demonstrates that the biochemical events in the muscle can be regulated by the mechanical constraints imposed on the muscle.

If we define efficiency as the ratio of the mechanical power to the total rate of energy release by the muscle, then the efficiency of muscle plotted against load has the form shown in Fig. 2.18b. Whereas maximum power output is obtained at a load of about $0.3P_0$, maximum efficiency is obtained at a load of about $0.5P_0$. The efficiency curve has a broad top, and only falls below 90% of the peak efficiency for loads less than $0.24P_0$ and greater than $0.7P_0$.

2.4.4 *How do cross bridges work?*

Much of our understanding of the way that cross bridges work originates from the experimenters and arguments of A. F. Huxley and his collaborators. Many of the points below are discussed in Huxley (1974).

(i) *XB act as independent force generators*

This can be concluded from the finding that the force produced by a muscle (Fig. 2.17a) is directly proportional to overlap of the filaments, and also from the finding that the maximal velocity of contraction is independent of the degree of overlap (Gordon, Huxley and Julian, 1966b). The concept of a tug-of-war team, made up of identical members, is useful here. The tension such a team can exert is proportional to the number of members of that team, but if the rope between two teams is cut, then the maximum rate at which the team can move is the speed with which any individual members can run.

46

(ii) *XB act in a cyclical manner*

The total shortening obtainable in a contraction of, say, frog muscle is from a sarcomere length of about 3.5 μm to one of about 1.5 μm. Each half-sarcomere has thus shortened about 1 μm. Since the total length of an XB is about 50 nm, the overall length change must be produced by many individual cross bridge actions. We can think of a complete cross bridge cycle being at least the four steps *attach – change shape and so produce force – detach – recover* as shown in Fig. 2.19.

Whether muscle shortening occurs or not depends upon the magnitude of the load on the muscle. If the combined action of all the cross bridges active at any instant of time is to produce a force greater than that of the load, then

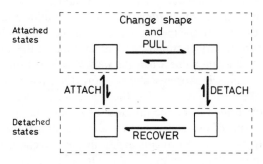

Fig. 2.19 Suggested cycle of cross bridge activity.

the filaments move relative to one another so as to produce shortening. After detaching from an actin monomer, the cross bridge can then attach to a new one further down the A filament, because in the time before the attachment occurs the filaments have moved. Note, of course, that in this case the filaments will be moving whilst any particular cross bridge is remaining attached, so that this attached cross bridge will be distorted by the movement, and the tension that it exerts will almost certainly change. A striking film of these ideas using annotated drawings has been made by Tregear (1971).

If the load on the muscle just balances the force produced by all the cross bridges which are attached and exerting tension at that instant then there will be no relative movement of the filaments; in this case a cross bridge will probably attach to the same actin monomer from which it previously detached. The muscle is then using up energy to stay still. Likewise, if the load on the muscle is more than the cross bridges can bear, then the filaments will slide so as to produce an extension of the sarcomere, thereby distorting the cross bridges in the opposite direction to that produced during shortening. The cross bridge will then attach to an actin monomer further up the filament.

(iii) *XB hydrolyse ATP in a cyclical manner*

An assumption commonly made, but for which there is as yet no more than circumstantial evidence, is that one (or possibly more than one) ATP molecule is hydrolysed for each cycle of mechanical activity. It is supposed that the cycle of activity undergone by the actomyosin acting as an ATPase is one and the same as the cycle of activity of the actomyosin acting as a mechanical force producer. If this is the case, then the rate of energy utilization by the muscle is proportional to the rate of cycling of the bridges.

Figure 2.19 showed the processes undergone in our hypothetical cycle. We imagine that the boxes represent distinct biochemical and mechanical states of the actin-myosin (as mentioned in Section 2.4.2 and in paragraph (v) of this section), and that the transitions between these states can be characterized in terms of normal biochemical rate constants.

In order to account for the finding that the rate of utilization of energy by the muscle increases when the muscle shortens, and decreases when the muscle is extended, it is necessary to suppose that at least one of the rate constants in the cycle has a moderate value when the muscle is held isometric, increases its value when the muscle shortens, and so allows the cross bridge to complete its working stroke, and decreases when the muscle is extended thereby causing the action of the bridge to be the opposite of its 'normal' operation during shortening. In terms of our concept of cross bridge distortion, we expect the rate constant to be increased for distortion in one direction, and decreased for distortion in the other direction.

(iv) *How is force produced?*

The cycle of Fig. 2.19 suggests, rather glibly, that the cross bridge attaches to the actin in one state, and that it then proceeds to a new state, changing shape in the process, thereby generating a force between its two attachment points on the two filaments.

Huxley and Simmons (1971, 1973; see also Huxley, 1974) have performed a series of experiments on isometrically contracting frog semitendinosus muscle, in which they have applied very rapid step changes of length to the muscle, and followed the time course of the changes in tension. They have been able to show that the initial, very rapid, changes of tension that occur are due to the activity of attached bridges before significant detachment or attachment has had time to take place, and they interpret their results in terms of a re-arrangement of the attached cross bridge population between two, or probably more, attached states; their model is this just the attached part of the cycle of Fig. 2.19.

Figures 2.20a, b show two sets of responses of frog muscle to such step changes of length. Figure 2.20a is on a slow time scale. The initial rise of tension

48

is that of the muscle developing tension when it is stimulated. The muscle is held isometric until the sudden tension change occurs, at which time the length has been changed. The upper records are of step increases of length,

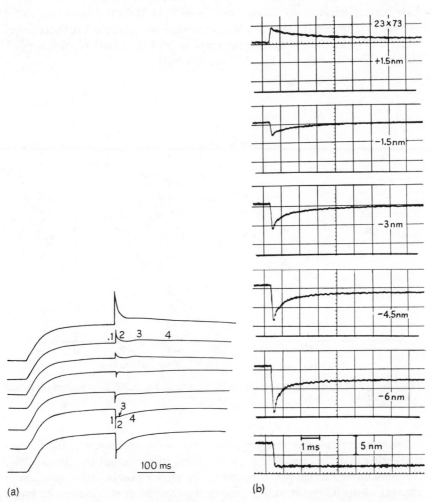

(a)   (b)

Fig. 2.20 Effect of applying step changes of length to an isometrically contracting frog semitendinosus fibre, using the spot-follower apparatus to make measurements on the central region of the fibre only. (a) Records on a slow time scale (Huxley and Simmons (1973)). (b) Records on a much faster time scale (Huxley, 1974).

and the lower records of step decreases of length. The magnitudes of the length changes progress smoothly from top to bottom. Figure 2.20b is of a similar set of records, but on a very much faster time scale, showing the tension records only at the step. Two of the traces in Fig. 2.20a have the numbers 1 to 4 drawn

49

on them, to indicate the four phases of the response that Huxley and Simmons distinguish. It will be shown below that phases 1 (which occurs at the same time as the length change and so is said to be an 'elastic' change) and 2 (a 'recovery' phase towards the isometric tension before the step) are too rapid for there to have been appreciable attachment and detachment. If the tensions at the end of the step (T_1), before any recovery has taken place, and at the end of the recovery phase 2 (T_2) are plotted against the magnitude of the length change applied, then the records shown in Fig. 2.21 are obtained.

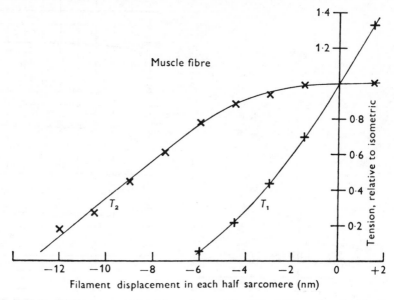

Fig. 2.21 T_1 and T_2 curves obtained from the records of Fig. 2.20b as described in the text (Huxley, 1974).

It is possible that there are large contributions to the responses from structures other than the cross bridges. If the filaments are compliant, for example, then sudden length changes to the ends of the muscle may be taken up as length changes in the filaments, rather than in the cross bridges. This can be tested experimentally by repeating the experiment at two different sarcomere lengths. If the responses seen are due to cross bridge activity, then the magnitude of the responses should be directly proportional to the degree of overlap (see Fig. 2.22a). If the responses are due to the filaments or Z discs stretching, or to other causes, then there will not be this simple proportionality. Huxley and Simmons (1971b) found that the responses of phases 1 and 2 were directly proportional in magnitude to the isometric tension developed at that sarcomere length (Fig. 2.22b), and this to the extent of overlap, establishing that at least about 80% of the response was due to the cross bridges.

It has been stated above that phases 1 and 2 of the transient response occur too rapidly for bridge attachment or detachment to have taken place to any appreciable extent. An alternative hypothesis is that the rate constant for

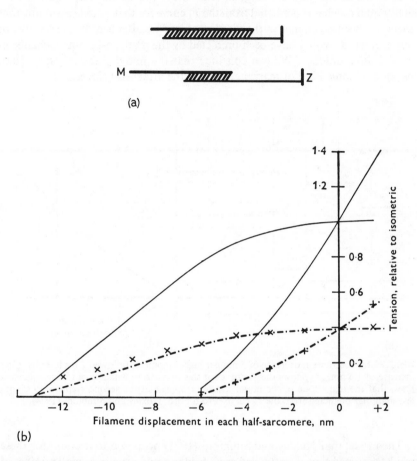

Fig. 2.22 (a) Effect of changing sarcomere length on number of cross-bridges overlapped by an I filament. (b) T_1 and T_2 curves at two sarcomere lengths. The full lines are redrawn from Fig. 2.21 and were measured at a sarcomere length of 2.2 μm. The crosses are the experimental points at a sarcomere length of 3.1 μm. At this sarcomere length the isometric tension was 39% of its value at 2.2 μm. The interrupted curves are the solid curves scaled down by 0.39. From Huxley (1974).

attachment is sufficiently high that the recovery phase 2 following a step decrease of length is accounted for in terms of new population of bridges attaching very rapidly. Such a model has been worked out quantitatively (Podolsky and Nolan, 1973). These two models (accounting for the transients in terms of (1) a rearrangement of bridges between different attached states in which the bridges generate different tensions and (2) a change in the number of attached bridges) can be tested by applying a second, test step to the muscle during the recovery process. Such an experiment is shown in Fig. 2.23. The magnitude of the elastic change in tension for the second step is almost exactly equal to what is predicted from the T_1 curve for that muscle assuming that there has been no change in the number of bridges attached. It is certainly not any greater, as would have been predicted by the model assuming a change in cross bridge number. We can conclude that the initial transients take place because of some kind of rearrangement of the attached bridges.

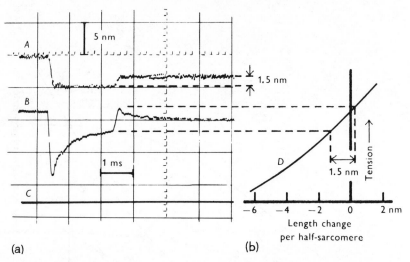

(a)   (b)

Fig. 2.23 (a) Step length decrease of 5 nm per half-sarcomere, followed 2 ms later by a step increase of 1.5 nm. (b) Shows the T_1 curve for this fibre, from which it can be seen (dashed lines) that the magnitude of the initial, elastic, response at the second step is very slightly less than that predicted from the T_1 curve. From Ford, Huxley and Simmons (1974).

The model that Huxley and Simmons (1971) proposed to account for these results assumed that the attached cross bridge could exist in a small number of different mechanical states (maybe corresponding to the attached states of Fig. 2.19), and that the effect of stretching the muscle thereby applying strains (or what I have called 'distortions') to the cross bridges, is to change the equilibrium between these states because the energy given to the cross bridge in stretching it changes the activation energy for the transfers from one state to the next. This model was able to account both for the shapes of the T_2 and T_1 curves, and also for the different times the tension took to reach the plateau level of phase 3 (from Fig. 2.20 it will be seen that the response is much more rapid for large decreases of length applied to the muscle than for small decreases of length). Much fuller explanations of these results and the model can be found in Huxley and Simmons (1971, 1973), Thorson and White (1973) and Simmons and Jewell (1974). The main point to be understood is that *rearrangement between attached states* of the cross bridge can account for the transients observed, and this model is a model for the way in which cross bridge force is developed—by a series of transitions between attached states in which each succeeding state strains the cross bridge more, exerting a sliding force between the two sets of filaments.

(v) *The XB cycle*

From the mechanical experiments the present level of understanding of cross bridges is thus in terms of a cycle of activity in which there is a fast equilibrium set up between two or more attached states of the cross bridge, and in which the rate constants for attachment and detachment of the bridges from the I filament are rather slower. We also expect these attachment and detachment rate constants to be dependent upon the precise alignment of the cross bridge to the actin monomer.

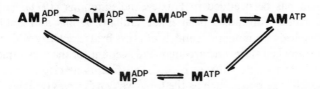

Fig. 2.24 Possible simplified biochemical cycle of activity of actomyosin during the hydrolysis of ATP. The arguments for this sequence are given by White and Thorson (1973). The cycle normally proceeds clockwise with the usual concentrations of nucleotides found in muscle. The two AM.ADP.P states have different conformations.

The biochemical events underlying the hydrolysis of ATP by both myosin and by actomyosin (i.e. myosin in the presence of actin) have been much studied. A simplified version of the likely sequence of events is shown in Fig. 2.24. An interesting point about the hydrolysis of ATP by myosin in the presence of actin is that the actin is associated and dissociated once each cycle of hydrolysis. This is different from what is normally thought to be the mechanism for an enzyme activator to work (it is normally thought to remain bound to an 'activating site' throughout the cycle, and indeed for many cycles). Of course this association and dissociation is exactly what is required to fit in with the mechanical cycle, and it is therefore commonly believed that the mechanical cycle and the biochemical cycle are one and the same. The cycle shown in Fig. 2.24 is certainly not complete. This area of the experimental study of muscle is probably the fastest developing area at the moment. White and Thorson (1973) and Taylor (1973) give reviews of the biochemical work. More recent papers by those doing this work are Bagshaw and Trentham (1974), Bagshaw *et al.* (1974), and Trentham, Eccleston and Bagshaw (1976).

2.4.5 *Some comments on the cross bridge cycle*

In the preceding section, we discussed the evidence for the cyclical behaviour of the cross bridge, and showed how its mechanics, down to a time resolution of better than 1 ms, can be accounted for in terms of a cycle of the form of Fig. 2.19. From the transient experiments we can conclude that the rate constants for the transitions between attached states of the cross bridge are much faster than those of the attachment or detachment processes; this means that if we are considering events that occur more slowly than those of the initial phases of the transient response we can consider that the attached cross bridges have taken up their epuilibrium distribution between the attached states.

It seems likely that in the not too distant future it will be possible to build a model of cross bridge behaviour which will be able to acount for the mechanical results of muscle down to the time resolution of the fastest events that have been recorded. An attempt to do this, in which the capability of attachment and detachment has been added to Huxley and Simmons (1971) model of the attached cross bridge, has been made (Julian, Sollins and Sollins, 1974). There have been a number of models of cross bridge activity which are based upon one attached state, and are unable to account for the very rapid transient response, but which nonetheless provide satisfactory descriptions of the slower events which take place, such as the force–velocity curves (Huxley, 1957 is the basis for these models).

These models have certain features in common which are of interest to us here.

(1) As the shortening velocity of the muscle is increased, the likelihood of a detached cross bridge moving past an actin monomer without having time to attach to it and produce its working stroke is increased. Thus, the greater the shortening speed the smaller the fraction of bridges which is attached at any time.

(2) At high shortening speeds it becomes increasingly likely that an attached bridge will be distorted to such an extent before it has time to detach that it starts to push instead of pull. The maximum obtainable shortening velocity is then obtained when the contribution of all the cross bridges which are pushing exactly balances the contribution from those which are pulling. To produce higher shortening velocities it is necessary to increase the rate of detachment of bridges in the pushing regions of distortion.

(3) The probability of detachment for bridges which have been distorted by sarcomere extension is less than for those with no distortion; the detachment probability increases for bridges distorted by sarcomere shortening. Thus, if the bridge can use the mechanical energy stored within it by helping to produce sliding it detaches more readily than if this does not happen. In particular,

note that if the muscle is extended and the energy of this extension is stored in the 'lengthening' distortions of cross bridges, then this energy is only available for further use for so long as those cross bridges remain attached. If they detach the energy is lost as heat, whilst if they remain attached the energy remains stored as in an elastic element and can be returned to the load if subsequent shortening occurs. The reduction in the detachment rate of bridges distorted by sarcomere extension obviously helps to store this energy.

References

Aidley, D.J. (1971) *The Physiology of Excitable Cells*, London: Cambridge University Press.

Alexander, R.McN. (1968) *Animal Mechanics*, London: Sidgwick and Jackson.

Auber, J. and Couteaux, R. (1963) Ultrastructure de la strie Z dans des muscles de dipteres. *J. Microscopie* **2**, 309–324.

Aubert, X. (1956) Le couplage energetique de la contraction musculaire. These d'Agregation, Universite Catholique de Louvain.

Bagshaw, D.R. and Trentham, D.R. (1974) The characterization of myosin-product complexes and of product-release steps during the magnesium-ion-dependent ATPase reaction. *Biochem. J.* **141**, 331–350.

Bagshaw, C.R., Eccleston, J.F., Eckstein, F., Goody, R.S., Gutfreund, H. and Trentham, D.R. (1974) The magnesium-ion-dependent ATPase of myosin. Two-step processes of ATP association and ADP dissociation. *Biochem. J.* **141**, 351–364.

Carlson, F.D. (1975) Structural fluctuations in the steady state of muscular contraction. *Biophys. J.* **15**, 633–649.

Close, R. (1967) Properties of motor units in fast and slow muscles of the rat. *J. Physiol.* **193**, 45–55.

Davies, R.E. (1964) Adenosine triphosphate breakdown during single muscle contractions. *Proc. R. Soc., Ser. B* **160**, 480–484.

Dudel, J. and Kuffler, S.W. (1961) Mechanism of facilitation at the crayfish neuromuscular junction. *J. Physiol.* **155**, 530–542.

Fenn, W. (1924) A quantitative comparison between the energy liberated and the work performed by the isolated sartorius muscle of the frog. *J. Physiol.* **58**, 175–203.

Ford, L.E., Huxley, A.F. and Simmons, R.M. (1974) Mechanism of early tension recovery after a quick release in tetanised muscle fibres. *J. Physiol.* **240**, 42P–43P.

Franzini-Armstrong, Clara (1970) Studies of the triad: I. Structure of the junction in frog twitch fibres. *J. Cell. Biol.* **47**, 488–499.

Gordon, A.M., Huxley, A.F. and Julian, F.J. (1966a) Tension development in highly stretched vertebrate muscle fibres. *J. Physiol.* **184**, 143–169.

Gordon, A.M., Huxley, A.F. and Julian, F.J. (1966b) The variation in isometric tension with sarcomere length in vertebrate muscle fibres. *J. Physiol.* **184**, 170–192.

Hanson, J. and Lowy, J. (1960) Structure and function of the contractile apparatus in the muscles of invertebrate animals. In *Muscle*, Volume I, ed. Bourne, G.H. London, New York: Academic Press.

Haslegrove, J. (1975) X-ray evidence for conformational changes in the myosin filaments of vertebrate striated muscle fibres. *J. mol. Biol.* **92**, 113–144.

Heistracher, P. and Hunt, C.C. (1969) The relation of membrane changes to contraction in twitch muscle fibres. *J. Physiol.* **201**, 589–611.

Hellam, D.C. and Podolsky, R.J. (1969) Force measurements in skinned fibres. *J. Physiol.* **200**, 807–819.

Hill, A.V. (1964) The efficiency of mechanical power development during muscular shortening. *Proc. R. Soc. Ser. B.* **159**, 319–324.

Hill, D.K. (1968) Tension due to interaction between the sliding filaments in resting muscle. The effect of stimulation. *J. Physiol.* **199**, 637–684.

Huxley, A.F. (1957) Muscle structure and theories of contraction. *Prog. Biophys. mol. Biol.* **7**, 255–318.

Huxley, A.F. (1974) Muscular Contraction. *J. Physiol.* **243**, 1–43.

Mechanics and energetics of animal locomotion

Huxley, A.F. and Niedergerke, R. (1954) Structural changes in muscle during contraction. *Nature* **173**, 971–973.

Huxley, A.F. and Simmons, R.M. (1971a) Proposed mechanism of force generation in striated muscle. *Nature* **233**, 533–538.

Huxley, A.F. and Simmons, R.M. (1971b) Mechanical properties of the cross bridges of frog striated muscle. *J. Physiol.* **218**, 59P–60P.

Huxley, A.F. and Simmons, R.M. (1973) Mechanical transients and the origin of muscular force. *Cold Spring Harbor Symp. Quant. Biol.* **37**, 669–680.

Huxley, H.E. (1971) The structural basis of muscle contraction. *Proc. R. Soc., Ser. B* **178**, 131–149.

Huxley, H.E. and Hanson, J. (1954) Changes in the cross-striations of muscle during contraction and stretch, and their structural interpretation. *Nature* **173**, 973–976.

Julian, F.J., Sollins, K.R. and Sollins, M.R. (1974) A model for the transient and steady-state mechanical behaviour of contracting muscle. *Biophys. J.* **14**, 546–562.

Katz, B. (1969) *The release of neural transmitter substances.* Liverpool: Liverpool University Press.

Lang, F., Sutterlin, A. and Prosser, C.L. (1970) Electrical and mechanical properties of the closer muscle of the Alaskan king crab, *Paralithodes camtschatica*. *Comp. Physiol. Biochem.* **32**, 615–628.

Lehman, W. and Szent-Gyorgyi, A.G. (1975) Regulation of muscular contraction: Distribution of actin control and myosin control in the animal kingdom. *J. gen. Physiol.* **66**, 1–30.

Marston, S.B., Rodger, C.D. and Tregear, R.T. (1976) Changes in muscle crossbridges when β,γ-imido ATP binds to myosin. *J. mol. Biol.* **104**, 263–276.

Offer, G. (1974) The molecular basis of muscular contraction. In: *Companion to Biochemistry: Selected Topics for Further Study*, ed. Bull, A.T., Lagnado, J.R., Thomas, J.O. and Tipton, K.F. London: Longman.

Podolsky, R.J. and Nolan, A.C. (1973) Muscle contraction transients, cross-bridge kinetics and the Fenn effect. *Cold Spring Harbor Symp. Quant. Biol.* **37**, 661–668.

Pringle, J.W.S. (1967) The contractile mechanism of insect fibrillar muscle. *Prog. Biophys. mol. Biol.* **17**, 1–60.

Prosser, C.L. (ed.) (1973) *Comparative Animal Physiology.* Philadelphia: Saunders.

Rack, P.H.M. and Westbury, D.R. (1969) The effects of length and stimulus rate on tension in the isometric cat soleus muscle. *J. Physiol., Lond.* **204**, 443–460.

Reedy, M.K., Holmes, K.C. and Tregear, R.T. (1965) Induced changes in orientation of the cross bridges of glycerinated insect flight muscle. *Nature* **207**, 1276–1280.

Ridgway, E.B. and Gordon, A.M. (1975) Muscle activation: Effects of small length changes on calcium release in single fibres. *Science* **189**, 881–883.

Simmons, R.M. (1976) *Muscle*, Outline studies in Biology; London: Chapman and Hall.

Simmons, R.M. and Jewell, B.R. (1974) Mechanics and models of muscular contraction. In: *Recent Advances in Physiology*, **9**, 87–147.

Squire, J.M. (1973) General models of myosin filament structure. III. Molecular packing arrangements in myosin filaments. *J. mol. Biol.* **77**, 291–323.

Squire, J.M. (1974) Symmetry and three-dimensional arrangement of filaments in vertebrate striated muscle. *J. mol. Biol.* **90**, 153–160.

Taylor, E.W. (1973) Mechanism of actomyosin ATPase and the problem of muscle contraction. *Current Topics in Bioenergetics*, **5**, 201–231.

Tregear, R.T. (1971) What makes muscles pull: The structural basis of contraction. *Wiley 16 mm Physiology Film Series.* London: John Wiley.

Trentham, D.R., Eccleston, J.F. and Bagshaw, C.R. (1976) Kinetic analysis of ATPase mechanisms. *Quart. Rev. Biophys.* **9**, 217–281.

Trombitas, K. and Tigyi-Sebes, A. (1974) Direct evidence for C-filaments in flight muscle of honey bee. *Acta biochemica biophysica hungarica* **9**, 243–254.

Wakabayashi, T., Huxley, H.E., Amos, L.A. and Klug, A. (1975) Three dimensional image reconstruction of actintropomyosin complex and actin-tropomyosin-troponin I-troponin T complex. *J. mol. Biol.* **93**, 477–497.

White, D.C.S. and Thorson, J. (1973) The kinetics of muscle contraction. *Prog. Biophys. mol. Biol.* **27**, 173–255. Reprinted in Pergamon Series in the Life sciences under the same title.

3 Muscle energetics and animal locomotion

G. Goldspink

3.1 Introduction

Perhaps the most fascinating aspect of muscle is its ability to convert chemical energy into mechanical work. Certainly this subject of muscle has attracted the attention of many physiologists and biochemists, and there is now a considerable amount of knowledge available concerning the mechanism of muscular contraction although the detailed molecular mechanism involved in the generation of force still eludes us. In producing work from chemical energy, muscle, like any other machine for producing work, complies with the laws of thermodynamics.

The basic bioenergetic concepts that will be used throughout this chapter are efficiency and economy and it is important that the distinction between these two terms should be made at the onset. The term *efficiency* will be used in the thermodynamic sense in that the efficiency values give the amount of energy converted into external work from the total amount of energy available. The amount of energy used in producing work is sometimes determined from biochemical measurements rather than heat measurements. Therefore, strictly speaking, we cannot talk about thermodynamic efficiency, we have instead to use the term biochemical or mechanochemical efficiency, although for most purposes they can be regarded as meaning the same.

The term *economy* has to be introduced because muscles have other functions besides contracting isotonically and producing work. In certain circumstances muscles develop force without any appreciable shortening in length taking place: this is called isometric contraction. Indeed, most muscles in terrestrial animals are almost continually engaged in maintaining posture. This applies not only when the animal is standing but also to locomotion when energy is being stored in the elastic elements of the limb and the muscles are required to maintain a steady tension. This will be discussed more fully later in the chapter. When muscles are contracting isometrically they are not doing external work, however, we know that energy is still being used and we need some way of expressing this, hence we use the term economy. Economy is usually obtained by dividing the time–tension integral by the amount of energy, in other words by dividing the area under the contraction curve by the number of joules of energy or in some cases the number of high energy phosphates used.

3.2 Fuel for muscular activity

Carbohydrates, lipids and proteins may all be used to provide energy for muscular contraction. As a general rule, carbohydrates, principally glycogen, act as short-term energy supplies. Glycogen is used for short bursts of intensive activity. It is also used at the onset of any activity to allow time for the lipids to be mobilized. Lipids on the other hand act as the main fuel for sustained activities particularly those associated with migration. As an energy store, lipid has several advantages over either glycogen or protein; the main advantage being that lipid has twice the number of calories per gram than either anhydrous protein or glycogen. The deposition of glycogen is associated with a considerable amount of water and only about 1/8th of the energy available per unit weight when glycogen rather than lipid is used as the fuel (Weis-Fogh, 1966). The oxidation of lipid also yields more metabolic water than either glycogen or protein and this may be of particular importance in long range aerial or terrestrial migration. There is also the point that lipid is less dense than water and therefore it increases the buoyancy of aquatic animals. However, the breakdown of lipid to yield energy is an aerobic process and it is therefore not a suitable fuel for periods of very vigorous activity when the oxygen supplies of the tissue may be depleted. Under these conditions the rate of glycolysis is increased and the lactate that is produced enters the blood stream and exerts an inhibitory effect on mobilization of the lipid deposits.

Although glycolysis may proceed in the absence of oxygen this results in an accumulation of lactate which causes a drop in blood and muscle pH, eventually leading to fatigue. Some invertebrates and vertebrates, which are adapted to living under aerobic conditions, have the ability to use various anaerobic pathways as well as glycolysis and this results in the accumulation of end products such as succinate and alanine in addition to lactate and pyruvate (Hochachka *et al.*, 1973). This sort of mechanism is found in fish living at considerable depths and in lakes that are covered in ice for long periods of time. They also operate in diving mammals and reptiles such as diving turtles which may remain submerged for periods of longer than 24 h. Catabolism of carbohydrate by the alternative pathways is still incomplete and hence inefficient, however it does enable the animal to occupy an environmental niche which otherwise would not be possible.

Accumulation of fat reserves before migration has been studied by several authors who have attempted to predict the distance that can be covered for a given amount of fuel. This has been carried out for bird migration by Pennycuick (1969) who calculated that a 400 g pigeon would be able to fly twelve kilometres per gram of fat whereas a 3.7 g ruby-throated hummingbird would cover 900 kilometres per gram of fat. Pennycuick's calculations are based on the weight of the bird, aerodynamic parameters and power output. Some birds such as the Arctic tern migrate over distances of 2000 miles or more without eating. Nevertheless, this sort of distance can be accounted for in terms of deposited fat. Migrating birds may in fact deposit as much as their own fat-free body weight in fat before commencing their migration. Indeed, it has been shown that birds can quite accurately judge the amount of fat deposited before they commence their migration. However, it is not known what mechanism is involved.

The extent to which carbohydrate or lipid is used for muscular contraction varies from species to species and probably from season to season. For example, some insects use mainly lipids whilst others use carbohydrates (principally the blood sugar trehalose) as the main fluid for flight. It is not clear why some insects such as Diptera and Hymenoptera should use carbohydrate as the fuel for flight whilst others such as the Orthoptera and Lepidoptera should primarily depend on lipids. Those that use lipids have in fact to convert most of their ingested carbohydrate into lipid before it can be used for sustained flight and in any case all insects still have to use carbohydrate for the first few seconds of flight to allow time for the lipid to be mobilized. Interestingly, some insects can also catabolize amino acids particularly those which feed on blood or other material with a high protein content. Perhaps the best known example is the tsetse fly which catabolizes considerable quantities of proline (Bursell, 1963).

It is known that different types of muscle fibres have different abilities to catabolize lipids and to catabolize carbohydrates. In general the slow red fibres which are specialized to produce slow isotonic movements or sustained isometric contractions catabolize lipids. The white fibres, on the other hand, rely mainly on glycolysis because the oxidative mechanisms cannot possibly keep up with their rapid rate of energy utilization. For this reason, the amount of stored glycogen is often much higher in the fast white fibres. However, it is perhaps misleading to refer to these fibres as glycolytic and the slow fibres as oxidative because the slow fibres, as well as having a higher oxidation capacity, can usually also metabolize glycogen more rapidly than the fast white fibres. This is borne out by the study of Patterson and Goldspink (1973) who showed that both red and white myotomal muscle of fish will oxidize pyruvate (glycolytic intermediate) more rapidly than octonoate (lipolysis intermediate). However, the red muscle oxidized both substrates twice as rapidly as the white muscle.

It is generally believed that proteins are not used to provide energy for muscular activity except as a last resort. The utilization of proteins or indeed the utilization of lipids in very high quantities can result in the build-up of ketone bodies in the blood causing acidosis. The ketone bodies originate from the liver and are metabolized in the other tissues of the body. The capacity of these tissues to metabolize ketone bodies is limited and therefore any extensive catabolism of protein results in their build-up in the blood. Nevertheless, many species of animals can be subjected to quite long periods of inanition when their carbohydrate supplies and even their lipid stores are depleted. These animals have to resort to protein as the source of energy. The susceptibility to ketosis seems to vary very considerably from species to species. Therefore, their ability to utilize protein also varies considerably. In this respect, man is very susceptible although eskimos who live on a high protein and lipid diet show far less tendency to develop ketosis than people from the more southern latitudes. Indeed, carnivores in general are less susceptible than herbivores. Fish seem to be the least susceptible as they can apparently rely on protein for 100 % of their energy demands for long periods of time and this is no doubt why they can exist for periods up to one year without being fed. It should be borne in mind that when proteins and lipids are utilized the ultimate energy-yielding pathways, i.e., those which produce the ATP, are essentially the same as those used for carbohydrate catabolism. This means that as well as being converted to pyruvate they may also be used in synthetic pathways which result in the production of glycogen.

Thus, even in cases where no dietary carbohydrate is available, it is still possible for the tissues, particularly liver, to manufacture glycogen. This is called glyconeogenesis. Until recently, it was thought that skeletal muscle did not have the ability to produce glycogen except from glucose (glycogenesis). However, work by Bendall and Taylor (1970) and Dyson *et al.* (1975) has shown that there is little doubt that skeletal muscle has the capacity for glyconeogenesis. In addition to metabolizing carbohydrate and lipid and in some circumstances proteins, muscle, particularly red muscle, can also metabolize intermediates such as lactate. These can be used either to produce energy straight away or to synthesize glycogen. The systems involved in energy production and storage are therefore really quite versatile.

3.3 Methods of measuring the energy used during contraction

At the present time there is, unfortunately, no method that is suitable for measuring the energy used by the contractile system or even by the muscles as a whole during locomotion. Therefore our knowledge of what is happening during running, swimming, etc., has to be extrapolated from energetic measurements made on muscles studied *in vitro* and from metabolic measurements made on the whole animal. It is possible to carry out some measurements on muscle *in situ* undergoing isometric contraction; but, unfortunately, it is always necessary to occlude the blood supply (Edwards *et al.*, 1972) and therefore this approach is of limited use in locomotion studies. It is quite possible that in the next few years, however, some of the methods will be adapted for making *in situ* measurements during various kinds of locomotion.

One of the oldest and still one of the most widely used methods is that of heat measurement. The energy that is not converted into work by the muscle is degraded into heat. The heat production of the muscle is measured using a very sensitive thermopile or, in the case of *in situ* measurements, by thermocouple or thermistor. For very precise measurements it is necessary to accurately control the environmental temperature and the only temperature at which this can really be conveniently achieved is 0°C. For this reason most of the precise measurements have been carried out on the frog sartorius which is thin and strap-like muscle which will contract at 0°C. The frog sartorius is in some ways a rather unusual muscle. Furthermore, very few muscles, including frog muscles, ever operate at 0°C and caution should be exercised in extrapolating all of the findings for this muscle to other muscles.

There are also several technical difficulties associated with heat measurements including that of keeping the area of contact between the muscle and the thermopile constant throughout the contraction. This is particularly difficult for isotonic contractions. There are also the problems of time lag in the transference of heat from the muscle to the recording apparatus. Nevertheless, work by A. V. Hill and others using this system has contributed much to our understanding, not only of muscle energetics, but of the contractile process itself. The other problem of heat production measurements is their interpretation as they include entropy changes as well as free energy changes. The energy available for doing work is the Gibbs free energy (G). This is described by the second law of thermodynamics as:

$$\Delta G = \Delta H - T\Delta S + pV$$

where $H =$ total heat or enthalpy, T the temperature in degrees absolute, S as the change in entropy and pV as the pressure volume work (as muscle may be considered to be an isothermal system with negligible volume changes the latter term can be omitted). Since some of the energy change is due to changes in the entropy of the system it is not possible to derive absolute efficiencies from heat and work measurements alone without knowing something about the entropy values. In practice, this means that heat measurements must really be combined with chemical measurements. One great advantage of heat measurements is that they are non-destructive. Data can be obtained almost instantaneously and an experiment can be repeated several times on the same muscle. Chemical methods on the other hand are very time-consuming because one is usually obliged to carry out assays on many samples or muscle pairs. One type of chemical method that is non-destructive is that of oxygen consumption measurements. However, most muscles have the power to respire anaerobically to a very considerable extent and therefore oxygen consumption measurements have to be accompanied by measurements of lactate levels. These can also be misleading because part of the oxygen debt may not be associated with anaerobic glycolysis. This has been termed the alactacid O_2 debt (Margaria *et al.*, 1933). We also have the added complication of not knowing how much energy is being derived from lipids or from glycogen, although a good indication of this can be obtained from RQ measurements. The other problem if one is interested in contractile efficiency, is that it is not possible to know how much of the oxygen consumed is being used in the general metabolism of the cell and is therefore not used in actually producing the work. The best one can do is to subtract the oxygen consumption measurements for resting muscle from those of the exercising muscle, but this is based on the assumption that the general metabolism does not change and that all the extra oxygen is being used to produce energy for the contractile apparatus. In brief, the problem with oxygen consumption measurements is that the measurements are too remote from the actual process of transduction of energy into work. A less remote method, also non-destructive, is that of measuring the levels of $NADH_2$. This involves an ingenious spectrofluorometric technique developed by Jobsis and Chance (1957). With this method the ratio of $NAD/NADH_2$ is measured on intact muscles using a sensitive dual wavelength fluorimeter. Although this method has certain drawbacks, it is perhaps potentially the most useful as a means of measuring the energy demand of muscles during locomotion. *In vivo* measurements could possibly be made by using fibre optic 'light pipes' connected to the light source, monochromator, photomultiplier, etc. The instrumentation could be miniaturized and carried on the animal's back with light pipe inserted in a particular muscle. The signal may then be transmitted using telemetry. The main drawback of this method is that it measures the position of equilibrium between NAD and $NADH_2$ rather than the through-put of energy. It is also still fairly remote from the contractile process and it is non-specific in that it will measure $NAD/NADH_2$ change associated with other processes in addition to the contractile process.

In order to investigate the energy requirements of the contractile process itself, it is necessary to use a more direct method, either by chemical or heat measurement or both. It is also usually necessary to isolate the reactions involved in the contractile process and the immediate energy supply. These reactions are shown in Fig. 3.1.

The central reaction of the contractile process is the hydrolysis of ATP by the myosin ATPase. For many years, all the attempts to detect a breakdown in ATP during contraction failed because the ATP that is hydrolysed is immediately rephosphorylated by phosphorylcreatine (PC) which is a second high energy store within the muscle and which in quantity usually exceeds the ATP level by 3–5 times. In mammalian fast muscle there is about 4 μmol g^{-1} ATP and about 20 μmol g^{-1} PC. In slow muscle the figures are somewhat less. It should be mentioned at this juncture that in invertebrate muscle arginine phosphate is usually used instead of phosphorylcreatine. In some worm muscles, the phosphogen is formed from the glycocyamine molecule. The reaction by which the ATP formed is rephosphorylated to ATP is a very rapid reaction and thus ATP is reconstituted as quickly as it is broken down. In 1962, Cain and Davies were able to inhibit the creatine phosphotransferase reaction in frog muscles by using fluoro-2,4-dinitrobenzene (FDNB). This inhibitor, better known as Sanger's reagent, is a fairly non-specific inhibitor which inhibits many of the reactions in the muscle fibres, but fortunately it leaves the contractile apparatus in a fully functional state. Cain and Davies (1962) using FDNB-treated muscles were able to show that the amount of ATP broken down was closely related to the amount of work performed. At about the same time, Mommaerts *et al.* (1962) were also successful in measuring the amount of energy used during contraction by measuring the breakdown of

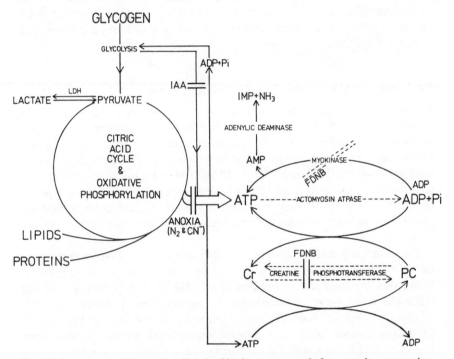

Fig. 3.1 A summary of the reactions involved in the energy supply for muscular contraction. Also shown is the site of action of the inhibitors used in biochemical energetics to prevent the replenishment of the immediate energy supplies, i.e. ATP and phosphoryl creatine (PC). FDNB = 2,4-fluoro dinitrobenzene. IAA = iodoacetate.

phosphocreatine after inhibiting glycolysis and oxidative phosphorylation with iodoacetate and nitrogen.

When measuring the change in levels of either ATP or PC during contraction, the usual experimental procedure is to compare the stimulated muscle with its contralateral muscle which acts as the control. There is always a certain amount of variation between right and left muscles or even samples taken from the same muscle, therefore one has to carry out assays on many samples or pairs of muscles. Although biochemical measurements are very time-consuming, they do represent one of the best methods of estimating the efficiency or economy of the contractile apparatus under different conditions. However, the values include the energy involved in activation as well as that used by the cross bridges.

Many workers have measured changes in muscle metabolite levels during exercise, particularly in man. These types of data are of little value in muscle energetics as the relationships with the contractile activity are not stoichiometric. Nevertheless, these sorts of measurements have yielded much useful data concerning the source of energy supply (e.g. carbohydrate or lipid) and in locating which muscles and even which type of fibres are being used during different kinds of locomotion.

3.4 Economy and efficiency of muscles working *in situ*

The cost of muscular work in man has been studied extensively under many different conditions. For an isometric contraction it seems that the energy expenditure as measured by oxygen consumption, is proportional to the time of the contraction, providing the tension developed by the muscle is kept constant (Lupton, 1923 and Muller, 1930). However, using heat production measurements Edwards *et al.* (1975) have shown that the rate of energy expenditure decreases with time. This latter finding is in agreement with chemical energy measurements carried out on isolated mammalian muscles (Awan and Goldspink, 1972).

The situation regarding efficiency in doing work, that is to say, in performing isotonic contractions is perhaps a more interesting one. A. V. Hill (1938) found that the oxygen consumption of an athletic man pedalling a bicycle ergometer under optimum conditions corresponded to an efficiency of 28 %. Similar, but slightly lower, efficiencies were recently reported by Pugh (1975) for racing cyclists riding in a normal unrestrained manner. Interestingly, Pugh found that the mechanical efficiency did not vary much with work rate or pedalling frequency. Gaesser and Brooks (1975) also found that the efficiency was fairly constant with speed although there was a tendency for it to drop at the higher speeds. Constancy of efficiency over a wide range of work rates and frequencies has also been noted by Bannister and Jackson (1967) on an Olympic oarsman during rowing.

Higher mechanical efficiency values of about 30 % have been obtained on isolated (A. V. Hill, 1950) and partially isolated muscle (Di Prampero *et al.*, 1969). In exercise studies it is to be expected that the mechanical efficiencies will be lower than for isolated muscles because in exercise, synergetic muscles are involved and there is more internal resistance in the intact limb which will tend to reduce the work output. The main difference between the results on exercising subjects and isolated muscle is that the latter do not show the same constancy of mechanical efficiency with work rate. The mechanical efficiency of isolated muscle is very much dependent on the velocity of shortening. This discrepancy can be understood if one considers that the muscles are composed of different types of fibres each of which has an optimum velocity of shortening. As work rate increases during exercise, different muscles and indeed different fibres populations become involved in the action and for this reason the efficiency can be kept reasonably constant. In isolated muscle preparations, all the muscle fibres are stimulated and therefore the maximum efficiency will be attained at the optimum rate of shortening of the predominant fibre type.

A mechanical efficiency of 20–30 % for muscles working *in situ* may seem to be low, however it is worth pointing out that internal combustion engines have efficiencies considerably lower than this and indeed the efficiency of steam engines is only of the order of 1 %. Electrical motors may have a high efficiency of the order of 90 %. However, the efficiency of converting fuel into electricity is only about 35 % hence in this case the comparable efficiency value is about the same as muscle. The figures that have been given for muscle are overall efficiency values and do not give us any real idea of the efficiency of the contraction process *per se*. The processes involved in producing ATP from glycogen or lipid are known to have an efficiency of approximately 50 %, therefore the value of the contractile process itself must presumably be at least double that for the overall efficiency of the muscle.

In order to measure the efficiency of the contractile apparatus, it is necessary to carry out measurements under conditions where there is no replenishment of the immediate energy supplies. These are described in the next section.

3.5 Energetics of the contractile process *per se*

In the laboratory, the muscle physiologist can make muscle contract in several different ways. Muscle can be made to develop tension without shortening (isometric contractions), it can be made to shorten at different velocities (isotonic contractions) and it can be forcibly lengthened whilst developing tension (negative work contraction). All of these conditions occur during locomotion but not necessarily in the same species or in the same muscle. The energy demand of muscle does differ very considerably according to the kinds of conditions that it is working under. Some types of muscle fibres are more efficient in working under certain conditions than other types of muscle fibre. It is perhaps not surprising, therefore, to find that there is a division of labour between different muscles or even between different fibres within the same muscle. The vertebrate limb possesses more muscle fibres and more muscles than it needs to perform the basic movements. It must be assumed that the diversity of muscles and muscle fibres enables the animal to finely regulate its movements and to carry them out with reasonably high efficiency. A description of fast and slow muscle fibres has been given in Chapter 1 and only the energetic aspects will be discussed here.

3.5.1 Isometric contraction

The rate of heat production of frog muscle during isometric tension has been measured by several workers. Abbott (1951) found that the rate of heat production was higher during the first few seconds. After this, the rate of heat production decreased to a constant level. The relationship between initial heat and the maintenance or stable heat is shown in Fig. 3.2. Aubert in a series of papers (Aubert, 1956, 1964, 1968) found that the extra heat produced during

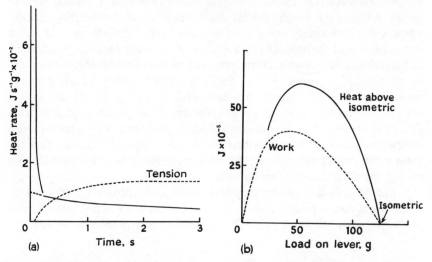

Fig. 3.2 Heat production during isometric and isotonic contractions. (a) The rate of heat production during an isometric contraction of toad muscle at 0°C. Initially the heat production (energy utilization) is high, however it drops to a fairly constant level after 0.25 s or so. The high initial heat is usually referred to as activation heat and the steady level as the maintenance heat. The course of tension development is shown as a dotted line. From Woledge (1971). (b) Shows the Fenn effect, that is to say more heat is emitted (more energy used) when a muscle is shortening than during an isometric contraction. This extra heat is often referred to as shortening heat.

the initial part of a tetanic isometric contraction (labile heat) could be lowered by increasing the frequency of stimulation and also by making the muscle contract several times prior to taking the measurements. He found that the basic rate of heat production (stable heat) could be decreased by increasing the muscle length and by using FDNB. It seems that isometric contractions of short duration are always less economical than those of longer duration. Indeed this point is illustrated by the results for isometric contraction economies for mammalian muscle given in Table 3.1. From the table it will be seen that the rate of energy expenditure is much greater during the first second or so than it is during the rest of the contraction. The extra energy that is associated with the initial phase of an isometric contraction is believed to be due to internal work that is involved in taking up the slack in the series elastic elements. Using laser beam diffraction methods, it has been shown that even though the contraction is considered to be isometric, the mean sarcomere length decreases considerably during the initial phase of a contraction (Larson *et al.*, 1968; Goldspink *et al.*, 1970a). The shortening of the sarcomeres would be expected to be associated with a greater energy turnover as isotonic contractions are usually more expensive than isometric contractions.

Table 3.1 The expenditure and economy of developing and maintaining isometric tension by the hamster biceps brachii (fast) and soleus (slow) muscles

	Length of contraction	*1.2 s*	*30 s*	*60 s*	
Expenditure (μmol high energy phosphate g^{-1})	Fast	1.5 ± 0.06	2.2 ± 0.05	2.5 ± 0.05	*in vitro*
	Slow	1.1 ± 0.03	1.5 ± 0.02	1.8 ± 0.01	
	Fast	—	—	2.2 ± 0.01	*in vivo*
	Slow	—	—	1.1 ± 0.01	
Economy (g.s μmol^{-1} high-energy phosphate)	Fast	99 ± 7	2483 ± 110	3635 ± 180	*in vitro*
	Slow	143 ± 8	6377 ± 250	9012 ± 230	
	Fast	—	—	4997 ± 431	*in vivo*
	Slow	—	—	9963 ± 560	

The amount of tension developed and maintained is measured by integrating the area under the tension curves and is expressed as gram seconds (g.s).
In vitro muscles were inhibited using iodoacetate and nitrogen. In all cases the amount of high energy used was estimated from the breakdown of phosphoryl creatine. From Awan and Goldspink (1972).

3.5.2 Isometric economy of fast and slow muscles

When one thinks about isometric contraction of muscles one usually thinks of their postural role. However, after reading Chapter 7 on terrestrial locomotion, it will be appreciated that isometric contraction of muscles is also very important in the dynamics of limb movement. Much of the muscle activity in running is associated with tensioning of the tendons which store energy. This tensioning or re-winding of the tendons is done with very little length change in the muscle fibres themselves. The economy of muscle fibres in developing and maintaining isometric tension is therefore of considerable importance in the dynamic as well as the static aspect of muscle design.

Slow muscles are often associated with a postural function. The reason for this is that slow muscle fibres are considerably more economical at developing and maintaining isometric tension than the fast fibres. Table 3.1 shows some measurements carried out on the fast and slow muscles of the hamster (Awan and Goldspink, 1972). From this table it will be seen that, for a contraction of 60 s duration, the economy of the slow muscle is almost three times greater than that of the fast muscle. This greater economy of the slow muscle can be fairly easily explained when one considers what is happening at the cross bridge level. In the slow muscle the cross bridge cycle is longer; this means that each cross bridge will have to be reprimed with ATP fewer times per second than is the case in fast muscle fibres. ATP is thus used only when the cross bridge detaches, whilst the cross bridges are engaged and pulling they are not using ATP. Hence, the longer the engagement time of the cross bridges the more economically they can maintain tension. Some tonic muscles such as the chicken anterior latissimus dorsi, which is the muscle which holds the wings back against the body can maintain tension for long periods of time with little usage of energy (Goldspink *et al.*, 1970b; Goldspink, 1975; Matsumoto *et al.*, 1973). However, these muscles are extremely slow contracting tonic muscles and it is unlikely that they are ever used for any function other than a strictly postural one.

3.5.3 Isotonic contractions

In isotonic contractions, the muscles are shortening and hence doing work. This type of activity is seen in terrestrial animals in swinging the limbs forwards and backwards, in fish in flexing the trunk from side to side and in birds in moving the wings up and down. It is, in fact, the most obvious type of muscle activity during locomotion. Fenn (1923, 1924) and A. V. Hill (1938) reported that heat production during an isotonic contraction was greater than for an isometric contraction. This is sometimes referred to as the Fenn effect. The extra heat Hill termed shortening heat, whereas the heat liberated during an isometric contraction he termed maintenance heat. The amount of extra heat used is dependent mainly on the amount of shortening (Hill, 1964). Cain and Davies (1962), using chemical measurements, found that the amount of chemical energy used was, however, related to the work done rather than to the distance shortened. In fact, the amount of ATP or PC broken down is not sufficient to explain all of the heat liberated during a single isotonic contraction (Davies *et al.*, 1967; Kushmerick and Davies, 1969). This discrepancy is more noticeable in a single lightly loaded contraction and may be due to entropy changes connected with conformational changes in the cross bridges (Davies *et al.*, 1967). The relationship between the heat produced and the chemical changes is still being actively studied by D. R. Wilkie, R. Woledge, Nancy Curtin and their co-workers at University College, London, who feel that the discrepancy may be due to the presence of an unknown chemical reaction or reactions rather than an entropy change (Curtin, Gilbert, Kretzschmar and Wilkie, 1974; Curtin and Woledge, 1974).

As muscle is a machine capable of doing work, the question of its efficiency naturally arises. Contractile efficiency measurements are, in practice, very difficult to make because there are so many other parameters that must be standardized, particularly if one is trying to measure maximum values. Also, as already mentioned, it is necessary to define clearly what part of the system is

68

being studied. Kushmerick and Davies (1969) found the efficiency of FDNB-treated frog muscle working at optimum velocity to be over 66%. This figure refers to the efficiency of the contractile process plus the ion pumping involved in the activation of contraction. These authors considered that an appropriate value for the contractile process itself might be of the order of 80–90% if the energy used for the ion pumping was subtracted from the total. It seems therefore that the transduction of energy at the cross bridge level is probably extremely efficient given the right conditions. An even higher value of 77% for the contractile efficiency plus activation has been given for tortoise muscle by Woledge (1968). However, he found that the overall efficiency which includes the replenishment reactions, to be about 35%. Tortoise muscle, therefore, appears to be considerably more efficient than frog muscle. This raises the interesting possibility that during the course of evolution it has, in some cases, been possible to sacrifice speed for efficiency.

3.5.4 *Efficiency of the contractile system in mammals and birds*

Most terrestrial vertebrate muscles are made up of motor units of different kinds. However, some muscles consist predominantly of slow units. The author and co-workers have studied the energetics of fast and slow muscle fibre by choosing suitable mammalian and avian muscles in which there is a preponderance of one fibre type. There are many difficulties in handling mammalian and avian muscles. They are, for example, more sensitive to anoxia and changes in temperature. Consequently, the procedures developed for frog muscle usually have to be modified before they can be applied to mammalian and avian muscles.

In order to find the maximum efficiency of fast and slow muscles, experiments have been carried out in which iodoacetate-/cyanide-treated muscles were made to contract isotonically whilst loaded with different weights. The energy used was estimated by measuring the amount of phosphoryl creatine used and the mechanochemical efficiency was calculated from the work done and the amount of high energy phosphate used. The efficiency of the fast and slow muscles is then plotted over a range of shortening velocities as shown in the curves in Fig. 3.3. In addition to plotting efficiency values it is also useful to plot the rate of high energy phosphate usage over the same range of shortening velocities (Fig. 3.4). From Fig. 3.3, it will be seen that the slow muscle is working most efficiently at shortening velocities of about one muscle length per second, whilst the fast muscle is working most efficiently at a shortening velocity of 5 muscle lengths per second. It will also be seen that, at its optimum rate of shortening, the slow muscle is more efficient than the fast muscle. It has already been stated that the slow tortoise muscle is more efficient than the faster frog muscle, thus the same situation seems to be true of the slow and fast twitch muscles found within the same animal.

The reason why slow muscles have a higher maximum efficiency is not known. It is known that slow muscles require fewer impulses per second in order to produce a smooth tetanus. Therefore part of the increased efficiency of the slow muscle may be due to their lower energy requirements for ion pumping. However, this is unlikely to be the full story.

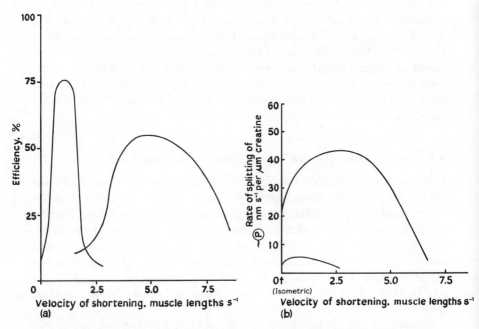

Fig. 3.3 The efficiency and rate of high energy phosphate utilization by mammalian fast phasic muscle (biceps brachii) and slow phasic muscle (soleus) whilst shortening at different velocities. (a) Efficiency expressed as a percentage against the rate of shortening in muscle lengths per second. The efficiency value is based on the assumption that each mole of phosphoryl creatine split is equivalent to 468.6 kJ. (b) The rate of splitting of high energy phosphate is nmol s^{-1} μmol of total creatine (total creatine is considered to be a better parameter than muscle weight). In both cases, the different velocities of shortening were obtained by loading the muscle to different extents.

It is important that a distinction should be made between the slow twitch muscles and the true slow or slow tonic muscles. It has already been stated that the slow tonic muscles are extremely economical when they contract isometrically. However, when they are required to shorten and do work their efficiency is low as compared with either the slow or the fast twitch muscles. It seems that this type of slow muscle is fully adapted for producing isometric tension rather than doing work, whereas slow twitch muscles are able to carry out both functions economically and efficiently.

70

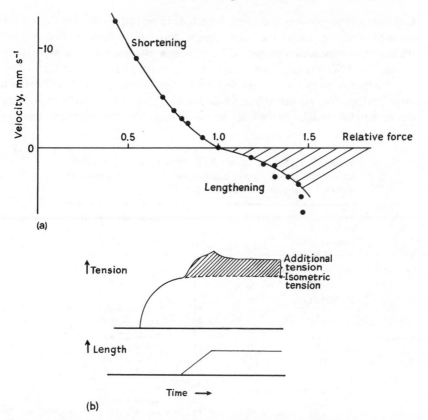

Fig 3.4 Negative work contractions. (a) The extension of the typical length–tension plot into the negative work region. (b) The sort of traces that are obtained when the muscle is stretched whilst being stimulated. The muscles are usually stretched using a Levin–Wyman ergometer or similar apparatus which is fitted with an isometric transducer. The hatched regions of both diagrams represent the extra tension produced in negative work.

3.5.5 Negative work contractions

The situation often arises during locomotion that muscles are stretched whilst still developing tension. When this is the case, the muscles usually develop more tension than they would do during an isometric contraction. This is predictable if we extend the A. V. Hill force–velocity curve as shown in Fig. 3.4a. From the energetics point of view, this extra tension is believed to be associated with no extra energy cost and this has important implications as far as locomotion is concerned. It means that we can descend a flight of stairs with less effort than it takes to ascend, even though our muscles still have to develop considerable force in order to prevent us from falling. The fact that negative work requires less total energy than positive work was demonstrated in a very interesting way by Abbott *et al.* (1952). They arranged to have two

71

bicycle ergometers connected back to back. One person pedalled in the normal forward direction whilst the other person pedalled backwards and resisted the action to a measured extent. A. V. Hill demonstrated the same experiment to the Royal Society in a more dramatic way when he showed that a woman doing negative work on the double bicycle could easily resist the efforts of a large healthy male athlete doing positive work. Thus, there seems to be little doubt therefore that negative work is considerably less expensive than positive

Table 3.2 Rates of total ATP breakdown by FDNB-treated muscles during stretching and shortening (negative work). From Curtin and Davies (1975)

Velocity $l_0 s^{-1}$	n	Mean rate of total chemical change $\mu mol\, g^{-1}\, s^{-1}$
A. Isometric contraction		
0.00	7	$+0.72 \pm 0.09$
B. Stretching		
0.13	11	$+0.18 \pm 0.042$
0.18	10	$+0.18 \pm 0.042$
0.33	7	$+0.22 \pm 0.143$
0.66	18	$+0.49 \pm 0.163$
1.33	10	$+0.45 \pm 0.173$
2.00	18	$+0.84 \pm 0.351$
C. Shortening		
0.20	10	$+0.69 \pm 0.035$
0.33	13	$+0.75 \pm 0.072$
0.66	16	$+1.08 \pm 0.105$
0.91	9	$+1.41 \pm 0.099$
1.33	9	$+0.73 \pm 0.065$
2.10	10	$+0.94 \pm 0.149$

work. Margaria (1968) demonstrated the importance of negative work contractions in human locomotion by measuring the energy cost of walking uphill and down hill. He concluded that in walking on the level that negative work was an important component but that it required relatively little energy.

This ability of muscle to do negative work relatively cheaply can be demonstrated on muscles working *in vitro* and indeed it is one of the inherent characteristics of the sliding filament/cross bridge type of mechanism. Recently, Curtin and Davies (1975) have shown that when the frog sartorius muscle is stretched at about 0.18 muscle lengths s^{-1} whilst it is developing force, the rate of ATP usage related to tension is only about 25% of that during an isometric contraction (Table 3.2). The reason why less energy is used per given amount of tension developed seems to be that extra tension is obtained by the pulling out of the cross bridges. Certainly, it is known that there is quite a lot of compliance in the cross bridges from the stretch studies described in the preceding chapter. Therefore, one can easily visualize parts of the cross bridge, in particular the S_2 part, acting like a spring. Extra tension will arise as these springs are forcibly extended. The matter is not quite as simple as it perhaps seems at first sight because the degree of extension of the muscle will usually be such that the cross bridges will be pulled off their original active site before they have gone through the complete cycle. This means that they still have the ability to combine with actin without requiring further ATP, and they presumably slide from one active site to the next. The molecular mechanism of negative work is not fully understood. However, it seems fortunate from the point of view of locomotion, particularly terrestrial locomotion, that muscle can develop extra tension when stretched in the activated state and that the tension produced during negative work is very inexpensive.

3.6 Implications of muscle energetics in locomotion

When the energetic characteristic of the different kinds of striated muscle are examined, we can begin to understand the reason for diversity of fibre type and the sort of division of labour that occurs during locomotion. To recapitulate, it has been shown that slow muscles are more economical than fast twitch muscles in developing isometric tension. Therefore, when there is an isometric component to the locomotion, we might expect to find that slow muscle fibres are involved. Also slow twitch muscles are more efficient in shortening and producing work, providing they are shortening at their optimum velocity. Therefore, when slow prowling or slow cruising movements are required, one might again expect to find that it is the slow twitch fibres that are involved. When more rapid movements are required, it is of course necessary to use fast contracting fibres as the slow fibres are not mechanically effective at higher shortening velocities and they also become very inefficient once their optimum rate of shortening is exceeded (Fig. 3.3). By the same contention it is not surprising to find that many muscles contain more than one type of fibre. In other words, muscles may, as it were, possess a two- or three-'geared' system. This enables them to contract efficiently over a wide range of shortening velocities by progressively recruiting different fibre populations.

3.6.1 Bird flight

The muscle fibre composition of the flight muscles is fairly homogeneous. It seems that the power requirements of flight are such that the bird cannot apparently indulge in a two- or three-geared system. The pectoral muscles in many birds make up 40% of the body weight and any diversity in the fibre population would add to this weight. Therefore in general we find that each species of bird has a particular flapping frequency and that this frequency does not vary to any great extent during flight. Alterations in speed during flight are achieved by changing the force and shape of the wing beat but more importantly by alternating between flapping and gliding or in the case of small birds by indulging in intermittent (swooping) flight. In order to avoid any confusion it should be pointed out that the two types of fibres found in the pigeon pectoralis muscle do not represent a two-geared system. The small fibres which are well supplied with mitochondria are probably involved in long sustained flight whereas the large fibres which are somewhat deficient in mitochondria are probably involved only in rapid take off when extra power is required. The pigeon can in fact take off very rapidly as compared with a gull or other kinds of medium-sized bird.

The myosin ATPase and hence the intrinsic speed of shortening of all the fibres is therefore matched to the basic flapping frequency of the bird. The flapping frequency of different species is related to body size (Greenewalt, 1962). In general, this means that the larger the bird the slower its muscle fibres. During gliding flight the muscle action is basically isometric. Therefore, as far as gliding is concerned, slow muscles will be a distinct advantage as they can produce isometric tension more economically than fast muscle fibres. This is probably one of the reasons why large birds have been able to use gliding as a means of energy saving. There are also some aerodynamic reasons why only large birds have been able to exploit gliding as a means of locomotion. However, it is a happy coincidence that their pectoral muscle fibres should be suitable for this as well as flapping flight.

The energy saving during gliding has been shown to be very considerable (Baudinette and Schmidt Nielsen, 1974). The reason for this is that the force required during gliding is only that required to maintain the posture of the bird in the air and so only a relatively few muscle fibres (motor units) need to be recruited at any one time. The pectoral muscles have to develop a force that is equal to the lift in order to keep the wings horizontal or slightly above horizontal and the force required is appreciable although it is not anything like as great as in flapping flight. The other important factor in energy saving is that muscle activity in gliding flight is predominantly isometric and the energy turnover by the contractile system will be low.

3.6.2 Fish swimming

In contrast to bird flight, fish swimming involves changes in the frequency of beat. As the speed of swimming increases there is an increase in both the frequency and the aptitude of the tail beat (Bainbridge, 1958). It is not surprising therefore to find that there are three basic types of muscle fibre in fish muscle and this provides the fish with a three-geared system. Fig. 3.5 shows the general diagram of the red, pink and white fibres of the fish. Studies using electromyography (Bone, 1966; Hudson, 1972; Davison *et al.*, 1976) and biochemical studies (Bone, 1966; Johnston and Goldspink, 1973) have strongly indicated that the red muscle fibres, usually found near to the lateral line, are the ones which are responsible for slow cruising movements. Measurements of the specific activity of the myosin ATPase of the different fibre types have shown these fibres to be slow contracting fibres (Johnston, Frearson and Goldspink, 1972). Energetically, this makes sense because the slow fibres will work with high efficiency at the sort of speeds involved in slow cruising. The slow muscle fibres may only account for a few per cent of the total musculature.

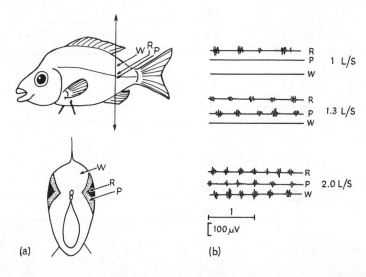

Fig. 3.5 (a) Diagram of a fish such as the carp showing the usual distribution of the red, white and pink muscle fibres. The arrows labelled R (red), P (pink) and W (white) show the positions at which EMG electrodes may be inserted to detect action potentials in the three layers during swimming at different speeds. (b) EMG records of the three layers at different swimming speeds. A switching mechanism was used to record from all three layers of the same fish whilst it was swimming at a given speed. Data from Davison *et al.* (1976).

However, the power requirement for slow swimming is very low (see Chapter 6). The white muscle fibres have a fast myosin ATPase and these are believed to be recruited only when rapid bursts of activity are required. The white fibres account for 70–80% of the musculature and it seems paradoxical that all this muscle is inactive most of the time. However, unlike the situation in the bird, this additional weight is no great problem to the fish because of the buoyancy of its aqueous environment. The precise function of the pink muscle fibre is not known; however, from biochemical studies (Davison *et al.*, 1975) it seems that this muscle has an intrinsic speed of shortening that is intermediate between that of the red and the white fibres. Thus, it may be assumed that the pink muscle fibres are recruited for swimming speeds which are slightly faster than slow cruising. These three kinds of fibres with different optimum rates of shortening enable the fish to swim at a range of speeds without a marked drop in efficiency.

Table 3.3a The concentrations of myoglobin, glycogen, lipid and certain glycolytic intermediates in the red, pink and white muscles of the carp

Compound	Muscle type		
Mean ± S.E.	Red	Pink	White
Myoglobin (mg g^{-1} dry wt.)	12.7 ± 1.1	5.4 ± 0.4	2.8 ± 0.4
Glycogen (mg g^{-1} dry wt.)	5.6 ± 1.2	3.8 ± 0.6	0.66 ± 0.15
Lipid (mg g^{-1} dry wt.)	20.2 ± 1.2	8.3 ± 1.1	5.8 ± 0.4
Pyruvate (μmol g^{-1} dry wt.)	0.60 ± 0.17	2.2 ± 0.7	1.4 ± 0.08
Lactate (μmol g^{-1} dry wt.)	35.6 ± 2.9	63.0 ± 11.9	29.1 ± 1.3

Table 3.3b Enzyme levels in the red, pink and white muscles of the carp

Enzyme	Muscle type		
	Red	Pink	White
Myofibrillar ATPase	0.25 ± 0.03	0.62 ± 0.07	1.09 ± 0.14
Creatine phosphokinase	1986 ± 241	2803 ± 288	4242 ± 659
Phosphorylase	18.9 ± 3.0	19.0 ± 0.8	12.7 ± 1.9
Hexokinase	3.33 ± 0.46	2.17 ± 0.28	0.76 ± 0.10
Pyruvate kinase	227 ± 21	464 ± 65	299 ± 51
Lactate dehydrogenase	299 ± 20	784 ± 74	440 ± 41
Malate dehydrogenase	1591 ± 231	1053 ± 100	686 ± 90
Succinic dehydrogenase	1332 ± 101	440 ± 99	199 ± 60
Cytochrome oxidase	4645 ± 717	1685 ± 242	609 ± 131

All enzyme activities are expressed as μmol substrate utilized per g dry weight per min except for myofibrillar ATPase activity which is in units of μmol Pi released per mg myofibrillar protein per min. Data in Tables 3.3a and 3.3b taken from Johnston, Davison and Goldspink (1977).

The involvement of the different kinds of muscle fibre does not require a complicated recruitment system by the central nervous system. The red fibres will no doubt be recruited first, the pink fibres will be next and if the speed of swimming has to be increased further then the white muscle fibres will become involved. At high speed the pink and particularly the red fibres will be working inefficiently. However, there is no need for these to be turned off by the central nervous system because they become mechanically ineffective at the higher speeds and they in fact split very little ATP (see Fig. 3.3b), since when muscle fibres are attempting to contract at speeds greater than their optimum velocity, they do not use much ATP. As the active filaments move more and more rapidly there is apparently less time for the cross bridges to go through the cycle of force production before they are pulled off the active site. When they are pulled off they are still primed with ATP, thus they can attempt to make contact with other active sites. The faster the filaments are moving the less probability there is of them making contact. Therefore at the higher swimming speeds the contractile apparatus of both the red and pink fibres will be moved passively.

The red muscle fibres are well supplied with mitochondrial enzymes and obviously have a predominantly oxidative metabolism. The pink fibres also have fairly high levels of oxidative enzymes but they are also particularly well supplied with glycolytic enzymes (Table 3.3). The red muscle fibres most probably always work in the steady state and never experience any oxygen debt. The pink muscle fibres are reasonably fast fibres and therefore have a more rapid energy turnover and they apparently do experience some oxygen debt. However, because the pink fibres are recruited reasonably frequently they also require the ability to recover rapidly, thus it is seen that they are generally very metabolically active. White muscle fibres, on the other hand, will consume energy at a very rapid rate and there is no way that the fish could get sufficient oxygen to this tissue to maintain it in a steady state during full activity. There is little point, therefore, in these fibres having high levels of oxidative enzymes. Because these white fibres fatigue very quickly, they are only used for short periods of time and most of the replenishment of immediate energy supplies and the oxidation of the lactate has to take place after the cessation of activity, as shown in Fig. 6.3.

Fish therefore afford a good example of the way that the energetic requirements for locomotion can be met using different types of muscle systems.

3.6.3 Terrestrial locomotion of mammals

Analysis of muscle action during terrestrial locomotion is much more complex. In running for example, there is a considerable amount of isotonic contraction involved in swinging the limbs backwards and forwards. Surprisingly, however, this is not the primary action. Certainly, it is believed that this action does not account for much of the energy consumed because the energetic cost of running in animals with light limbs is about the same as for ones with heavier limbs (Taylor *et al.*, 1974). This is because much of the energy involved is stored from step to step, as elastic strain energy in the tendons. This involves a considerable change in the length of tendons, but not in the muscle itself. A more detailed description of the elastic storage of energy during running and jumping will be found in Chapter 7. The important aspect of terrestrial locomotion from an energetics point of view is that the forces are derived mainly from isometric contractions. This means a decreased energy expenditure because isometric contractions are considerably less expensive than isotonic contractions. In this context, muscle efficiency measurements are therefore not really very relevant. What we really need to talk about is the overall energy expenditure involved in this mode of locomotion and this is obviously much lower than it would be in a simple non-compliant mode of locomotion.

It is well known that the cost of locomotion for large animals is considerably less than that for small animals (Taylor *et al.*, 1970; Schmidt-Nielsen, 1971). There are probably several factors that contribute to this decreased cost of transport. The most obvious one is that muscles in large animals will not need to be activated as often as those in small animals. Although the energy required for activation of muscle is not very great as compared with that used in contraction, the initial part of a contraction is very definitely the most expensive part of the contraction as this is associated with a lot of internal work in the muscle. The fact that the muscles in smaller animals have to develop force more times a second will mean that they have a higher cost of transport. In other words, developing tension is expensive, maintaining it once it is developed is relatively cheap. Thus animals with a longer stride duration should be able to cover ground much more economically.

Last, but not necessarily least, is the fact that the muscles of large animals have more slow fibres than fast fibres. Slow twitch fibres are both more economical and more efficient than fast fibres, therefore this must represent a considerable energy saving. We may assume that in all types of terrestrial locomotion there is the same sort of progressive recruitment of fibres that is found in fish swimming. These fibre types have different efficiency maxima and the progressive recruitment of fibre types enables the muscle to work efficiently over a wide range of shortening velocities. However, large animals presumably need fewer fast fibres because they have longer muscles and hence their overall rate of shortening is high in any case and they probably also have a longer period of time in which to tension the tendons.

Although, at first sight, it seems that a lot of information is available about muscle action during locomotion it is true to say that there is still much to be learned about muscle energetics and locomotion, particularly with regard to the division of labour between different types of muscle fibres. Many of the current questions will most probably be answered within the next few years as several groups of research workers are employing elegant methods to investigate muscle action *in situ* and we shall no longer have to rely solely on extrapolating from information gained from isolated muscles.

Acknowledgements

This review was written whilst the author was in receipt of a Science Research Council Grant for a study of the bioenergetics of different vertebrate muscles.

References

Abbott, B.C. (1951) The heat production associated with the maintenance of a prolonged contraction and the extra heat produced during large shortening. *J. Physiol., Lond.* **112**, 438–445.

Abbott, B.C., Bigland, B. and Ritchie, J.M. (1952) The physiological cost of negative work. *J. Physiol., Lond.* **117**, 380–390.

Aubert, X. (1956) *Le Couplage Energetique de la Contraction Musculaire.* Brussels: *Editions Arscia.*

Aubert, X. (1964) Tension and heat production of frog muscle tetanized after intoxication with fluoro-2,4-dinitro-benzene, FDNB. *Pfluegers Arch. ges. Physical.* **281**, 13 pp.

Aubert, X. (1968) In: *Symposium on Muscle, Budapest,* Ernst, E. and Straub, F.B., eds., Budapest: Akatemiai Kiado, pp. 187–189.

Awan, M.Z. and Goldspink, G. (1972) Energetics of the development and maintenance of isometric tension by mammalian fast and slow muscle. *J. Mechanochem. Cell Motility* **1**, 97–108.

Bainbridge, R. (1958) The speed of swimming of fish as related to size and the frequency and amplitude of the tail beat. *J. exp. Biol.* **35**, 109–133.

Bannister, E.W. and Jackson, R.C. (1967) The effect of speed and load changes on oxygen intake for equivalent power outputs during bicycle ergometry. *Int. Z. angew. Physiol.* **24**, 284–290.

Baudinette, R.V. and Schmidt-Nielsen, K. (1974) Energy cost of gliding flight in herring gulls. *Nature* **248**, 83–84.

Bendell, J.R. and Taylor, A.A. (1970) A meyerhof quotient and the synthesis of glycogen from lactate in frog and rabbit muscle: A reinvestigation. *Biochem. J.* **118**, 887–893.

Bone, Q. (1966) On the function of the two types of myotomal muscle fibres in elasmo-branch fish. *J. Mar. Biol. Ass. U.K.* **46**, 321–349.

Bursell, E. (1963) Aspects of the metabolism of aminoacids in the tsetse fly, *Glossina,* (Diptera). *J. Insect Physiol.* **9**, 439–452.

Cain, D.F. and Davies, R.E. (1962) Breakdown of triphosphate during a single contraction of working muscle. *Biochem. Biophys. Res. Commun.* **8**, 361–366.

Curtin, N.A. and Davies, R.E. (1975) Very high tension with little ATP breakdown by active skeletal muscle. *J. Mechanochem. Cell Motility* **3**, 147–154.

Curtin, N.A., Gilbert, C., Kretzschmar, K.M. and Wilkie, D.R. (1974) The effect of the performance of work on total energy output and metabolism during muscular contraction. *J. Physiol., Lond.* **238**, 455–446.

Curtin, N.A. and Woledge, R.C. (1974) Energetics of relaxation in frog muscle. *J. Physiol.* **238**, 437–446.

Davies, R.E., Kushmerick, M.J. and Larson, R.E. (1967) ATP, activation and the heat of shortening of muscle. *Nature* **214**, 148–151.

Davison, W., Goldspink, G. and Johnston, I.A. (1976) Division of labour between fish myotomal muscles during swimming. *J. Physiol., Lond.* **263**, 185P.

Di Prampero, P.E., Cerretelli, P. and Piiper, J. (1969) Energy cost of isotonic tetanic contractions of varied force and duration in mammalian skeletal muscle. *Pflugers Arubine* **305**, 279–291.

Dyson, R.D., Cardenos, J.M. and Barsottii, R.J. (1975) The reversibility of skeletal muscle pymurate kinase and an assessment of its capacity to support glyconeogenesis. *J. Biol. Chem.* **250**, 3316–3321.

Edwards, R.H.T., Hill, D.K. and Jones, D.A. (1972) Effect of fatigue on the time course of relaxation from isometric contractions of skeletal muscle in man. *J. Physiol.* **227**, 26P.

Edwards, R.H.T., Hill, D.K. and Jones, D.A. (1975) Heat production and chemical changes during isometric contractions of the human quadriceps muscle. *J. Physiol., Lond.* **251**, 303–315.

Fenn, W.O. (1923) A quantitative comparison between the energy liberated and the work performed by an isolated sartorius of the frog. *J. Physiol., Lond.* **58**, 175–208.

Fenn, W.O. (1924) The relation between the work performed and the energy liberated in muscular contraction. *J. Physiol., Lond.* **58**, 373–395.

Gaesser, G.A. and Brooks, G.A. (1975) Muscular efficiency during steady-rate exercise: effects of speed and work rate. *J. appl. Physiol.* **38**, 1132–1139.

Goldspink, G. (1975) Biochemical energetics for fast and slow muscle. In: *Comparative Physiology: Functional Aspects of Structural Materials.* Bolis, L., Maddrell, S.H.P. and Schmidt-Nielsen, K., eds., Amsterdam: North-Holland.

Goldspink, G., Larson, R.E. and Davies, R.E. (1970a) Fluctuations in sarcomere length in the chick posterior latissimus dorsi muscle during isometric contraction. *Experientia* **26**, 16–18.

Goldspink, G., Larson, R.E. and Davies, R.E. (1970b) Thermodynamic efficiency and physiological characteristics of the chick anterior latissimus dorsi muscle. *Z. vergl. Physiol.* **227**, 848–850.

Greenewalt, C.H. (1962) Dimensional relationships for flying animals. *Smithson. misc. Coll.* **144**, 2.

Hill, A.V. (1938) The heat of shortening and the dynamic constants of muscle. *Proc. R. Soc., Ser. B,* **126**, 136–195.

Hill, A.V. (1964) The efficiency of mechanical power development during muscular shortening and its relation to load. *Proc. R. Soc., Ser. B,* **159**, 319–324.

Hochachka, P.W., Fields, J. and Mustafa, T. (1973) Animal life without oxygen: Basic biochemical mechanisms. *Am. Zool.* **13**, 543–555.

Hudson, R.C.L. (1972) On the function of the white muscles in teleosts at intermediate swimming speeds. *J. exp. Biol.* **58**, 509–522.

Jobsis, F.F. and Chance, B. (1957) Time relations between muscular contraction and response of cytochrome chain. *Fed. Proc.* **16**, 68.

Johnson, I.A., Davison, W. and Goldspink, G. (1977) Energy metabolism of carp swimming muscle. *J. comp. Physiol.* B (In press).

Kushmerick, M.J. and Davies, R.E. (1969) The chemical energetics of muscle contraction. II. The chemistry, efficiency and power of maximally working sartorius muscles. *Proc. R. Soc., Ser. B* **174**, 315–354.

Larson, R.E., Kushmerick, M.J., Haynes, D.H. and Davies, R.E. (1968) Internal work during maintained tension of isometric tetanics. *Biophys. Abs.* **8**, (7) M.A.4.

Lupton, H. (1923) An analysis of the effects of speed on the mechanical efficiency of human muscular movement. *J. Physiol., Lond.* **57**, 337–349.

Margaria, R. (1968) Positive and negative work performances and their efficiencies in human locomotion. *Int. Z. angew. Physiol. einschl. Arbeitsphysiol.* **25**, 339–351.

Margaria, R., Edwards, H.T. and Dill, D.B. (1933) The possible mechanisms of contracting and paying the oxygen debt and the rate of lactic acid in muscular contraction. *Am. J. Physiol.* **106**, 687–715.

Matsumoto, Y., Hoekman, T. and Abbott, B.C. (1973) Heat measurements associated with isometric contraction in fast and slow muscles of the chicken. *Comp. Biochem. Physiol.* **46A**, 785–797.

Mommaerts, W.F.H.M., Seradarian, K. and Marechal, G. (1962) Work and chemical change in isotonic muscular contractions. *Biochim. biophys. Acta* **57**, 1–12.

Muller, E.A. (1930) Der Einfluss der Koutraktionsgeschwindigkut auf den Energieverbrauch bei einer statischten Arbeit. *Arbeitsphysiologie* **3**, 298–308.

Paterson, S. and Goldspink, G. (1973) The oxidation of pyruvate and octonoate by red and white myotomal muscles of the crucian carp (*Carassius carassius*). *Experientia* **29**, 629–630.

Pennycuick, C. (1969) The mechanics of bird migration. *Ibis* **111**, 525–556.

Pugh, L.G.C.E. (1975) The relation of oxygen intake and speed in competition cycling and comparative observations on the bicycle ergometer. *J. Physiol., Lond.* **241**, 795–808.

Schmidt-Nielsen, K. (1971) Locomotion: Energy cost of swimming, flying and running. *Science* **177**, 222–226.

Seaherman, H.J., Taylor, C.R. and Maloiy, G.M.O. (1976) Maximum aerobic power and anaerobic glycolysis diving, running in lions, horses and dogs. (In press).

Taylor, C.R., Schmidt-Nielsen, K. and Raab, J.L. (1970) Scaling of energetic cost of running to body size in mammals. *Am. J. Physiol.* **219**, 1104–1107.

Taylor, C.R., Shkolnik, A., Dmiel, R., Baharar, D. and Borat, A. (1974) Running in, cheetahs, gazelles and goats: energy cost and limb configuration. *Am. J. Physiol.* **227**, 848–850.

Weiss-Fogh, T. (1966) Metabolism aod weight economy in migrating animals, particularly birds and insects. *Nutrition and Physical Activity. Symp. Swedish Nutrition Foundation.*

Woledge, R.C. (1971) Heat production and chemical change in muscle. *Prog. Biophys. mol. Biol.* **22**, 37–74.

4 Co-ordination of invertebrate locomotion

F. Delcomyn

4.1 The problem of co-ordination

Animals are capable of locomotion because of their ability actively to move their bodies or appendages. In order for these movements to be successful in propelling the animal from place to place, they clearly must occur in specific sequences rather than at random; that is, they must be co-ordinated. The ability of an animal to co-ordinate its movements during locomotion therefore requires that it be able to choose what movement to make and when to make it, and the biological problem is to determine what mechanisms allow the animal to make these choices. Now, since the active movement of any part of an animal's body usually requires the contraction of one or more muscles, and since the contraction of a muscle usually requires its excitation by a motor nerve, the question of how co-ordination, that is, the selection and timing of movements, occurs becomes a question of how the nervous system is able to programme a spatially and temporally meaningful pattern of impulses in motor neurons. Seen from this point of view it should be obvious that the specific problem of how animals co-ordinate movements during locomotion is really a part of the broader problem of how animals co-ordinate movements during any behaviour, and this is the context within which the subject of this chapter will be discussed. (See Hoyle (1975) for a vigorous statement of this view.)

The subject is clearly vast enough for a book of its own. In this chapter, therefore, I can only briefly survey our knowledge of co-ordinating mechanisms in the invertebrates. Nevertheless, I will also try to give the reader something of the flavour of current research, an understanding of the major questions being asked today, through a more detailed discussion of some topics.

In order to conserve space, I have been selective in the use of references, citing mainly recent or historically important work, and reviews. Readers interested in more detail should consult appropriate reviews or the proceedings of two recent symposia on locomotion (Stein *et al.*, 1973; Herman *et al.*, 1976).

rdinating mechanisms

:tems

neurobiologists as justification for working with
imals have nervous systems which are considerably
rates, and are therefore much more amenable to
se neural mechanisms which underlie behaviour.
of diffuse neural elements rather than a distinct,
elenterates has not induced all neurobiologists
here has nevertheless been considerable interest
ization of these primitive animals. This interest
r simplicity of organization, but also from the
umed to be at an early stage in the evolution
es to the course of evolution of more complex
result has been a considerable body of work,
in recent reviews (Bullock and Horridge,
osephson, 1974a, b; Muscatine and Lenhoff,
on gained by reading this literature is that
siderably greater degree of behavioural
one might have expected by considering
ry scale.
the locomotion of many coelenterates.
lyfish is one good example. It was long
rhythmic pulsations of the bell (umbra) are
net of neurons at its margin, the frequency of beating being
determined by one of the marginal ganglia acting as a pacemaker. While it
was known that some jellyfish had more than one nerve net, the concept of
functional specialization in these nets had not yet taken hold. Recent work,
however, has revealed previously unsuspected complexity. In most coelenterates, including jellyfish, there may be several specialized conducting systems,
each, in many cases, driven by its own system of pacemakers and often controlling different behavioural acts. Mackie (1975) has reported that in the
jellyfish *Stomotoca atra*, for example, there are at least six such distinct
systems. Tentacular movements, swimming and 'crumpling' (a kind of withdrawal response) are only three of the behaviours which are controlled, each
by a separate system. Colonial hydroids such as *Tubularia* also may have complex hierarchies of pacemaker and conducting systems (Josephson, 1974a).

Control of the vigorous swimming exhibited by some sea anemones appears
even more complex, since this behaviour involves a well-co-ordinated sequence
of actions rather than isolated, simple movements such as those discussed
above. Robson has suggested that both the side to side bending of *Stomphia*
(Robson, 1961) and the tentacular lashing of *Gonactinia* (Robson, 1971)

during swimming are controlled by separate pacemaker systems. But little is really known of the co-ordinating mechanisms which underlie these activities, and how individual components of the entire behavioural sequence are integrated is still a mystery.

Further work on complex behaviours like swimming in sea anemones should yield exceptionally interesting results, because they have all the characteristics of fixed action patterns of higher animals, being released by specific stimuli and progressing through a stereotyped sequence of actions to a definite end. Ross (1973) has speculated that a separate nerve net may be responsible for each separate, complex behavioural act. (The different conducting systems probably represent functional specializations of a single net.) This interesting suggestion raises the possibility that the co-ordinating mechanisms necessary for the production of a complex behaviour evolved *before* the evolution of a central nervous system (CNS) as such, and that the CNS may have developed partly as a morphological consolidation of several semi-independent nerve nets, each of which already controlled a complex behaviour.

4.2.2 The emergence of the central nervous system

Lost in the glare of the scientific attention directed at the coelenterates on the one hand and the complex and large annelids, molluscs and arthropods on the other, the flatworms, round worms and various minor coelomate phyla occupy a kind of neurobiological no-man's-land. This is especially unfortunate in the present context because these phyla can be viewed as representing a series of natural experiments in nervous system structure and function. Detailed comparative studies of even the relatively simple locomotion exhibited by these groups ought to yield a rich harvest in understanding of the many mechanisms by which co-ordination of locomotion might take place, and perhaps even clues as to the advantages inherent in those mechanisms actually used in higher organisms.

Of these neglected groups, flatworms are probably the best known. As can be seen from Bullock's comprehensive review (Bullock and Horridge, 1965), most of the work has been done on the Turbellaria. This is partly because flatworms in this class show much greater variability in neural structure than either the flukes or tapeworms, ranging from virtually brainless species with extensive nerve nets to those with a well-developed brain, ladder-like, paired nerve cords and little in the way of a peripheral nerve plexus or net. Early work had shown that the brain, while not absolutely necessary for some types of locomotion to occur, apparently was for other kinds. So, for example, undulating swimming could be elicited in some species of brainless Turbellaria but not in others. While spontaneous creeping was apparently not affected by lack of a brain (Bullock and Horridge, 1965), in *Planocera*, removal of the brain seemed to abolish any locomotion which involved co-ordinated movements (such as swimming or rapid crawling), but not slow gliding progression

(Gruber and Ewer, 1962). More recent work has shown that in at least one flatworm, *Notoplana acticola*, the peripheral nerve plexus is primarily sensory, and that locomotion is normally initiated via activation of this system (Koopowitz, 1973). Control of the individual muscles of the body, on the other hand, involves direct connections between the muscle and the brain. Cutting these connections destroys the link and leaves the animal unable to make voluntary movements (see Koopowitz, 1974, for review).

To date we have no knowledge of the role played by various parts of the brain or peripheral nervous system in co-ordinating the muscular (or ciliary) movements necessary for locomotion. The recent demonstration that electrophysiological techniques can be applied successfully to Turbellaria, and that basic integrative phenomena such as habituation can be studied (Koopowitz, 1975), perhaps means that we will soon have a better understanding of these animals.

We know even less about co-ordinating mechanisms in roundworms and other minor groups than we do about them in the flatworms. A few studies have been done on swimming in nemertineans, which have close affinities with the flatworms. These have shown similar kinds of effects of brain removal in some groups: cessation of spontaneous locomotion, but the resumption of normal movements upon mild electrical stimulation (Bullock and Horridge, 1965, Chapter 10). However, Bullock (Bullock and Horridge, 1965) points out that no adequate controls were run, and it is quite possible that the cessation of locomotion was due only to the effects of the trauma of the operation.

Some interesting studies have been done on nematodes, which have the peculiar structural feature of having the muscles send long projections to the nerve cord rather than vice versa (DeBell, 1965). Intracellular studies have shown (Jarman, 1959) that the individual muscle cells in a given body region contract rhythmically and in a co-ordinated fashion. The rhythm appears to be myogenic, and co-ordination apparently takes place via current spread between the fingers of the muscular projections, not by neural mechanisms (DeBell, 1965). Shishov (1968) has shown by ablation experiments that one function of the brain is to regulate rather than co-ordinate locomotion, as *Ascaris* shows continuous rather than periodic undulations after the brain is removed. Croll (1975) suggests that there may be a separate control centre for rearward movement.

4.3 The co-ordination of whole body movements

4.3.1 Creeping and looping movements

A wide variety of invertebrates, such as snails, earthworms and caterpillars, progress by creeping along, as described in Chapter 8. While the gliding movement of molluscs has not been investigated electrophysiologically, we do have some information on the co-ordinating mechanisms which underlie peristalsis in some of the annelids and the larval insects. Technical difficulties presented by the presence of a 'hydraulic skeleton' in these animals have caused most modern workers to bypass these groups, so that most of our information is old. For that reason if no other, these animals would probably repay study using current techniques.

Peristaltic progression in both annelids and caterpillars seems to be due to a combination of reflex effects from segment to segment as the peristaltic wave passes, and some kind of neural pattern generated within the CNS. Spontaneous locomotion depends on the presence of the subesophageal ganglion in caterpillars, and the forward wave of contraction will pass over up to three adjacent, denervated segments (von Holst, 1934) from which one might infer that reflex feedback in these segments is not required for propagation of the peristaltic wave. Weevers (1966) has demonstrated a variety of reflex effects mediated by muscle receptor organs in the body wall of caterpillars. However, since his studies showed that these reflexes elicited a pattern of muscular response quite different from that which can be observed during crawling, he concluded that the role of these reflexes was to maintain a desired body movement in the face of varying external forces, not to aid in the co-ordination of the peristaltic wave. Bullock (Bullock and Horridge, 1965, Chapter 14) suggests, largely on circumstantial evidence, that the rearward peristaltic wave of annelids is generated primarily by the CNS.

On the other hand, no one has demonstrated the presence of an appropriately timed pattern of output in an isolated nerve cord in either annelids or caterpillars. In earthworms, it is possible to initiate contraction of circular muscle by applying a stretch to the body wall (Gray, Lissmann and Pumphrey, 1938), and Friedländer (1894, cited by Bullock and Horridge, 1965) claimed to have obtained propagation of peristaltic waves across a gap separating two halves of an earthworm as long as these were linked mechanically by pieces of thread tied to the body wall. Gardner (1976), in his recent review, concludes that in earthworms the origin of the peristaltic wave is still completely unknown.

The looping locomotion of geometrid moth caterpillars ('inchworms') and leeches out of water may seem rather different from peristaltic creeping, but as Hughes and Mill (1974) have pointed out, it may be viewed as peristalsis in which the part of the body which is moving forward but is not attached to the substrate encompasses most of the length of the body. Nothing is known about the co-ordination of this movement in caterpillars, but in leeches it appears to be largely the result of the action of a chain of reflexes. Gray, Lissmann and Pumphrey (1938) showed that each of the actions of the leech during its progression: attachment of one sucker, release of another, contraction or extension of the body, etc., can be elicited by the action which normally precedes it, thereby forming a classical closed reflex chain. The CNS may only participate in the starting or stopping of locomotion.

While creeping in polychaetes is accomplished by parapodial movements

86

sometimes combined with sinuous bends of the body rather than by peristalsis, it may profitably be discussed here. Lawry's (1970) recent study, and his review (Lawry, 1973), provide most of the known details. He reported two main findings: (1) Removal of one or more parapodia had no effect on the sequence of movement of the remaining appendages. (2) If the parapodial nerve running from the CNS to the parapodia were cut, or if the parapodial ganglion were removed with the parapodium, waves of movements of the parapodia in the unoperated segments would not pass the site of the lesion or ablation. Instead, new waves would be set up on the other side of the cut. Apparently, an intact nerve cord, parapodial nerve and parapodial ganglion are required in each segment for normal propagation of waves of parapodial movements.

While we seem to have a better understanding of the neural basis of locomotion in annelids and caterpillars than we do for any of the other groups already discussed, we are still a long way from answering even the most basic questions about co-ordination in these groups. Although the data for earthworms and leeches suggest a prominent role of reflexes in co-ordinating movements in these animals, while caterpillar and polychaete locomotion may be more centrally organized, these conclusions should be considered tentative at best. In the first case, it is not sufficient to demonstrate that reflexes which might initiate various movements are present, since these reflexes may well only serve a role in refining movements whose timing is established entirely by interactions within the CNS (see Section 4.4.1). In the second case, noting that severing the nerve cord or ablating certain ganglia results in disruption of a locomotor pattern does not eliminate the possibility of an important role for reflexes in co-ordination since such operations may disrupt *intersegmental* reflex pathways. Lawry (1973) has pointed out that no one has yet recorded patterned output from a de-afferented polychaete nerve cord. The conclusion drawn from this should be that there is a great deal more to be done before we can begin to comprehend the complexity of co-ordinating mechanisms of these movements.

4.3.2 Swimming

Leeches

While swimming in some annelids is accomplished by a series of side to side twists of the body, in leeches it consists of a series of vertical rather than horizontal waves. Perhaps surprisingly, we know quite a lot more about the regulation of this movement than we do about the movements already discussed, due partly to the fact that the medicinal leech, *Hirudo medicinalis*, has become a favourite subject for some cellular neurophysiological research.

Leeches swim by flattening the body and entering into a series of vertical undulations (Gray, Lissmann and Pumphrey, 1938; Kristan, 1974; Kristan,

Stent and Ort, 1974a). These undulations are powered by alternating, sequential contractions of the dorsal and ventral longitudinal muscles. This has been shown by Kristan *et al.* (1974a, b) by direct recordings from nerves during swimming in a semi-intact preparation. Given this information, two main questions arise: (1) How is the alternation of the contractions of the dorsal and ventral longitudinal muscles in one segment maintained? (2) How is the proper delay between the contraction of muscles in one segment and those in adjacent segments maintained?

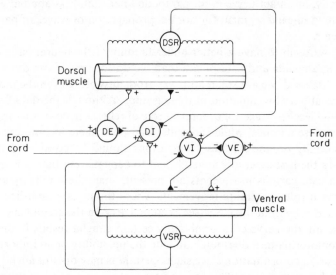

Fig. 4.1 A diagram of connections among ventral and dorsal motoneurons and hypothetical stretch receptors in a leech. Note that proper reciprocal activation of the dorsal and ventral longitudinal muscles depends on the presumed sensory input. See text for further discussion. Symbols: DE, VE – dorsal and ventral excitatory motoneurons; DI, VI – dorsal and ventral inhibitory motoneurons; DSR, VSR – presumptive dorsal and ventral stretch receptors. Each circle may represent more than one neuron. Pathways shown may be polysynaptic. (From Kristan, 1974.)

Studies of the interactions between motoneurons in a single segment have established the source of their rhythmic output (Kristan, 1974; Kristan *et al.*, 1974b). The most important known and presumed interactions are presented diagrammatically in Fig. 4.1. Two aspects of this network should especially be noted: both the dorsal and ventral longitudinal excitatory motor neurons are under tonic excitation from other neurons within the nerve cord, and, there is no central connectivity between these sets of neurons sufficient to establish alternating bursts of activity. Instead, stretch receptors which presumably exist in or near the animal's skin, and which would be active when the body was bending away from them, could send impulses into the CNS to influence (via inhibitory neurons, as shown) the activity of both dorsal and ventral longitudinal muscles. Thus, stretch of the ventral stretch receptor as the animal bent dorsally would result in eventual inhibition of the dorsal muscles via the dorsal inhibitors, and activation of the ventral muscles by the on-going tonic excitation as it was released from the inhibition of the ventral inhibitor. (But see p. 114.)

While the evidence for the action of input from stretch receptors described above is largely indirect (someone has yet to record from the sense organs), evidence for the source of co-ordination between adjacent ganglia is much more direct (Kristan *et al.*, 1974b). Recordings from the nerve cord connecting two

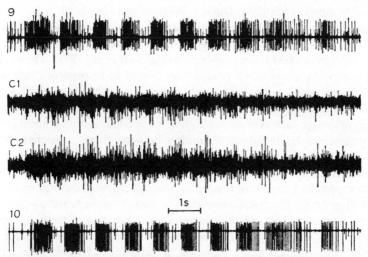

Fig. 4.2 Records of rhythmic neural activity in motor nerves (9, 10) and interganglionic connectives (C1, C2) during swimming. The motor nerve records are from adjacent ganglia, while the connective records are both from the connective joining the two ganglia. Note the presence of large units in records C1 and C2 which are phase-locked with the bursting motor activity. (From Kristan *et al.*, 1974b.)

ganglia have revealed activity which is phase-locked with the motor bursts generated by the CNS (Fig. 4.2). This activity may well represent co-ordinating information travelling from one ganglion to the next to ensure the proper timing of the muscles in that segment, similar to the 'co-ordination' fibres described in crayfish (Stein, 1974). Such fibres may be able to override at least some of the influence of the sensory feedback loop described above.

Molluscs

While several kinds of molluscs, such as octopus, scallops, sea butterflies, sea hares and others are capable of swimming, the neural basis of this behaviour has been investigated in very few. Probably the best known example is that of the nudibranch, *Tritonia*. The swimming behaviour, which can be elicited by contact with certain predaceous starfish, consists of several discrete actions performed in a fixed sequence (Willows, Dorsett and Hoyle, 1973): local reflex withdrawal of the branchial tufts and rhinophores, elongation of the animal and flattening of the oral veil, a series of about five cycles of vigorous alternating dorsal and ventral flexions, and finally, relaxation. Only minor variations in this behaviour (such as in the number of cycles of dorsal and ventral flexions) are ever observed – the sequence itself is always the same. By recording intra-cellularly from cells in the brain of an otherwise intact animal during swimming (a tethered preparation), it was shown that the stereotyped behaviour was reflected in stereotyped activity in a number of cell groups in the brain of the animal (Willows *et al.*, 1973; Dorsett, Willows and Hoyle, 1973).

Further study showed that the pattern of activity exhibited by these cell groups in tethered preparations could also be elicited by appropriate electrical stimulation in isolated brain preparations (Dorsett *et al.*, 1973). The sequence and pattern of activity of the groups thus seemed to be entirely a function of neural connections within the brain, requiring no external input to help establish the timing. Experiments also suggested that the alternation of the activity of groups such as those controlling the dorsal and ventral flexor muscles was maintained by reciprocal inhibition between them, a mechanism also postulated for alternation of muscle groups during flight (Wilson, 1964a; see Section 4.4.1).

One especially interesting feature of this system was a group of neurons called trigger cells. Electrical stimulation of the part of the brain in which these cells are located always induced swimming, and bursting of these cells always preceded it. It was therefore thought that these cells were responsible for initiating or triggering the entire sequence of behaviour (Willows *et al.*, 1973). This was an important concept, since it appeared to provide a much-needed physiological analogue of the ethologists' sign or releasing stimulus, a brief cue whose presence could initiate an entire sequence of behaviour. Examples are known in other invertebrates of neurons which can drive co-ordinated behaviour (e.g., Bowerman and Larimer, 1974), but in these cases the neuron(s) must be active continuously during the behaviour.

However, recent work by Getting (1975) has undermined this view. Getting reported that in isolated brains he was unable to initiate swimming by intra-cellular stimulation of several of the trigger cells, even though he could induce spikes in them, and that he could elicit the swimming pattern even when some of the trigger cells were hyperpolarized strongly enough to prevent bursting not only in the hyperpolarized cells but in others in the trigger group as well. Interestingly, Getting's results have not proved to be repeatable on the northern (Puget Sound) population of *Tritonia* (Willows, personal communication). Getting obtained his animals from Los Angeles, and although they appeared indistinguishable morphologically (Getting, personal communication), this difference may be a factor.

The resolution of the problem must await the results of further experiments. There seems to be no evidence of any sustained neural activity which is on during the entire swimming cycle. A functional 'trigger' to initiate the behaviour would therefore still seem to be necessary, although this function could be served by the initial sensory input itself. If the cells we have been calling the trigger group do not in fact have a role in initiating the behaviour, then their function must also be discovered. Work on *Tritonia* may therefore be expected to attract special attention for some years to come.

4.3.3 Locomotion in echinoderms

While echinoderms are of a convenient size and readily available from the seas of the world, their calcarious skeletons and exceedingly small neurons (axons about 0.1 μm) have combined to discourage work on the group. In consequence, we have only a poor understanding of the means by which locomotion in these animals is co-ordinated. This is especially unfortunate because the problems in co-ordinating the activity of the hundreds of tube feet used for walking by most echinoderms are quite interesting.

There have been two main theories of how movements of the tube feet are controlled in starfish, one of the most popular experimental animals (see e.g., Kerkut, 1954). The first states that the direction of stepping in each arm is a function of the combination of forces acting on the feet of that arm – i.e., the arm which exerts the strongest pull induces the others to move in the same direction. The other invokes a system of central control over the arms to ensure that all the tube feet step in approximately the same direction. There are certainly indications of the independent nature of the stepping of individual feet, such as the fact that there are no fixed phase relationships between the stepping of feet in any one arm. Some feet also may for a while even move at different rates from others. But it is equally true that the feet of all arms tend to step in the same direction, and when that direction changes, as it frequently does, it changes simultaneously in all arms. These and other observations by Kerkut (1954), Smith (1966) and others, especially on the effects of lesions in the radial or circumoral nerves, lead most modern workers to the conclusion that there are neural 'centres' at the junction of the circumoral nerve ring and each radial nerve which are essential to normal co-ordination between arms. (See reviews by Smith in Bullock and Horridge (1965), Chapter 26, and Binyon (1972).)

4.4 The co-ordination of arthropod locomotion – flight

4.4.1 Arthropods and the central programming of locomotion

Partly because they possess hard exoskeletons and jointed appendages which force them to make movements that are more restricted and easier to analyse than the movements of soft-bodied invertebrates, adult arthropods have long been especially popular subjects for research on mechanisms underlying co-ordination, and a great deal of work in this field has been done on them. The result of this work has often been concepts which are relevant not just to the arthropods, but to all animal groups. Indeed, one of the most important modern concepts of the neural control of behaviour, that of 'central programming' of behaviour, was first clearly demonstrated in an arthropod. The idea of central programming (also called central control or central patterning) is that the central nervous system is capable of producing a pattern and sequence of motor nerve activity appropriate to a rhythmic behaviour without using any of the sensory information generated by the ongoing behaviour for timing cues. This is in contrast to the older idea that the CNS could only generate a patterned, repetitive motor output by using cues from the sensory signals produced by the ongoing activity. The idea of central control has now come to dominate the entire field of 'neuroethology' research, and the mechanism is thought to underlie not only the repetitive movements of wings and legs in insects (Wilson, 1961; Pearson and Iles, 1970), but such diverse behaviours as swimming in nudibranch molluscs (Willows *et al.*, 1973) and grooming in mice (Fentress, 1973), as well as some vertebrate locomotion (see Chapter 5).

While a certain amount of work is still being done to determine whether particular behavioural acts are co-ordinated via central control, many investigators have turned to the problem of what arrangement of and interactions between neurons allow the CNS to generate a complex spatial and temporal pattern of motor impulses without relying on any external cues for the main features of that pattern. Since our knowledge of the neuroanatomy of any invertebrate has lagged far behind our knowledge of neural physiology, until the last 2 or 3 years we could really only tackle this problem in a theoretical way, by building neural models. Neural models represent networks of living neurons, and may be constructed of electronic analogues of neurons or be simulated on a computer. When properly done, studies of such models may contribute significantly to the progress of physiological work by allowing us quickly to test a variety of theoretical possibilities, and by providing us with quantified, testable predictions about the operation of the living system (Wilson, 1966b).

One of the most influential of the modern models of central control mechanisms was proposed by Wilson (1964a) on the basis of his own work on the locust flight system (Wilson, 1961; Wilson and Weis-Fogh, 1962) and the computer simulation studies of Reiss (1962). The model, shown in Fig. 4.3, uses a pair of reciprocally innervated neurons driven by common input from a continuously active driver neuron to generate a pattern of alternating activity in the two output lines. Wilson's conception was that there is a network of neurons with properties approximated by the model for each set of antagonistic muscles controlling the movement of each appendage. These networks, called 'oscillators' by Wilson, would then be coupled together in some way to ensure the proper timing of movements of one appendage relative to another. Wilson (1966b) and Wilson and Waldron (1968) performed extensive simulations of this and similar models. The tested models unfortunately were often relatively unstable, in that in some cases parameters of the model neurons had to be set within rather narrow limits in order for the net-

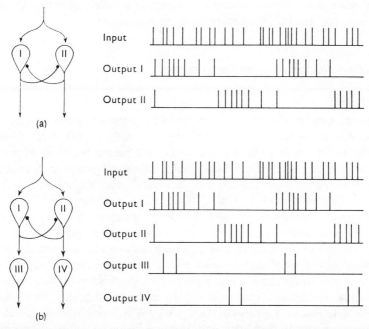

Fig. 4.3 Hypothetical models of neural 'oscillators', networks of connections by which the alternating bursts of activity in antagonistic muscle pairs during rhythmic behaviour could be generated. The accompanying diagrams of input and output patterns show that continual input would be converted to appropriately timed bursts of output. Note that if I and II are to represent motoneurons (a), the reciprocal bursts must overlap. This constraint may be eliminated by allowing I and II to represent interneurons (b). In flying or other rhythmic behaviour, each appendage is thought to be controlled by its own oscillator. (From Wilson, 1966b).

work to function properly (Wilson, 1966b). Wilson and Waldron (1968) suggested that perhaps in the living animal there were several networks whose interaction increased the stability of the output. They, and later Perkel and Mulloney (1974), also showed that networks in which post-inhibitory rebound was an important factor were significantly more stable than those in which it was not. Recently, Dagan, Vernon and Hoyle (1975) have developed a model system which is both reciprocating and self-excitatory, being initiated by only a single trigger pulse. This sort of network has obvious relevance to systems such as that controlling swimming in *Tritonia* (Willows *et al.*, 1973; see Section 4.3.2).

In the end, modelling can only have biological significance in so far as it generates models the properties of which are accurate reflections of those of living systems. Modelling of neural networks has been considerably retarded by lack of adequate neuroanatomical information. Now that new techniques of staining (see Kater and Nicholson, 1973) have begun to bear fruit, we may begin to see more rigorous testing via models of a variety of networks which have been proposed for the generation of motor patterns.

4.4.2 Co-ordination via central programmes

Locust flight

The mechanism by which the movements of the wings of locusts are co-ordinated during flying was the subject of Wilson's early studies, and of all the behaviours for which we believe control is via a central programme, it is the most thoroughly studied (Wilson, 1961, 1964a; Wilson and Weis-Fogh, 1962; Wilson and Wyman, 1965) and the most thoroughly reviewed (Wilson, 1964b, 1968a; Huber, 1974). During flight, a locust moves each wing up and down with its contralateral partner, the hindwings about 10% earlier than the front wings. In a series of elegant experiments, Wilson showed that the alternating activity of the elevator and depressor muscles for each wing, the synchronous movement of each wing with its partner, and even the lag between front and rear wings could all be produced by the CNS without it requiring any timing information from sense organs in the moving wings or elsewhere (see Fig. 4.4). Since the reviews of this work by Wilson and others (cited above) discuss in some detail the evidence for this conclusion, as well as some proper-

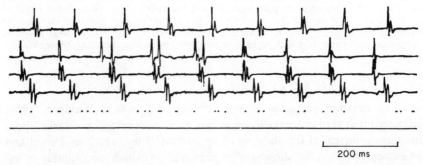

200 ms

Fig. 4.4 Rhythmic activity in four flight muscles of a wingless locust during random stimulation (monitored on the bottom trace). Note that the rhythmic output is maintained even though the stimulus is random, and the CNS is receiving no phasic sensory feedback from moving wings. (From Wilson and Wyman, 1965. Reproduced from *Biophys. J.*, 1965, p. 134, by copyright permission of the Rockefeller University Press.)

ties of the neurons whose activity underlies the generation of flight output, I will not repeat it again here.

However, actually finding in the living ganglion the cells responsible for generating the flight pattern, and describing the mechanism by which the pattern is produced, has proved to be exceptionally difficult. Page (1970), for example, reported complete lack of success in his attempts to locate and record from interneurons which participated in the regulation of flight. Recent advances in both intracellular recording techniques (Hoyle and Burrows, 1973) and neuroanatomical knowledge of ganglionic structure (Tyrer and Altman, 1974) have led the way to more successful approaches. Burrows (1973) has exploited these techniques to demonstrate the existence of a variety of indirect pathways linking a motoneuron with its ipsilateral antagonist and its contralateral homologue. Appropriate delays were found in the linking pathways to ensure that all the motoneurons were excited at the appropriate time during the cycle of wing movements. While there is considerable evidence that coupling between motoneurons cannot alone be responsible for generation of the pattern of flight output (Wilson, 1961, 1964a; Burrows, 1973), these findings of Burrows clearly indicate that the final description of the complete flight system will have to include such pathways to be complete.

In addition, while it is clear that the central nervous system is capable of generating a properly timed motor pattern without using any rhythmic sensory feedback, recent work has shown that it may in fact use such feedback. Waldron (1968) showed that a light flashing at a frequency close to that of the normal wing beat was capable of entraining that beat. Wendler (1974) showed that forcibly moving a wing over a narrow range of frequencies close to the normal wing beat frequency, could also entrain the beating of the free wings. And Burrows (1975) showed that stretch receptors at the bases of the wings synapse directly on the flight muscle motoneurons. Input from these receptors was shown to be able to influence the timing of the spikes produced in the motor nerves. There is no doubt that the isolated CNS can produce a properly patterned motor output closely similar to the pattern observed during flying in intact locusts. However, these papers make it clear that during straight flight rhythmic sensory feedback may influence the timing of the pattern to a small extent.

Flight in myogenic fliers

Locusts have two pairs of wings to control during flying, and since the pairs do not beat in synchrony, the insect must have mechanisms for ensuring that proper phasing between them is maintained, as discussed in the preceding section. More specialized insects such as bees and flies, however, do not have the same problem, because they either fly with the front and rear wings physically coupled together (bees) or they only have one pair of wings. In these insects the problem of how wing movements can be properly timed and co-ordinated is obviously simplified. But there is a feature of the physiology of the flight muscles of these insects which transforms the problem entirely.

The muscles which move the wings in locusts and other unspecialized insects, such as crickets, cockroaches and dragonflies, are neurogenic. That is, they contract only when one or more nerve impulses is delivered to them. In order for the muscles to be made to contract in a particular pattern or sequence, the nerve impulses delivered to those muscles must also be patterned and sequenced, as I have discussed. However, the muscles which move the wings of flies and bees are myogenic. Once activated by a nerve impulse, such muscles can for a short time contract merely in response to a mechanical stretch. As long as nerve impulses continue to arrive at the muscles, they will continue to respond to stretch by contracting. Since the muscles which power wing movements are mechanically linked to the wings, the beating of the wings will cause a quick stretch to be applied alternately to the wing levator and depressor muscles. Therefore, once the wings begin moving, the nervous system does not need to deliver a carefully patterned sequence of motor impulses to the muscles; it need only deliver impulses frequently enough to keep the muscles activated so they continue to respond to the stretch repetitively imposed on them.

Studies of the actual pattern of activation of the main flight muscles in flies, bees and bugs have shown that motor impulses to wing levator and depressor muscles indeed lack the closely controlled pattern which can be seen in locusts (Wyman, 1966; Mulloney, 1970). In fact, Wilson and Wyman (1963) have argued that in flies only the average frequency of motor nerve impulses is important to the muscle. Close study showed (Wyman, 1966) that over periods of many seconds one could detect weak, intermittent relationships between the activity of one unit and that of another, but in general, tight coupling seemed to be absent.

However, recent work has made it clear that absence of *tight* coupling does not mean absence of *all* coupling. In honey bees, it is clear that action potentials in the dorso-ventral muscles (those responsible for wing elevation) have a strong tendency to occur at the start of the upstroke (Bastian and Esch, 1970; Esch, Nachtigall and Kogge, 1975). The potentials in these myogenic muscles do not occur with every wing beat, nor do they occur only at the preferred phase of the wing beat cycle, but they do occur at that preferred time most frequently. Examination of long records of activity in the muscles responsible for depression of the wing showed that while these muscles occasionally exhibit a preferred phase, they often do not. In flies, a study in which phase relationships between muscle activity and wing movements were sought by statistical analysis of long records has shown that in these insects there are often (but not always) clear preferred relationships (Heide, 1974), although these are not obvious in short records. Heide (1974) has suggested that some 'wingbeat-synchronous feedback' (i.e., phasic sensory input) was important in establishing the observed phase preferences. However, the important question of what role such feedback may actually play in the establishment of such preferred phases is one which no investigator has yet rigorously attempted to answer.

In contrast to these studies of inter-muscle co-ordination, an interesting study has recently been carried out by Harcombe and Wyman (1977) on intra-muscle co-ordination in the fruit fly *Drosophila*. Their main finding was that the output of the motoneurons innervating the individual fibres of each single flight muscle was highly patterned. The motoneurons had a strong tendency to fire in a cyclic, repetitive pattern, such that each neuron fired once per cycle, seriatim, and all the neurons divided the cycle into approximately equal intervals. (The motoneurons of each separate muscle had their own character-istic firing frequency, so there was no apparent coupling between muscles.) The results of a variety of experiments led Harcombe and Wyman to conclude that strong, inhibitory connections among all the motoneurons of each muscle was the primary cause of the observed pattern. At this level of co-ordination, and in *Drosophila*, then, it appears that central connections between moto-neurons can explain completely the observed pattern of output. Nevertheless, we still know considerably less about the neural basis of flight co-ordination in flies, bees and similar insects than we do in locusts.

4.4.3 The effects of sensory input on flight

If sensory input is not required for the production of normal flight patterns (at least in locusts), what role does it play? The answer, which holds for every insect studied in detail, is that it enables the insect to detect and respond to unintended deviations from its flight path, and to regulate the speed and direc-tion of its flight.

Flight stability

In order to maintain stable flight, an insect must be able to correct deviations from its preferred orientation in three planes: the horizontal (deviation in which is yaw), the vertical along the body axis (pitch) and the vertical at right angles to the body (roll). Insects can easily make such corrections.

Roll correction seems to be primarily a visual response. Wilson (1968b) and others have shown that when a tethered flying locust is allowed to roll in the dark, it flies in no preferred position. However, it always flies dorsal side toward the light if light is available. If an artificial horizon is rotated in front of the locust, it will roll to retain its preferred orientation to the light-dark boundary. The primary orientation is actually of the head, which selects its preferred position first. Proprioceptors at the neck detect deviations from the normal position and their output in turn causes differential twisting of the wings, which then rotates the body (Camhi, 1970a).

Yaw and changes in pitch can be detected by sensing changes in the direction of wind impinging on the insect's head, and several wind-detecting sense organs have been described. In locusts, there are a number of short hairs scattered about the front of the head. These hairs are grouped into sets, each set having a unique orientation on the head of the insect. Each set responds differentially to wind from different directions, and each therefore acts as a sensitive indicator of the presence of wind from a particular direction (Camhi, 1969a). The primary sensory neurons synapse with interneurons the response patterns of which tend to sharpen discriminations between wind directions (Camhi, 1969b). In flies, but apparently not in locusts, the antennae are used to help correct yaw during flight (Gewecke, 1974).

Other sense organs may also be used. Pringle (1948), Schneider (1953) and others have shown that the halteres are important organs in the regulation of flight stability in flies. Many insects also use visual cues to detect yaw.

Given that an insect is capable of regulating its orientation in space, what is the means by which this may be accomplished? The aerodynamic forces which cause a change in orientation may be the result of the rudder-like use of legs or the abdomen (locusts, Dugard, 1967; Camhi, 1970a, b; bugs, Govind and Burton, 1970), or of differential wing twisting or other changes in the stroke (locusts, Dugard, 1967; Wilson, 1968b; moths, Kammer and Nachtigall, 1973; bugs, Govind and Burton, 1970; flies, Schneider, 1953; beetles, Burton, 1971). But our interest here lies more in the question of what co-ordinating mechanisms underlie turning and other changes in orientation. In locusts, present evidence suggests that sensory input from various sense organs is superimposed on the endogenously generated rhythm delivered to the flight muscle motoneurons (Wilson, 1968b; Waldron, 1967). When the input (e.g., from wind receptors on the head) is asymmetrical, the effect on the muscles is also asymmetrical, resulting in a differential effect on the movements of the right and left wings (Dugard, 1967). Sensory input may impinge on non-flight muscles as well, in which case its effects may be switched on by activation of the flight control system. For example, a flying locust will swing out its legs and abdomen in response to air directed at it from one side of the head or the other, but will exhibit no such response when it is not flying (Camhi and Hinkle, 1974).

In some insects, different muscles control the basic up and down movements of the wings (the power-producing movements) than control twisting and other aspects of wing attitude during the stroke (flies, Nachtigall and Wilson, 1967; moths, Kammer, 1971). In such insects, the sensory input which initiates a corrective manoeuvre may therefore bypass the main flight muscles and influence only the others. In several moths, for example (the wings of which are powered by indirect flight muscles), during turns one can see significant phase shifts in the relative timing of muscle potentials in some of the muscles which adjust wing attitude during the stroke, but not in the indirect muscles themselves (Kammer, 1971; Kammer and Nachtigall, 1973). Schneider (1953) and Gewecke (1974) suggest that in flies asymmetric sensory input from the antennae, generated by a sudden yaw, acts asymmetrically on the direct flight muscles which control wing twisting, so as to produce a turning tendency which counteracts the original yaw stimulus. Nevertheless, the possibility of the presence of such a mechanism does not seem to exclude direct effects on the main flight muscles. In beetles, differential changes in wing amplitude are the result of changes in the average frequency of impulses delivered to the indirect flight muscles on one side of the body only (Burton, 1971). Wyman (1970) suggested that the sudden changes in power output to different flight muscles he observed over long periods of time in certain flies was the basis of the well-known erratic flight path of these flies.

Flight speed

Several receptors have been shown to exert significant control over flight speed. Sensory hairs on the heads of locusts provide important input to the flight oscillator, since covering these hairs with wax causes a marked reduction in the speed of flight (Gewecke, 1974). Due to their sensitivity to changes in wind velocity, they also play an important role in reflexes which allow the locust to avoid stalls due to an angle of attack which is too large (Camhi, 1969c). In some insects, antennae seem to have an opposite effect on flight speed. In both blowflies and locusts, immobilizing the antennae results in an increase in speed of flight (Gewecke, 1974). However, the antennae of some beetles seem to play no role in the regulation of speed (Schneider and Krämer, 1974).

Receptors at the bases of the wings in locusts have a general excitatory effect on wing beat frequency, and therefore also on speed of flight (Wilson, 1964b). Destruction of these receptors caused a reduction in the frequency of wing beating proportional to the number of sense organs destroyed. Destruction of all four resulted in a reduction of about 50 %, destruction of any two of 25 %, and so forth. It had been thought that the specific sense organ involved was a stretch receptor at the base of each wing. Since later work (Burrows, 1975) showed that these receptors synapsed directly on major flight muscle motoneurons, it has been suggested that other receptors might be better candidates to provide the general tonic effect which can be observed.

4.5 The co-ordination of arthropod locomotion – swimming

4.5.1 Insects

Although the vast majority of insects are entirely terrestrial, there are a number of important groups which spend all or part of their lives in or on water. Unfortunately, very little work has been done on mechanisms of co-ordination of swimming. While some insects move pairs of legs alternately in the water, as do walking insects on land, others move them synchronously (see Nachtigall, 1974). Studies of these insects on land and in water would be especially interesting, because movements of the appendages during walking and during swimming may be entirely different (Walcott, 1968), suggesting that entirely separate neural programmes may be responsible for co-ordination in the two cases.

4.5.2 *Crustacean swimming and the control of swimmeret beating*

While many small crustaceans are entirely free-swimming throughout their lives, most of the larger ones, such as crabs, crayfish, lobsters and the like, spend their adult lives walking about on the bottom rather than swimming. As a consequence there has been little work on the co-ordination of swimming movements in these animals. Spirito (1972) has studied the leg movements of one swimming crab, *Callinectes*, and concluded on the basis of amputation experiments that leg movements are probably centrally co-ordinated. Other crustacea swim by moving their tails or tail appendages. The sand crab *Emerita*, for example, beats its uropods (appendages of the last abdominal segment) in order either to swim or to tread water. The results of ablation and de-afferentation experiments (Paul, 1971, 1976) suggested that swimming was entirely centrally programmed, although reflexes might help adjust the finer features of the beat. Treading water, however, required a different pattern of motor activity, and this pattern could only be generated when proprioceptive feedback from sense organs in the base of each uropod was present. The crab must therefore be able to ignore or respond to this input at will, depending on whether it wanted to swim or tread water.

Although crayfish and lobsters usually walk, swimming, driven by powerful tail flips, is well known in these animals. This 'escape' swimming can be elicited by stimulation of giant axons in the nerve cord, and has therefore usually been investigated in the context of studies of these cells. However, Schrameck (1970) has clearly shown that crayfish (and presumably lobsters as well) are quite capable of swimming without ever using the giant axons, and that therefore co-ordination of swimming does not require involvement of these axons at all. We are left with the surprising fact that in spite of the extent of our knowledge of the neural basis of crayfish behaviour, we know virtually nothing about how swimming is co-ordinated in this animal!

Many small crustacea swim by beating the ventral abdominal appendages called swimmerets. The young of large animals such as lobsters also use these paddle-like structures for swimming. The neural mechanisms underlying the co-ordination of the beat of swimmerets has been studied extensively in crayfish and lobster adults (in which they persist). Since these mechanisms may be presumed to be similar in young lobsters or crayfish, or in small, related crustacea, I will discuss them here. It has been shown that the basic metachronal rhythm of swimmeret beating, which starts with the movement of the most posterior pair and sweeps forward, is generated entirely by the CNS, without any need for timing cues from sensory feedback (Ikeda and Wiersma, 1964; Davis, 1969). As in the locust flight system, reflexes serve to adjust the finer aspects of the beat by affecting, for example, power output (Davis, 1969).

The basic unit of the system is also thought of in a similar manner, as a neuronal 'oscillator' or pacemaker (Ikeda and Wiersma, 1964) controlling the individual movements of each swimmeret, with the entire system being co-ordinated by appropriate endogenous coupling between the oscillators.

It is further known that the operation of the swimmeret system is driven by 'command' interneurons located in the ventral nerve cord (Wiersma and Ikeda, 1964). The detailed work of Davis and Kennedy (1972a, b, c) has shown that

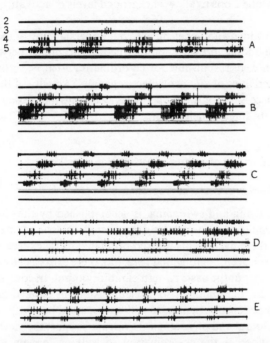

Fig. 4.5 Recordings from the nerves innervating the power stroke muscles on the left side in four segments of the abdomen, showing the effects of stimulating five different command interneurons. Note the wide differences in the response of the nerves in each segment following the stimulation of the different command fibres. The numerals 2–5 refer to the segment from which each trace was recorded. (From Davis and Kennedy, 1972a.)

individual command fibres affect two main aspects of motor output delivered to the swimmerets: the period between successive bursts (i.e., the frequency of repetition of bursts), and the population of motoneurons activated (Davis and Kennedy, 1972a, b). When stimulated, each excitatory command fibre would produce bursts at a period within the range 300–800 ms, with higher frequencies of stimulation usually producing shorter periods. However, no single fibre was found which would encompass the entire range. In addition, some command neurons would activate most of the powerstroke muscles in the anterior segments and few of those in the posterior ones, and vice versa (Fig. 4.5). Interactions of several command fibres were additive, so it appeared that in the intact animal more than one command fibre would normally be active.

Quantitative studies of the timing of bursts produced by command inter-neuron stimulation (Davis and Kennedy, 1972c) combined with the other work led to several conclusions, the main one being that the output of the command fibres impinged on the postulated neuronal oscillators in each half ganglion (the oscillators being separate from the motoneurons) and also sent information directly to motoneurons to allow background levels of activation or inhibition to be set.

Stein (1974) added another element to our understanding of the system when he demonstrated the presence of interneurons running between adjacent ganglia, which were responsible for co-ordinating, that is, setting the timing between the activities of any two adjacent ganglia. Activity of these fibres was shown to be necessary for the proper delay between synergistic muscles in one ganglion and its anterior neighbour.

4.6 The co-ordination of arthropod locomotion – walking

4.6.1 Co-ordination via central programmes

Walking, like flying and swimming, is accomplished by the rhythmic move-ments of appendages. Although long considered a classic example of a behav-iour in which these movements were co-ordinated by reflex feedback, evidence has recently been accumulating that here, too, there exists a central programme capable of laying out the basic pattern of motor output. In contrast to the flight system, however, the timing of some elements of this pattern can be strongly affected by some sensory signals, as will be discussed below.

Both behavioural and physiological evidence support the view that a central programme underlies the co-ordination of walking movements. Delcomyn (1971a, 1973) inferred that, in cockroaches, reflexes alone could not account for the sequence of leg movements because some parameters of these movements were extremely stable in the face of a variety of experimental manipula-tions. Clarac and Coulmance (1971) drew similar inferences concerning walking in crabs. Hoyle (1964) suggested that walking in locusts was at least initiated by a central command (his 'motor tape') since the pattern of subliminal activity in leg muscles was similar to the pattern observed during overt move-ments elicited by similar stimuli and no proprioceptors could be providing cues when no movements took place. This view would seem to hold for some crustacea as well, since Bowerman and Larimer (1974) and others found 'command' fibres in the circumesophageal connectives of crayfish, the stimu-lation of which would induce walking movements.

The obvious experiment to try, of course, would be the de-afferentation of all the walking legs. While technically difficult, the ideal of complete de-afferentation without damage to motor nerves has been approached in the cockroach. Pearson and Iles (1970) were able to show that rhythmic, alternating bursts of activity in a pair of antagonistic leg muscles in a cockroach persisted even after all sensory connections between each leg and the CNS were cut (Fig. 4.6), thereby depriving the nervous system of any timing cues possibly provided by sense organs in the legs. The main features of the bursts in such de-afferented preparations (duration, frequency, etc.) were very similar to the features of bursts in free-walking insects and consistent with the movements of the legs in intact animals (Pearson, 1972; Delcomyn and Usherwood, 1973). These findings strongly supported the claim that the bursts do represent

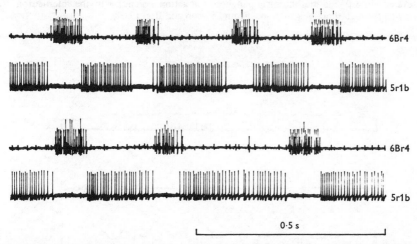

Fig. 4.6 Records of alternating bursts of activity in extensor (5r1b) and flexor (6Br4) leg motoneurons in a de-afferented cockroach preparation. Top and bottom records continuous. (From Pearson and Iles, 1970.)

part of the output of the locomotor control system. On the basis of this work there is little doubt that some system of central neurons, an oscillator, is capable of generating the properly timed alternation of antagonistic muscles in each leg without reference to incoming sensory signals.

Pearson and Iles (1973) have also investigated the question of whether individual oscillators are neurally linked to one another. They present two kinds of evidence to support such linkage. In one series of experiments, they found axons in the connectives between adjacent thoracic ganglia which fired in phase with the firing of some of the flexor motoneurons of the leg (Fig. 4.7). Since they were able to eliminate the possibility that these fibres were either primary sensory fibres or interneurons which followed sensory input, they suggested that the fibres were interneurons probably carrying information necessary for co-ordination from one ganglion to another. (Compare the co-ordination fibres described by Stein (1974) for the crayfish swimmeret system.

0.5 s

Fig. 4.7 Records of bursts in flexor leg motoneurons (top trace) from a metathoracic leg and an interneuron, in the ipsilateral meso-metathoracic connective. Since the connective was severed close to the mesothoracic ganglion, the action potentials in the interneuron are leaving the metathoracic ganglion. (From Pearson and Iles, 1973.)

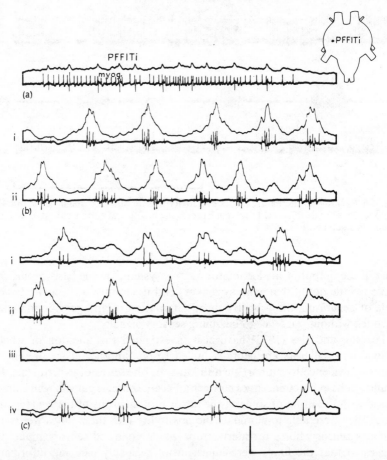

Fig. 4.8 Intracellular recordings from a leg flexor motoneuron (PFFITi, top traces), and electromyograms from the flexor muscle (myog., bottom traces). The records show rhythmic, spontaneous activity in the motor neuron which is similar to that seen during walking. Scale: 10 mV and 500 ms. (From Hoyle and Burrows, 1973.)

See Section 4.5.2.) In another series of experiments they showed that in a de-afferented preparation the output of one oscillator bore a fixed relationship to that of another, but that the exact phase between the two differed from that in intact insects. Their evidence thus strongly suggests that while mechanisms by which individual oscillators may be coupled together centrally do exist, these do not work to establish the phase relationships between the activity of leg muscles observed in intact animals. The normal relationship seems to be a function of some necessary sensory input, as discussed in the next section.

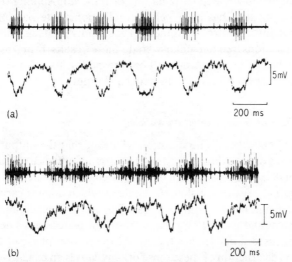

Fig. 4.9 Extracellular records from flexor motor neurons (top traces) and intracellular records from non-spiking interneurons (bottom traces) during rhythmic activity. (a) Record from an interneuron the depolarizations of which coincide with the intervals between flexor bursts. (b) Record from another interneuron which depolarizes as the flexor fires. (Modified from Pearson and Fourtner, 1975.)

Recent attempts to probe the properties of the oscillator itself have been made in insects by the use of intracellular recordings from individual neurons in the thoracic ganglia. Hoyle and Burrows (1973) successfully penetrated somata of a number of motoneurons in locusts, including several which innervated leg muscles. One of their most interesting findings was that these motoneurons would sometimes show strong waves of depolarization (Fig. 4.8), waves which produced rhythmic bursts of muscle potentials, and which the authors thought might be an expression of the activity of a locomotor control system. These and similar observations led them to support one of the main hypotheses regarding the walking oscillator, that the oscillator can not be composed only of coupled motoneurons, but must include several interneurons which drive the motor nerves in the appropriate pattern of activation. This conclusion was also reached by Burrows and Horridge (1974) in a later study.

105

The most interesting information regarding the probable properties of these driver interneurons (in cockroaches) was reported by Pearson and Fourtner (1975) when they presented good evidence that at least some of these drivers were non-spiking cells. They located four different types of such non-spiking interneurons in each half ganglion, each type distinguished from the others by its effects on specific motoneurons. Their evidence strongly suggests that depolarization of each type directly either excites or inhibits the flexor or extensor motoneurons innervating the ipsilateral leg (Fig. 4.9). On the basis of a number of experiments, including one in which they reset an ongoing rhythm by momentarily depolarizing one of these non-spiking cells, Pearson and Fourtner suggest that these cells actually constitute part of the neural oscillator thought to control movement of each individual leg. Although many important questions remain unanswered, these results bring us the closest we have yet been to understanding how the nervous system of any animal generates patterned output from un-patterned input.

4.6.2 *The effect of sensory input on walking*

On the other hand, there is no question whatever that the output of many of the sense organs of an arthropod leg has a profound influence on the movement of that leg, or at least on the pattern of motor signals delivered to it. Destruction of a chordotonal organ may alter the angle at which a leg is carried (locusts, Usherwood, Runion and Campbell, 1968) or alter the end point of flexion (crabs, Fourtner and Evoy, 1973). Input from hair plates, found near the joints of the legs of insects, determines the end of the flexion phase of the stepping movement, as destruction of these sense organs causes an exaggerated forward movement of the leg in stick insects (Wendler, 1964) and cockroaches (Wong and Pearson, 1976). Further, sense organs such as insect campaniform sensilla which can detect changes in cuticular stress (i.e., the load on the leg) seem to be responsible for adjusting either the firing frequency of the load-bearing muscles (cockroaches, Pearson, 1972; Delcomyn and Usherwood, 1973) or the duration of the power stroke (crabs, Fourtner and Evoy, 1973).

In addition, a number of other inter- or intraleg reflexes have been described. However, the importance of these reflexes to walking is not at all clear. Demonstrated inter-segmental reflexes are usually labile and weak (Wilson, 1965; Delcomyn, 1971b; Hoyle and Burrows, 1973; Burrows and Horridge, 1974), and therefore may not play an important role. Leg reflexes elicited by imposing movement on all or part of a leg usually resist the imposed movement, which would seem logically to limit their importance during natural movements. Barnes, Spirito and Evoy (1972) presented evidence that in crabs such resistance reflexes are turned off centrally during walking. While Burrows and Horridge (1974) report they could find no evidence of an active suppression of any reflex during natural or other movements in insects, they do point out

106

that reflexes which oppose any natural movement tend to be swamped by the massive synaptic activity driving or inhibiting the motoneurons during such movement. It may well be that central suppression is more important in animals such as crabs in which peripheral inhibition of muscles is an important element of normal neuromuscular control. In any case, the importance to co-ordination during walking of many well known leg reflexes may be minimal.

But this compendium of the actual or presumed effects of specific sense organs on walking fails, as do many of the papers themselves, to come to grips with the question so often stated to be of central interest: is any sensory input necessary for the maintenance of the observed timing relationships between the moving legs? While studies of specific known sense organs or reflexes have rarely been able to provide a positive answer to this question, other types of studies have. These are studies of perturbations of the normal gaits of arthropods.

A survey of the variety of gaits to be found among arthropods would be interesting in this context, but such a survey is outside the scope of this chapter. There are also several good recent reviews (Wilson, 1966a; Hughes and Mill, 1974) and original research papers (Barnes, 1975; Macmillan, 1975; Bowerman 1975a) to which the interested reader can turn.

In any case, the main question of interest can easily be stated: does interference with normal leg movements result in any change in gait? The answer is clearly, yes. Amputation of one or more legs in cockroaches (Delcomyn, 1971a), stick insects (Wendler, 1964), crabs (Barnes, 1975), crayfish (Grote, personal communication), scorpions (Bowerman, 1975b) and many other arthropods (see Hughes and Mill, 1974, for review of insect work) usually has an obvious and dramatic effect on the timing of the movements of the remaining legs relative to one another. One interpretation given to this result is that loss of input from the sense organs in the amputated leg(s) is the cause of the changed gait, and that therefore this sensory input must normally play an important role in determining the timing. Several criticisms of this interpretation have been voiced, such as, that amputees walk slower and therefore use a different but normally occurring gait, and that the effect may be due to changes in loading on the remaining legs, not missing input. The original inference has held up remarkably well, however. Detailed studies of the gaits of amputee cockroaches (Delcomyn, 1971a) and crabs (Barnes, 1975) have invalidated the first criticism. Although the fact that fitting an animal with an artificial leg in place of each amputated one restores normal timing in stick insects (Wendler, 1964) and crayfish (Grote, personal communication) would seem to support the second point, it actually does not. In amputation experiments the legs are usually removed just distal to the coxo-femoral joint. This operation leaves the coxa intact, but a femoral stump too short to touch the ground. Outfitting this stump with an artificial leg would allow sense organs in the coxa and the coxo-femoral joint to respond to the placement of the artificial limb on the ground, and therefore restore normal timing. Other experiments also support the hypothesis that the relevant sense organs are on the proximal part of the leg.

The results of experiments in which legs are tied up or fastened in a fixed position also support the view that specific input from the legs is required for normal timing, since in all arthropods in which this has been attempted, spiders (Wilson, 1967), crabs (Evoy and Fourtner, 1973), lobsters (Macmillan, 1975), cockroaches (Pearson and Iles, 1973) and others, the procedure has had

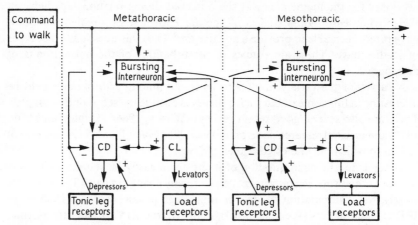

Fig. 4.10 A hypothetical model of neural connections which might underlie leg movements during walking in cockroaches. The command to walk provides tonic input to the bursting interneurons, and to the pool of depressor (=extensor) motoneurons (CD). The bursting interneuron in turn delivers excitatory bursts of impulses to the pool of levator (=flexor) motoneurons (CL) and inhibitory impulses to the extensors. This causes reciprocal output between extensor and flexor muscle groups in each leg. Bursting interneurons in adjacent ganglia are coupled by reciprocal inhibition. Feedback from leg receptors can set the firing of each burster, and, via the inhibitory connections between adjacent pairs, influence the relative timing between them. (From Burrows and Horridge, 1974. Reproduced by permission of the Royal Society.)

a profound effect on the walking gait. In some instances it has been shown that the method of fastening a leg, that is, whether the individual segments of the leg are free to move or not, makes a difference in the nature of the disturbance which results (e.g., Pearson and Iles, 1973).

Somewhat surprisingly, only Pearson and Iles (1973) appear to have approached the problem directly by comparing the pattern of output in meta-thoracic and mesothoracic leg motoneurons during normal walking with that to be observed in partly de-afferented preparations during rhythmic activity. As was pointed out above, they found that the pattern, that is, the timing of bursts in one segment relative to bursts in another, was quite different in the two cases, and that the pattern in de-afferented preparations closely resembled that which would be expected in animals with legs amputated. Further experiments suggested that it was lack of input from the campaniform sensilla located on the trochanter, at the coxo-tibial joints which was primarily responsible for the timing of bursts observed in de-afferented preparations.

The conclusion one may draw from all this work seems clear. The neural system responsible for co-ordination of leg movement during walking is similar in principle to that which regulates wing movements during flight, in that the movements of each appendage are controlled by a single, endogenous oscillator which can produce appropriately timed flexor and extensor activity without using any sensory signals as timing cues. The two systems differ in that while individual oscillators seem to be coupled centrally in both cases, the proper timing of the output of ipsilateral walking oscillators seems to require sensory input from the moving legs, while flight oscillators are coupled so as to produce the normal ouput pattern endogenously. The hypothetical mechanism by which the system may work is shown in Fig. 4.10, which incorporates the models of Pearson and Iles (1970, 1973) and Delcomyn (1971a). It should be emphasized that the details, and even the main hypothesis, are based largely on work on cockroaches. The ideas may be valid for other insects, but their applicability to the crustacea and other arthropods remains to be determined.

Acknowledgements

I thank the many people who allowed me to cite their unpublished work, and who commented upon earlier drafts of this article. The detailed comments of Drs Peter Hartline and Ann Kammer were especially helpful. Preparation of this chapter was supported in part by NIH grant NS12142.

References

Barnes, W.J.P. (1975) Leg co-ordination during walking in the crab, *Uca pugnax*. *J. comp. Physiol.* **96**, 237–256.

Barnes, W.J.P., Spirito, C.P. and Evoy, W.H. (1972) Nervous control of walking in the crab, *Cardisoma guanhumi*. II. Role of resistance reflexes in walking. *Z. vergl. Physiol.* **76**, 16–31.

Bastian, J. and Esch, H. (1970) The nervous control of the indirect flight muscles of the honey bee. *Z. vergl. Physiol.* **67**, 307–324.

Binyon, J. (1972) *Physiology of Echinoderms*. Oxford: Pergamon Press.

Bowerman, R.F. (1975a) The control of walking in the scorpion. I. Leg movements during normal walking. *J. comp. Physiol.* **100**, 183–196.

Bowerman, R.F. (1975b) The control of walking in the scorpion. II. Co-ordination modification as a consequence of appendage ablation. *J. comp. Physiol.* **100**, 197–209.

Bowerman, R.F. and Larimer, J.L. (1974) Command fibres in the circumoesophageal connectives of crayfish. II. Phasic fibres. *J. exp. Biol.* **60**, 119–134.

Bullock, T.H. and Horridge, G.A. (1965) *Structure and Function in the Nervous Systems of Invertebrates*. San Francisco: W. H. Freeman.

Burrows, M. (1973) The role of delayed excitation in the co-ordination of some metathoracic flight motoneurons of a locust. *J. comp. Physiol.* **83**, 135–164.

Burrows, M. (1975) Monosynaptic connexions between wing stretch receptors and flight motoneurons of the locust. *J. exp. Biol.* **62**, 189–219.

Burrows, M. and Horridge, G.A. (1974) The organization of inputs to motoneurons of the locust metathoracic leg. *Phil. Trans. R. Soc. Lond., Ser. B* **269**, 49–94.

Burton, A.J. (1971) Directional change in a flying beetle. *J. exp. Biol.* **54**, 575–585.

Camhi, J.M. (1969a) Locust wind receptors. I. Transducer mechanics and sensory response. *J. exp. Biol.* **50**, 335–348.

Camhi, J.M. (1969b) Locust wind receptors. II. Interneurons in the cervical connective. *J. exp. Biol.* **50**, 349–362.

Camhi, J.M. (1969c) Locust wind receptors. III. Contribution to flight initiation and lift control. *J. exp. Biol.* **50**, 363–373.

Camhi, J.M. (1970a) Yaw-correcting postural changes in locusts. *J. exp. Biol.* **52**, 519–531.

Camhi, J.M. (1970b) Sensory control of abdomen posture in flying locusts. *J. exp. Biol.* **52**, 533–537.

Camhi, J.M. and Hinkle, M. (1974) Response modification by the central flight oscillator of locusts. *J. exp. Biol.* **60**, 477–492.

Clarac, F. and Coulmance, M. (1971) La marche latérale du crabe (*Carcinus*). Coordination des mouvements articulaires et régulation proprioceptive. *Z. vergl. Physiol.* **73**, 408–438.

Croll, N.A. (1975) Components and patterns in the behaviour of the nematode *Caenorhabditis elegans*. *J. Zool. (Lond.)* **175**, 154–176.

Dagan, D., Vernon, L.H. and Hoyle, G. (1975) Neuromimes: self-exciting alternate firing pattern models. *Science* **188**, 1035–1036.

Davis, W.J. (1969) Reflex organization in the swimmeret system of the lobster. I. Intrasegmental reflexes. *J. exp. Biol.* **51**, 547–563.

Davis, W.J. and Kennedy, D. (1972a) Command interneurons controlling swimmeret movements in the lobster. I. Types of effects on motoneurons. *J. Neurophysiol.* **35**, 1–12.

Davis, W.J. and Kennedy, D. (1972b) Command interneurons controlling swimmeret movements in the lobster. II. Interaction of effects on motoneurons. *J. Neurophysiol.* **35**, 13–19.

Davis, W.J. and Kennedy, D. (1972c) Command interneurons controlling swimmeret movements in the lobster. III. Temporal relationships among bursts in different motoneurons. *J. Neurophysiol.* **35**, 20–29.

DeBell, J.T. (1965) A long look at neuromuscular junctions in nematodes. *Q. Rev. Biol.* **40**, 233–251.

Delcomyn, F. (1971a) The effect of limb amputation on locomotion in the cockroach, *Periplaneta americana*. *J. exp. Biol.* **54**, 453–469.

Delcomyn, F. (1971b) Computer-aided analysis of a locomotor leg reflex in the cockroach, *Periplaneta americana*. *Z. vergl. Physiol.* **74**, 427–445.

Delcomyn, F. (1973) Motor activity during walking in the cockroach *Periplaneta americana*. II. Tethered walking. *J. exp. Biol.* **59**, 643–654.

Delcomyn, F. and Usherwood, P.N.R. (1973) Motor activity during walking in the cockroach *Periplaneta americana*. I. Free walking. *J. exp. Biol.* **59**, 629–642.

Dorsett, D.A., Willows, A.O.D. and Hoyle, G. (1973) The neuronal basis of behavior in *Tritonia*. IV. The central origin of a fixed action pattern demonstrated in the isolated brain. *J. Neurobiol.* **4**, 287–300.

Dugard, J.J. (1967) Directional change in flying locusts. *J. Insect Physiol.* **13**, 1055–1063.

Esch, H., Nachtigall, W. and Kogge, S.N. (1975) Correlations between aerodynamic output, electrical activity in the indirect flight muscles and wing positions of bees flying in a servomechanically controlled wind tunnel. *J. comp. Physiol.* **100**, 147–159.

Evoy, W.H. and Fourtner, C.R. (1973) Nervous control of walking in the crab, *Cardisoma guanhumi*. III. Proprioceptive influences on intra- and intersegmental coordination. *J. comp. Physiol.* **83**, 303–318.

Fentress, J.C. (1973) Development of grooming in mice with amputated forelimbs. *Science* **179**, 704–705.

Fourtner, C.R. and Evoy, W.H. (1973) Nervous control of walking in the crab, *Cardisoma guanhumi*. IV. Effects of myochordotonal organ ablation. *J. comp. Physiol.* **83**, 319–329.

110

Gardner, C.R. (1976) The neuronal control of locomotion in the earthworm. *Biol. Rev.* **51**, 25–52.

Getting, P.A. (1975) *Tritonia* swimming: triggering of a fixed action pattern. *Brain Res.* **96**, 128–133.

Gewecke, M. (1974) The antennae of insects as air-current sense organs and their relationship to the control of flight. In: *Experimental Analysis of Insect Behaviour*, Barton Browne, L., Ed. pp. 100–113. New York: Springer-Verlag.

Govind, C.K. and Burton, A.J. (1970) Flight orientation in a coreid squash bug (Heteroptera). *Can. Ent.* **102**, 1002–1007.

Gray, J., Lissmann, H.W. and Pumphrey, R.J. (1938) The mechanism of locomotion in the leech (*Hirudo medicinalis* Ray). *J. exp. Biol.* **15**, 408–430.

Gruber, S.A. and Ewer, D.W. (1962) Observations on the myo-neural physiology of the polyclad, *Planocera gilchristi*. *J. exp. Biol.* **39**, 459–477.

Harcombe, E.S. and Wyman, R.J. (1977) Output pattern generation by *Drosophila* flight motoneurons. *J. Neurophysiol.* (In press).

Heide, G. (1974) The influence of wingbeat synchronous feedback on the motor output systems in flies. *Z. Naturforsch.* **29**, 739–744.

Herman, R.M., Grillner, S., Stein, P.S.G., and Stuart, D.G., Eds. (1976) *Proceedings of the International Conference on Neural Control of Locomotion*. New York: Plenum Press.

von Holst, E. (1934) Motorische und tonische Erregung und ihr Bahnenverlauf bei Lepidopterenlarven. *Z. vergl. Physiol.* **21**, 395–414.

Horridge, G.A. (1968) *Interneurons*. San Francisco: W. H. Freeman.

Hoyle, G. (1964) Exploration of neuronal mechanisms underlying behavior in insects. In: *Neural Theory and Modeling*, Reiss, R.R., Ed., pp. 346–376. Stanford: Stanford University Press.

Hoyle, G. (1975) Identified neurons and the future of neuroethology. *J. exp. Zool.* **194**, 51–73.

Hoyle, G. and Burrows, M. (1973) Neural mechanisms underlying behavior in the locust *Schistocerca gregaria*. II. Integrative activity in metathoracic neurons. *J. Neurobiol.* **4**, 43–67.

Huber, F. (1974) Neural integration (central nervous system). In: *The Physiology of Insecta*, 2nd edn., Rockstein, M., Ed. Vol. IV, Ch. 1, pp. 3–100. New York: Academic Press.

Hughes, G.M. and Mill, P.J. (1974) Locomotion: terrestrial. In: *The Physiology of Insecta*, 2nd edn., Rockstein, M., Ed. Vol. III, Ch. 5, pp. 335–379. New York: Academic Press.

Ikeda, K. and Wiersma, C.A.G. (1964) Autogenic rhythmicity in the abdominal ganglia of the crayfish: the control of swimmeret movements. *Comp. Biochem. Physiol.* **12**, 107–115.

Jarman, M. (1959) Electrical activity in the muscle cells of *Ascaris lumbricoides*. *Nature* **184**, 1244.

Josephson, R.K. (1974a) The strategies of behavioral control in a coelenterate. *Am. Zool.* **14**, 905–915.

Josephson, R.K. (1974b) Cnidarian neurobiology. In: *Coelenterate Biology. Reviews and New Perspectives*, Muscatine, L. and Lenhoff, H.M., Eds. Ch. VI, pp. 245–280. New York: Academic Press.

Kammer, A.E. (1971) The motor output during turning flight in a hawkmoth, *Manduca sexta*. *J. Insect Physiol.* **17**, 1073–1086.

Kammer, A.E. and Nachtigall, W. (1973) Changing phase relationships among motor units during flight in a saturniid moth. *J. comp. Physiol.* **83**, 17–24.

Kater, S.B. and Nicholson, C., Eds. (1973) *Intracellular Staining in Neurobiology*. New York: Springer-Verlag.

Kerkut, G.A. (1954) The mechanisms of coordination of the starfish tube feet. *Behaviour* **6**, 206–232.

Koopowitz, H. (1973) Primitive nervous systems. A sensory nerve-net in the polyclad flatworm *Notoplana acticola*. *Biol. Bull.* **145**, 352–359.

Koopowitz, H. (1974) Some aspects of the physiology and organization of the nerve plexus in polyclad flatworms. In: *Biology of the Turbellaria*, Riser, N.W. & Morse, M.P., Eds. pp. 198–212. New York: McGraw-Hill.

Koopowitz, H. (1975) Electrophysiology of the peripheral nerve net in the polyclad flatworm *Freemania litoricola. J. exp. Biol.* **62**, 469–479.

Kristan, W.B., Jr. (1974) Neural control of swimming in the leech. *Am. Zool.* **14**, 991–1001.

Kristan, W.B., Jr., Stent, G.S. and Ort, C.A. (1974a) Neuronal control of swimming in the medicinal leech. I. Dynamics of the swimming rhythm. *J. comp. Physiol.* **94**, 97–119.

Kristan, W.B., Jr., Stent, G.S. and Ort, C.A. (1974b) Neuronal control of swimming in the medicinal leech. III. Impulse patterns of the motor neurons. *J. comp. Physiol.* **94**, 155–176.

Lawry, J.V., Jr. (1970) Mechanisms of locomotion in the polycheate, *Harmothoë. Comp. Biochem. Physiol.* **37**, 167–179.

Lawry, J.V., Jr. (1973) Central pacemaker hierarchies and peripheral feedback in the movements of worms. In: *Neurobiology of Invertebrates.* Salánki, J., Ed. pp. 341–352. Budapest: Akadémiai Kiadó.

Mackie, G.O. (1975) Neurobiology of *Stomotoca.* II. Pacemakers and conduction pathways. *J. Neurobiol.* **6**, 357–378.

Macmillan, D.L. (1975) A physiological analysis of walking in the American lobster (*Homarus americanus*). *Phil. Trans. R. Soc. Lond., Ser. B* **270**, 1–59.

Mulloney, B. (1970) Impulse patterns in the flight motor neurones of *Bombus californicus* and *Oncopeltus fasciatus. J. exp. Biol.* **52**, 59–77.

Muscatine, L. and Lenhoff, H.M., Eds. (1974) *Coelenterate Biology. Reviews and New Perspectives.* New York: Academic Press.

Nachtigall, W. (1974) Locomotion: mechanics and hydrodynamics of swimming in aquatic insects. In: *The Physiology of Insecta*, 2nd edn., Rockstein, M., Ed. Vol. III, Ch. 6, pp. 381–432. New York: Academic Press.

Nachtigall, W. and Wilson, D.M. (1967) Neuro-muscular control of dipteran flight. *J. exp. Biol.* **47**, 77–97.

Page, C.H. (1970) Unit response in the metathoracic ganglion of the flying locust. *Comp. Biochem. Physiol.* **37**, 565–571.

Paul, D.H. (1971) Swimming behavior of the sand crab, *Emerita analoga* (Crustacea, Anomura). III. Neuronal organization of uropod beating. *Z. vergl. Physiol.* **75**, 286–302.

Paul, D.H. (1976) Proprioception from non-spiking sensory cells required for a swimming behavior in the sand crab, *Emerita analoga.* In: *Proc. Int. Conf. Neural Control Locomotion*, Herman, R.M., Grillner, S., Stein, P.S.G. and Stuart, D.G., Eds. pp. 785–787. New York: Plenum Press.

Pearson, K.G. (1972) Central programming and reflex control of walking in the cockroach. *J. exp. Biol.* **56**, 173–193.

Pearson, K.G. and Fourtner, C.R. (1975) Nonspiking interneurons in walking system of the cockroach. *J. Neurophysiol.* **38**, 33–52.

Pearson, K.G. and Iles, J.F. (1970) Discharge patterns of coxal levator and depressor motoneurones of the cockroach, *Periplaneta americana. J. exp. Biol.* **52**, 139–165.

Pearson, K.G. and Iles, J.F. (1973) Nervous mechanisms underlying intersegmental co-ordination of leg movements during walking in the cockroach. *J. exp. Biol.* **58**, 725–744.

Perkel, D.H. and Mulloney, B. (1974) Motor pattern production in reciprocally inhibitory neurons exhibiting postinhibitory rebound. *Science* **185**, 181–183.

Pringle, J.W.S. (1948) The gyroscopic mechanism of the halteres of Diptera. *Phil. Trans. R. Soc. Lond., Ser. B* **233**, 347–384.

Reiss, R.F. (1962) A theory and simulation of rhythmic behavior due to reciprocal inhibition in small nerve nets. *AFIPS Proceedings, Spring Joint Computer Conference*, pp. 171–194.

Robson, E.A. (1961) The swimming response and its pacemaker system in the anemone *Stomphia coccinea. J. exp. Biol.* **38**, 685–694.

Robson, E.A. (1971) The behaviour and neuromuscular system of *Gonactinia prolifera*, a swimming sea-anemone. *J. exp. Biol.* **55**, 611–640.

Ross, D.M. (1973) Some reflections on actinian behavior. *Publications of the Seto Marine Biological Laboratory, 20.* (*Proceedings of the Second International Symposium on Cnidaria*), 501–512.

Schneider, G. (1953) Die Halteren der Schmeissfliege (*Calliphora*) als Sinnesorgane und als mechanische Flugstabilisatoren. *Z. vergl. Physiol.* **35**, 416–458.

Schneider, P. and Krämer, B. (1974) Die Steuerung des Fluges beim Sandlaufkäfer (*Cicindela*) und beim Maikäfer (*Melolontha*). *J. comp. Physiol.* **91**, 377–386.

Schrameck, J.E. (1970) Crayfish swimming: alternating motor output and giant fiber activity. *Science* **169**, 698–700.

Shishov, B.A. (1968) Contribution to the study of the function of the nervous system in *Nematoda*. In: *Neurobiology of Invertebrates*, Salánke, J., Ed. pp. 487–492. Budapest: Akadémiai Kiadó.

Smith, J.E. (1966) The form and functions of the nervous system. In: *Physiology of Echinodermata*, Boolootian, R.A., Ed. Ch. 21, pp. 503–511. New York: Interscience.

Spirito, C.P. (1972) An analysis of swimming behavior in the portunid crab *Callinectes sapidus*. *Marine Behaviour and Physiology* **1**, 261–276.

Stein, P.S.G. (1974) Neural control of interappendage phase during locomotion. *Am. Zool.* **14**, 1003–1016.

Stein, R.B., Pearson, K.G., Smith, R.S. and Redford, J.B., Eds. (1973) *Control of Posture and Locomotion*. New York: Plenum Press.

Tyrer, N.M. and Altman, J.S. (1974) Motor and sensory flight neurons in a locust demonstrated using cobalt chloride. *J. comp. Neurol.* **157**, 117–138.

Usherwood, P.N.R., Runion, H. and Campbell, J.I. (1968) Structure and physiology of a chordotonal organ in the locust leg. *J. exp. Biol.* **48**, 305–323.

Walcott, B. (1968) The locomotion of giant water bugs. Ph.D. Thesis, University of Oregon.

Waldron, I. (1967) Neural mechanisms by which controlling inputs influence motor output on the flying locust. *J. exp. Biol.* **47**, 213–228.

Waldron, I. (1968) The mechanism of coupling of the locust flight oscillator to oscillatory inputs. *Z. vergl. Physiol.* **57**, 331–347.

Weevers, R.deG. (1966) The physiology of a lepidopteran muscle receptor. III. The stretch reflex. *J. exp. Biol.* **45**, 229–249.

Wendler, G. (1964) Laufen und Stehen der Stabheuschrecke *Carausius morosus*: Sinnesborstenfelder in den Beingelenken als Glieder von Regelkreisen. *Z. vergl. Physiol.* **48**, 198–250.

Wendler, G. (1974) The influence of proprioceptive feedback on locust flight co-ordination. *J. comp. Physiol.* **88**, 173–200.

Wiersma, C.A.G. and Ikeda, K. (1964) Interneurons commanding swimmeret movements in the crayfish, *Procambarus clarkii* (Girard). *Comp. Biochem. Physiol.* **12**, 509–525.

Willows, A.O.D., Dorsett, D.A. and Hoyle, G. (1973) The neuronal basis of behavior in *Tritonia*. III. Neuronal mechanism of a fixed action pattern. *J. Neurobiol.* **4**, 255–285.

Wilson, D.M. (1961) The central nervous control of flight in a locust. *J. exp. Biol.* **38**, 471–490.

Wilson, D.M. (1964a) Relative refractoriness and patterned discharge of locust flight motor neurons. *J. exp. Biol.* **41**, 191–205.

Wilson, D.M. (1964b) The origin of the flight-motor command in grasshoppers. In: *Neural Theory and Modeling*, Reiss, R.R., Ed. pp. 331–345. Stanford: Stanford University Press.

Wilson, D.M. (1965) Proprioceptive leg reflexes in cockroaches. *J. exp. Biol.* **43**, 397–409.

Wilson, D.M. (1966a) Insect walking. *A. Rev. Ent.* **11**, 103–122.

Wilson, D.M. (1966b) Central nervous mechanisms for the generation of rhythmic behaviour in arthropods. *Symp. Soc. Exp. Biol.* **20**, 119–228.

Wilson, D.M. (1967) Stepping patterns in tarantula spiders. *J. exp. Biol.* **47**, 133–151.

Wilson, D.M. (1968a) The nervous control of insect flight and related behaviour. *Adv. Insect Physiol.* **5**, 289–338.

Wilson, D.M. (1968b) Inherent asymmetry and reflex modulation of the locust flight motor pattern. *J. exp. Biol.* **48**, 631–641.

Wilson, D.M. and Waldon, I. (1968) Models for the generation of the motor output pattern in flying locusts. *Proc. I.E.E.E.* **56**, 1058–1064.

Wilson, D.M. and Weis-Fogh, T. (1962) Patterned activity of co-ordinated motor units, studied in flying locusts. *J. exp. Biol.* **39**, 643–667.

Wilson, D.M. and Wyman, R.J. (1963) Phasically unpatterned nervous control of dipteran flight. *J. Insect Physiol.* **9**, 859–865.

Wilson, D.M. and Wyman, R.J. (1965) Motor output patterns during random and rhythmic stimulation of locust thoracic ganglia. *Biophys. J.* **5**, 121–143.
Wong, R.K.S. and Pearson, K.G. (1976) Properties of the trochanteral hair plate and its function in the control of walking in the cockroach. *J. exp. Biol.* **64**, 233–249.
Wyman, R.J. (1966) Multistable firing patterns among several neurons. *J. Neurophysiol.* **29**, 807–833.
Wyman, R.J. (1970) Patterns of frequency variation in dipteran flight motor units. *Comp. Biochem. Physiol.* **35**, 1–16.

Note added in Proof

Recent work has shown that the swimming undulations of the leech (pp. 88–89) are not co-ordinated only by sensory feedback from stretch receptors in the body wall. Kristan and Calabrese (1976) showed that an isolated chain of at least 8–12 ganglia could be stimulated to produce rhythmic bursts of impulses in the motoneurons which resembled the bursts produced in semi-intact preparations during swimming, and Friesen, Poon and Stent (1976) have identified several interneurons which seemed to be the drivers of the motoneurons. The latter authors suggest that these interneurons in each ganglion comprise the neural oscillator driving swimming. This new work strongly supports the idea that a central oscillator (p. 92) underlies *all* rhythmic activities in invertebrates. In those cases in which the evidence in favour of this view is not clear, we may simply not yet have done the right experiments.

Friesen, W.O., Poon, M. and Stent, G. (1976) An oscillatory neuronal circuit generating a locomotory rhythm. *Proc. natn. Acad. Sci. U.S.A.* **73**, 3734–3738.
Kristan, W.B., Jr. and Calabrese, R.L. (1976) Rhythmic swimming activity in neurones of the isolated nerve cord of the leech. *J. exp. Biol.* **65**, 643–668.

5 Activation and co-ordination of vertebrate locomotion

Mary C. Wetzel and D.G. Stuart

5.1 Ensembles of locomotor movements

More information has been assembled about vertebrate stepping within the last few years than in two centuries of previous research. Detailed, documented stepping reviews have been made recently for the cat (Wetzel and Stuart, 1975) and more generally across vertebrates (Grillner, 1975), although not as yet for man. In this chapter we summarize the more commonly accepted terms by which stepping is defined and methods by which it is measured. We then try to extract the most important ideas about how the movements are co-ordinated by drawing freely from studies in neurophysiology, morphology, behavioural psychology, and theoretical engineering. The reader is directed, where necessary, to fuller expositions in these specialized areas of knowledge.

From the Greeks, by way of their word for motion, *kinesis*, we have inherited a variety of terms to describe locomotion and other animal movements. Two of these terms are especially useful. *Kinetics* refers to the forces and energy associated with movement, while *kinematics* refers to aspects of movement apart from considerations of mass and force. The kinds of sensory messages that are fed to the nervous system, the central interneuronal networks, and the types of outflow in motor neurons that will cause muscles to contract in the correct periodic order and at appropriate strength have all evolved in accordance with constraints set by kinetic and kinematic variables.

Vertebrate locomotion is a challenging behaviour to study because in their natural worlds these animals move over a wide range of speeds from walking through galloping. Vertebrates not only require a wide dynamic range of speeds, but they must also be able to change speed quickly in order to avoid obstacles, escape danger, or pursue food. Diversity of stepping movements is seen within a given species as animals meet their everyday goals, and also in the variants of gait (the manner of foot movements) across species as different as humans, lizards, cats, or elephants. Complicated neuronal circuitry is needed so that each animal can utilize information from its specialized environment and survive. The control machinery has evolved, however, so as to be able to generate stepping rhythms even without sensory feedback from the periphery. A schema in which stepping movements can occur in rudimentary form without peripheral input but are fully co-ordinated only when afferent channels are involved as well, is perhaps a primeval characteristic, since it is true of invertebrates as well as vertebrates (see, for example, review by Wilson, 1964).

Although man is still the exception, within the past few years an increasing quantity of research has established that the isolated vertebrate spinal cord can sustain relatively effective stepping movements. In animals, this machinery can be activated by several sources (principally a set of descending supra-spinal pathways), and it normally functions in close harmony with segmental afferent input. Much of our knowledge has come from quadrupeds, the cat in particular. Man's specialized use of his forelimbs has made his bipedal loco-motion rather different from most vertebrate stepping, although a crawling infant retains the basic four-footed pattern (Muybridge, 1957).

Before turning to the neural pathways for locomotion, the ways are de-scribed by which elements of movements—motor units, muscles, joints, and limbs—are joined to form co-ordinated ensembles. As measurements of stepping characteristics have become more accurate, it has also been necessary to continually revise and make more rigorous the language of kinetics and kinematics.

5.1.1 Co-ordination within the limb: motor unit to muscle to joint

Motor Unit to Motor Unit

The most informative experiments about the role of the individual moto-neuron (the large α cell) which, together with the muscle fibres (muscle unit) it innervates, comprise a *motor unit*, have come from the high decerebrate cat preparation under so-called controlled locomotion. Although the animal is deprived of volition, it can walk and even run on a moving treadmill belt, so long as a region under the inferior colliculus in the brainstem is activated by electrical stimulation (Shik, Severin and Orlovsky, 1966). It has been pos-sible to record in this preparation the activity of single α motoneurons in filaments of ventral root fibres that exit from the spinal cord and supply muscles that operate on the ankle joint. As might have been expected, fibres destined for flexors (to close the joint) and extensors (to open the joint) are active at different times during stepping (Severin, Shik and Orlovsky, 1967).

As it unfolds, the step cycle of a single limb includes one complete forward and backward movement, with respect to the body. Extensor motoneurons are active during the first half of the *stance* phase, the period during which the foot touches and thrusts forward from the surface, while flexor motoneurons are active during the early part of the *swing* phase, while the limb moves forward above the surface (Severin, Shik and Orlovsky, 1967). Both kinds of motoneurons discharge impulses at steady rates that are well below the fusion frequency for the muscle units they supply. The duration of the train of flexor motoneuron impulses is fairly constant as the cat moves faster, in accordance with the relative constancy of swing duration across velocities (Arshavsky *et al.*, 1965). The extensor motoneuron train becomes shorter, however, as the animal moves faster, as is reasonable, since the stance phase also shortens (Arshavsky *et al.*, 1965).

Only a few details are known about how activities of different motor units are co-ordinated during stepping. It would be of value to know, for example, if small motoneurons are recruited before larger ones, in accordance with Henneman's (1968) *size principle* of recruitment, which has been demonstrated and, despite many onslaughts, confirmed to quite a reasonable degree in a variety of situations in both anesthetized animals and conscious humans (Burke and Edgerton, 1975; Milner-Brown, Stein and Yemm, 1973; Freund, Büdinger and Dietz, 1975). The controlled locomotion preparation lends itself to study of the recruitment order and usage of motoneurons, but only preliminary observations were made in the Severin, Shik and Orlovsky (1967) analysis. In that study several examples were given of extensor motoneuron firing patterns during increased force and speed of controlled locomotion as was provided by increasing the intensity of brainstem stimulation. Individual motoneurons retained their particular firing patterns, and increased force was presumably accomplished by recruitment of additional motoneurons and their muscle units. A few examples were also given of simultaneous recordings from two α axons. In each case the order of motoneuron recruitment was fixed on repeated induction of controlled stepping. There was an unemphasized example of a smaller motoneuron (as adjudged by its lower spike height) being recruited before a larger one. What was emphasized, however, was that an extensor unit recruited at a relatively high strength of midbrain stimulation might fire earlier in the stance phase than a unit recruited at a lower strength of stimulation.

Of further value would be information on the relative usage of different types (FF = fast, fatiguing rapidly; FR = fast, resistant to fatigue; and S = slow) of motor units during locomotion of different gaits, speed, and power.

As already mentioned in Chapter 1, most mammalian muscles are an admixture of different fibre types, which can be conveniently lumped into three basic categories: FG, for fast twitch–glycolytic; FOG, fast twitch, oxidative–glycolytic; and SO, slow twitch, oxidative. Mutability between the fibre types is not excluded, of course, particularly between FG and FOG fibres (for review of this classification system and fibre typing in general see Chapter 1 and Burke and Edgerton, 1975). Recent glycogen depletion studies on cat gastrocnemius by Burke and his collaborators indicate that each α axon supplies muscle fibres of quite similar histochemical profiles, and that the mechanical properties of each muscle unit match its histochemical properties (for a short review, see Burke, 1975). As such, mixed mammalian muscles can be considered as having three motor unit types: FF for fast, rapidly fatiguing units composed, in our opinion, of FG fibres; FR for fast, fatigue-resistant units composed of FOG fibres; and S for slow fatigue-resistant units composed of SO fibres. Mutability between FF and FR units is emphasized for units composed of fibres intermediate between FG and FOG. Figure 5.1 is from Burke (1975) and shows further properties of these units, including their force developing capabilities, which are far greater for FF than FR and S units.

It is attractive to speculate that as the speed and power of locomotion increase, there is a progressive recruitment of S, FR, and finally FF units. Until there are further advances in the information that can be extracted from electromyograms it will remain difficult to test this arrangement with electrophysiological experiments. The reader is referred, however, to Burke and

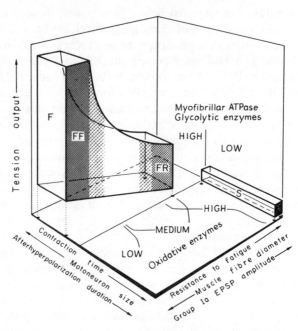

Fig. 5.1 Tri-dimensional histogram showing the association of motoneuron properties (size, after hyperpolarization duration and group Ia EPSP amplitude) with the mechanical (tension output, contraction time, resistance to fatigue) and histochemical (concentration of oxidative and glycolytic enzymes, affinity for myofibrillar APTase) properties of the muscle units they supply. Relative muscle fibre size is also indicated. FF, FR and S stand for motor unit types as defined in the text (from Burke, 1975, Fig. 1).

Edgerton's (1975) recent review which includes metabolic evidence based on histochemical profiles of selected muscle fibres taken from a variety of species, including man, immediately after various forms of work activity executed at different levels of intensity and endurance. In quite general terms, such studies support the hypothesis that the usage of motor units is directly correlated with the oxidative capacity and force generating capabilities of their muscle fibres.

118

Muscle to muscle

The co-ordination of motor unit usage within a single muscle during stepping is obviously influenced by the role it plays in the elaboration of each step. There are, for example, well over 30 cat hindlimb muscles available to participate in each step. Unfortunately, and despite the availability of appropriate technology, little is known concerning the precise timing, consistency, and relative intensity of even one such muscle's activity during four-legged stepping. Indeed, information of this kind is at a far more advanced level for the human, its relevance to prosthetics having been well recognized in the 1940s and 1950s (for review, see Eberhart, Inman and Bressler, 1964).

Despite these restrictions, promising beginnings have been made for both cat (Engberg and Lundberg, 1969; Gambaryan *et al.*, 1971) and dog (Tokuriki, 1973a, b; 1974). Figure 5.2 summarizes, for the hindquarters, Tokuriki's ambitious attempt to provide an electromyographic analysis of stepping during walking (1973a), trotting (1973b), and galloping (1974) gaits. While details are reviewed elsewhere (Wetzel and Stuart, 1975; Wetzel, Atwater and Stuart, 1975) note the subtle mutability and individuality of each muscle's EMG activity. There were differing degrees of usage in each gait and some muscles even became active in different phases of stepping, depending on whether their particular limb was leading or trailing its counterpart on the opposite side. It would be of great value to extend these promising beginnings and relate each muscle's histochemical profile to its role in locomotion and other movements. In the cat hindlimb, for example, vastus intermedius and soleus are atypical in being composed almost exclusively of SO fibres (Ariano, Armstrong and Edgerton, 1973). These muscles not only participate in stepping but presumably remain more active than their extensor synergists during maintenance of standing postures. Tensor fascia latae is also atypical, being composed almost exclusively of FG fibres. Note that Fig. 5.2 suggests it participates but minimally even in galloping. Most other hindlimb muscles studied by Ariano *et al.* (1973) contain FG, FOG, and SO fibres, except for caudofemoralis and pectineus, hip extensors with virtually no SO fibres. The locomotor role of all these muscles has yet to be determined, however.

As with individual motor units, probably whole muscles fulfil both a general function together with their synergists, and an additional and special role for particular tasks. The fast gastrocnemius, for example, can extend the ankle quite adequately during both posture and locomotion without assistance from its synergists (if the latter have been denervated, Wetzel *et al.*, 1973), yet in the intact cat it would be most useful for rapid, forceful movements. Similarly, the four digit flexor muscles (plantaris, flexor hallucis longus, flexor digitorum longus and brevis) in the cat presumably play somewhat different parts in fast, forceful than in slower, sustained flexion or protrusion of the claws, since their motor units have quite different contractile speeds and tensions (Goslow *et al.*, 1972).

Another co-ordinative feature of synergistic muscles is that their activity profiles are slightly displaced, one from another, presumably to smooth the cycle. Gastrocnemius, for example, may remain active longer during the stance phase than does soleus in the cat (Gambaryan *et al.*, 1971) and, in fact, there are subtle timing differences among the four ankle extensors: lateral gastrocnemius, medial gastrocnemius, plantaris, and soleus (Messinger, 1974).

119

WALK

		33	34	35	36	37	38	39	40	41	42	43	44	45	46	47	48	49	50	51	52
Swing phase	fore-swing	++	0	++	++	0	0	±	0	++	0	+	0	0	+	0	0	++	±	0	0
	mid-swing	0	±	++	++	0	0	±	±	0	0	++	0	0	++	0	0	+	++	0	0
	aft-swing	+	0	+	+	++	++	0	++	0	++	+	++	+	±	+	++	0	++	+++	++
Stance phase	fore-stance	0	±	++	++	+	++	0	+++	0	0	0	0	±	±	+++	+++	0	0	+++	+++
	mid-stance	0	+	++	+	+	0	0	0	0	0	0	0	0	+++	++	+++	0	0	0	++
	aft-stance	++	±	++	±	0	0	±	0	±	0	0	0	0	++	0	0	0	0	0	0

TROT

		33	34	35	36	37	38	39	40	41	42	43	44	45	46	47	48	49	50	51	52
Swing phase	fore-swing	±	0	±	±	0	0	±	±	±	0	+	0	0	++	0	0	++	±	0	0
	mid-swing	0	0	±	+	0	0	0	0	0	0	++	0	0	±	0	0	+	+	0	0
	aft-swing	±	0	±	+	±	+	0	++	0	+	0	+	±	±	+	++	±	±	+++	±
Stance phase	fore-stance	±	0	+	+	+	+	0	+++	0	±	0	±	0	0	++	++	±	0	++	+
	mid-stance	0	0	±	±	±	±	0	++	0	0	0	0	0	0	++	+++	0	0	0	++
	aft-stance	+	0	±	+	0	0	±	±	0	0	±	0	0	++	±	++	0	0	±	0

GALLOP

			33	34	35	36	37	38	39	40	41	42	43	44	45	46	47	48	49	50	51	52
Trail	Swing Phase	fore-swing	0	0	++	+	0	0	±	0	+	++	++	0	0	++	0	0	++	++	+	0
		mid-swing	0	±	++	+	0	0	0	+	0	0	+	++	0	+	±	±	++	++	0	0
		aft-swing	++	±	+	+	+++	+++	0	+++	+++	++	0	+++	+++	±	+	+++	+	++	+++	+
	Stance phase	fore-stance	+	0	++	++	+++	+++	±	+++	+	±	0	+	0	+	±	+++	+++	0	0	++
		mid-stance	++	0	++	+	0	0	0	0	0	0	±	0	0	+	+++	++	0	0	0	+++
		aft-stance	+	0	++	++	0	0	+	0	+	++	+	±	0	+++	0	0	+	±	+	0
Lead	Swing phase	fore-swing	0	0	++	+	0	0	±	0	0	++	+	0	0	++	0	0	++	+	±	0
		mid-swing	++	0	++	++	0	0	±	±	0	0	++	0	0	++	±	±	++	++	±	0
		aft-swing	+	0	+	+	+	++	+++	++	0	+++	++	0	++	+++	0	++	+++	+	++	+
	Stance phase	fore-stance	+	±	++	++	+++	+	±	+++	+	±	±	0	+	+	+++	+++	0	0	++	++
		mid-stance	0	+	++	±	0	0	±	++	0	0	±	0	0	+++	++	+++	0	0	0	++
		aft-stance	0	±	+	+	0	0	+	0	0	++	+	0	0	+++	0	0	+	±	0	0

Remarks. Nil (0), negligible (±), slight (+), moderate (++), marked (+++), and very marked (++++).

Fig. 5.2 An electromyographic analysis of 20 hindquarter muscles in a dog during walking, trotting, and galloping. Seven muscles could be recorded from simultaneously as the animal moved on a motor-driven treadmill. The numbers at the top of each record indicate the muscles: 33, sacrospinalis (L4) (longissimus dorsi); 34, obliquus externus abdominis; 35, rectus abdominus; 36, iliopsoas; 37, gluteus superficialis; 38, gluteus medius; 39, tensor fasciae latae; 40, biceps femoris; 41, semitendinosus; 42, semimembranosus; 43, sartorius; 44, gracilis; 45, adductores; 46, rectus femoris; 47, vastus lateralus; 48, vastus medialis; 49, tibialis anterior; 50, extensor digitorum longus; 51, gastrocnemius; 52, flexor digitorum profundus (from Tokuriki, 1973a, Table 1; 1973b, Table 1; 1974, Table 1).

In addition to co-ordination between muscles at a particular joint that have the same broad synergistic action (such as flexion or extension), there are complicated inter-relationships between antagonistic muscles at a joint (flexors versus extensors) and between muscles at different joints. Most of our information at present has come from experiments in which kinematic measurements (usually from movie film) have been compared with EMG activity of the muscles during different parts of the step cycle (see in particular Engberg and Lundberg, 1969; Gambaryan *et al.*, 1971; Goslow, Reinking and Stuart, 1973a). To a large extent the movements of flexion (closing) and extension (opening) at a joint are promoted by inertial forces that continue after active muscle contraction has ceased. As the limb swings forward, for example, the initial opening of knee and ankle joints, after flexion is completed, is almost passive under the pull of gravity, since the extensors remain relatively inactive until shortly before touchdown. There is also active co-operation between flexors and extensors to accomplish particular movements, as is signalled by their co-contraction. As the limb touches down, the flexor muscles are active, presumably to brake and stabilize the forward motion, while the extensors contract even prior to ground contact (see below) to provide for support and subsequent thrustoff.

Two-joint or bifunctional muscles provide continuity of movement across joints. As forward speed changes, so do their activity profiles. They may contract either to actually open or close a joint or else to increase the stability of a bone so that another bone may move. In the decerebrate cat, for example, the two-joint muscle, biceps femoris posterior, flexes the knee and extends the hip during walking, being active during the swing and also the stance phases (Gambaryan *et al.*, 1971). During galloping, the muscle extends the hip exclusively and is active only during the stance phase (Gambaryan *et al.*, 1971).

While the nervous system determines the order of muscle activation through the step cycle (see following sections), the bone–tendon–muscle systems in different animals are adapted for specialized co-ordinated movements. The cheetah has morphological ties between many muscles that probably serve its fast sprint toward prey. Muscles that may be one- or two-joint in the domestic cat have become two- or three-joint in the cheetah. Gastrocnemius is primarily an ankle extensor in the cat, although it arises partly from the femur, but in the cheetah the muscle is truly a two-joint knee flexor and ankle extensor (Gambaryan *et al.*, 1971).

It deserves emphasis that the visco-elastic properties of bone–tendon–muscle systems are also adapted for the execution of locomotion. Hip, knee, and ankle extensors become active before, rather than after the foot touches the ground (Engberg and Lundberg, 1969) to provide what has been termed 'elastic bounce' (Cavagna, 1970) on ground contact (for review see Goslow *et al.*, 1973a and Chapter 7). Grillner (1972) has developed a model that emphasizes the role of muscle stiffness (the force developed per length increase) in meeting the postural and locomotor requirements for force development by ankle extensors. His calculations reveal that much of the load compensation on ground contact is attributable to the visco-elastic properties of the muscles themselves and not to reflexly induced changes. The latter would be brought into play too slowly, particularly in high speed movement. Note finally Fig. 5.3 from Stephens, Reinking and Stuart (1975), which shows how well-attuned the length-tension properties of cat medial gastrocnemius are to its locomotor demands. The peak of active tetanic tension is near the longest lengths that are actually employed for heavy force requirements. In contrast, the declining limb of the curve is set at lengths beyond maximum *in situ* length and that are never attained in locomotion.

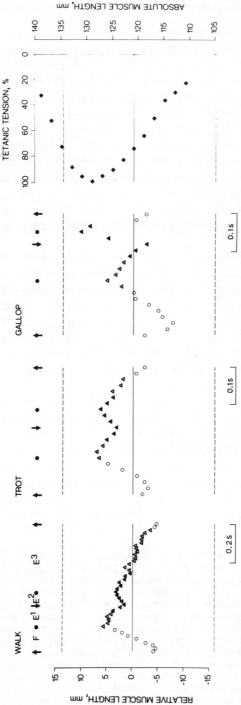

Fig. 5.3 Relationship of the length tension curve to the usage of medial gastrocnemius in locomotion. The four phases of the step cycle (F and E[1] for the swing phase and E[2] and E[3] for the stance phase) are shown in real time for representative walking (1.5 mph), trotting (3.5 mph), and high speed (16.3 mph) galloping gaits of an 'idealized' large (4 ± 0.35 kg) cat (for procedural details see Goslow *et al.*, 1973a,b). Instantaneous lengths (circles and triangles) of medial gastrocnemius are shown together with the length associated with quiet standing (thin solid line) and the maximum and minimum *in situ* length (broken lines). For the stepping curves, open circles indicate periods in which the muscle can be considered as passive and open triangles indicate points of relatively low level electrical activity. Filled triangles indicate periods of heightened electrical activity. Note that electrical activity increases abruptly ca 30 ms before the foot hits the ground (E[1]–E[2] junction) and ceases ca 30 ms before the end of the stance phase. The left side vertical calibration shows mm of lengthening (+) and shortening (–) relative to standing length (o). The right side of the figure shows the mean length-active tension curve in relation to the overall stepping changes. The right side length scale is absolute. The procedure for normalizing and fitting the length-tension curve to these natural movements is explained in the cited reference (from Stephens, Reinking and Stuart, 1975, Fig. 11).

Synthesis of movements in the whole limb: basic and conditional patterns

The basic sequencing and features of movements through an entire step cycle are seen at all forward speeds and in most natural and experimental settings. These basic patterns are described now. Later, some departures are noted from regular patterns that have been observed under different stepping conditions.

The terminology of limb movements has been revised from time to time and will probably undergo still more changes. Most emphasis has been placed on understanding the hindlimb. One of the most useful early schematics was that by Philippson (1905) in which joint angles at hip, knee, and ankle were referred to contact durations (whether the limb was up in the swing or down in the stance) through a complete hindlimb *step cycle*. Four epochs were defined and are diagrammed in Fig. 5.4: F (flexion at all joints, commencing at foot lift-off), E^1 (beginning extension at all joints, starting about midway through the swing phase), E^2 (an 'extension' epoch starting at touchdown, in which the knee and ankle but not the hip begin to yield or flex under the weight of the animal, even while the extensors are still contracting actively), and E^3 (final extension in which knee and ankle join the hip in extension prior to thrustoff). It will be noted from Fig. 5.4 that knee and ankle are in phase for much of the step cycle. Hip and toe (not shown in Fig. 5.4) angles shift from flexion to extension and vice versa rather differently from knee and ankle or each other. In this manner, the entire limb can show a smooth resultant action with two cycles, as is shown by the total direct hip-to-toe length (Fig. 5.4). In this whole limb measure the four epochs are still discernible but might better be considered as two foldings and two unfoldings than as flexion or extension.

While there are somewhat greater excursions of joint angles and hip-to-toe direct distances as speed increases, the general scheme of co-ordination is maintained irrespective of velocity, as was shown relatively early by Arshavsky *et al.* (1965) and confirmed by a variety of other work. The wavelike excursions that represent successive changes at one joint and between the different joints do not shift relative positions as speed increases. The principle change with forward speed is the decrease in stance duration. The excursions are compressed in time, so that thrustoff force is delivered more and more quickly. While the foot is off the surface and moving forward, in contrast, there is little change in swing duration.

Unfortunately, little is known about the details of net forces that are exerted by the limbs or at different joints through the step cycle, at least in experimental animals. In the cat one might attempt to calculate forces by the formula, force = mass × acceleration. Manter (1938) actually measured distance changes as the bones moved, differentiated twice with respect to time (to calculate acceleration), and also weighed segments of the limb to estimate their masses. The estimated forces compared favourably with forces measured during the stance phase as the cats walked across spring-supported platforms. As might be expected, vertical force was greatest at the peak of the yield

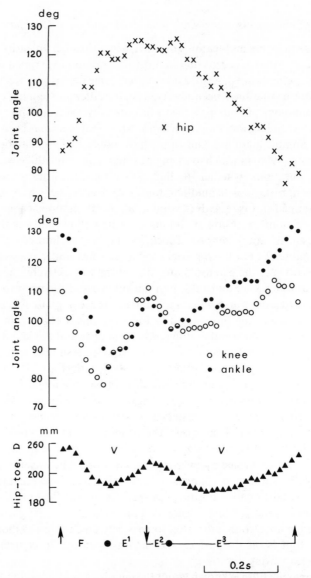

Fig. 5.4 The step cycle of a single hindlimb of the cat during moderately slow walking. Along the bottom appear the 4 epochs defined by Philippson (1905) as F (flexion), E¹, E², and E³ ('extensions'). Upward arrows indicate lift-off and the downward arrow, touch-down. Note that joint angles at the top of the figure show flexion of hip, knee, and ankle (the hip angle was measured with reference to the pelvis, so flexion appears as an increasing angle) during F, extension of all 3 joints during E¹, a yield or closure of only the knee and ankle joints during E², and extension of all 3 joints at the onset of E³ and until lift-off. The lower, hip–toe curve, shows concurrent changes in the total length of the limb (measured from the hip pivot point to the tip of the toe). The V's indicate the point at which the toe passes under the hip pivot. This whole-limb measure shows how the separate excursions of different joints are translated to 2 smooth cycles of motion (from Wetzel, Atwater and Stuart, 1975, Fig. 1).

(see also Grillner, 1972). Many substantial technical problems impede understanding of kinetics, however. The small size of the cat prevents accurate measurements of distances from movie film. Only a few data have been taken as yet from force (Yager, 1972) or length (Proschazka *et al.*, 1974) transducers implanted in the hindlimb. In the far larger human, it is far easier to measure kinetic data, by externally placed transducers (see Herman *et al.*, 1973).

Mathematical analyses of the limb movements during stepping has advanced recently for animals (and man as well) more from utilizing engineering theory than from kinetic information gained in the laboratory. Engineering principles have been applied to the construction of legged devices with a hip and knee, with the hip being able to move up and down with respect to the substrate (McGhee, 1967; Frank, 1968). This simplified leg can step quite efficiently. As is apparently true for biological systems, it has been possible to develop an automaton leg that will operate either with or without feedback. Control algorithms have been based on a simple *finite state* model, which assumes that a limb is in one or the other of two mutually exclusive and exhaustive categories or states; namely, on or off the surface of the ground. Control laws have been developed to drive the leg by preprogrammed sets of signals or by feedback-modified signals, and in some cases actual locomotor machines have been built (see Frank and McGhee, 1969 for more extended discussion). Control theory and laboratory data can already be brought together in a few instances. For example, the ideal trajectory of the body and its parts is a horizontal straight line in the forward direction, a motion that will minimize transfer between potential and kinetic energy. It is also known that the cat minimizes vertical excursions of its toe while it is walking (Wetzel, Atwater and Stuart, 1975).

In summarizing the whole limb's stepping behaviour, the main scheme of co-ordination from motor unit to muscle to joint to limb of the normal animal is probably similar at different speeds. There are rather minor adjustments in magnitudes of most timings and excursions, although stance duration can vary widely. These conclusions rest, however, on data that have been gained primarily under conditions of unperturbed stepping. When perturbations are introduced while the animal is running, or if the entire environment is altered (as from overground stepping to stepping on a motor-driven treadmill), single limb kinematics shift. Some measures shift more than others.

The feature of stepping that is most impervious to deformation is probably the duration of the swing phase. It changes only slightly in the dog if a brake is applied to a joint at different times (Orlovsky and Shik, 1965), or if the load upon the animal is increased (for a given velocity) by pulling back on a restraining harness or tilting the treadmill belt on which the animal runs (Orlovsky, Severin and Shik, 1966a). Stance duration, in contrast, is altered readily by such manipulations, just as it continuously changes with changes in forward speed. When stride duration and velocity are invariant, swing-stance durations can differ by times that are not large (less than 50 ms) but are stat-

125

istically significant for the cat, depending on the incentive for running. Performances that are maintained by food reward alone have been compared with performances when food was paired with aversive shock or an air jet applied to the hindquarters (Lockard, Traher and Wetzel, 1975). As is true for stance duration, the absolute value at an individual joint can vary over quite a wide range. This conclusion is reasonable from casual observations of slinking or strutting cats, as well as from laboratory findings (Lockard *et al.*, 1975).

5.1.2 Co-ordination of limb to limb in whole body progression

Although the single limb's step cycle seems to be relatively uniform across speed and conditions, a variety of footfall patterns may be established by the four limbs in various forms of progression. If any overall theme for the movements has appeared, it is one of exquisite responsiveness to external needs, given that mechanical and energy requirements are satisfied to maintain forward velocity. To an even greater extent than is true for the single limb's cycle, interlimb co-ordination depends upon both forward velocity and (when velocity is controlled) situational variables. The patterns of interlimb co-ordination can only be interpreted, moreover, by considering the different functions of the hind- and forelimbs as they move the body along.

Interlimb co-ordination across velocities

As described in Chapter 7, the universal slow gait of quadrupeds (and human infants) is the crawl whose footfall order is (if the sequence is said to start with the left hindlimb): left hind, left fore, right hind, and right fore (LH, LF, RH, RF). This footfall pattern appears in Fig. 5.5a in the form of a gait diagram that was developed by Hildebrand (1959). *Gait* is understood to mean a given pattern of footfalls or, more rigorously, 'a periodic sequence of binary k-tuples (where k = the number of legs) that represents the successive states of the legs' (McGhee, 1968, p. 69). It has been recognized for many years that this pattern provides a good deal of stability, since at very slow speeds only one foot leaves the surface at any given time, and the centre of gravity of the entire body always falls within a triangle of support marked by the positions of the three remaining limbs. The continuous support of the centre of gravity provides what is called *static stability* (McGhee and Frank, 1968). Recent work (McGhee and Frank, 1968) has shown that while five additional patterns might theoretically be used in crawling, the particular one in which the above order (LH, LF, RH, RF) occurs provides maximal static stability, since the centre of gravity is always farthest from the boundaries of the support triangle. It is not surprising, then, that the pattern is seen for all known quadruped species, and that the fit is so good between the mechanical demands on the terrestrial form and the goals of its neural control programme.

As speed increases, variants in gait across species become more evident. In general, there is a gradual change in phase relations between limbs, with increasing speed, toward the trotting pattern, in which diagonal limbs (LH–RF and RH–LF) descend almost simultaneously (Fig. 5.5b). Sometimes, however, the gait is one of a lateral support pattern, the pace, in which ipsilateral limb touchdowns are more or less synchronized (Fig. 5.5c). The stability properties

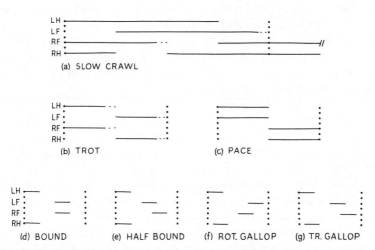

Fig. 5.5 Idealized gait diagrams for patterns of interlimb coordination that are commonly seen in vertebrates. The method of display is that suggested by Hildebrand (1959). While no time scale appears, swing durations (open spaces) are similar at all speeds, as is normal (see text). The 4 limbs in each of the 7 diagrams appear from top to bottom as LH (left hind), LF (left fore), RF (right fore), and RH (right hind). Contacts of each foot appear as horizontal lines (stance duration). Vertical rows of closed circles indicate the duration of a stride, as referred to 2 successive LH touchdowns. (a) Slow crawl. Dashed lines at the end of LF & RF stances indicate 'ideal' pattern (maximal static stability, McGhee and Frank, 1968) in which the forefoot touches in the print of its ipsilateral hindfoot just as that foot lifts off. In reality, of course, the forefoot usually lifts off a little earlier than the full 3/4 of stride duration (solid line for LF and RF). Otherwise the fore- and hindlimbs would be displaced from each other laterally. (b) Trot. A running footfall pattern in which diagonal limbs descend simultaneously. There may be periods in which all four feet are in the air (dashed lines show these options). (c) Pace. Ipsilateral limbs descend simultaneously. (d–g) Gallops used at high speeds. Additional patterns exist that are the mirror images of e, f and g.

of neither trotting nor pacing are well understood, but in trotting the striking characteristic is one of successive, interleaved falls by the two sets of diagonal limbs. In trotting, as in all more rapid gaits, there are periods of static instability in which total support by all limbs is relinquished in the interest of gathering speed. The concept of *dynamic stability* is used when progression continues during periods in which static stability is interrupted within the four-legged sequence of animals or the two-legged sequence of man (Frank and McGhee, 1969). While much of this work is in the computer-simulation or descriptive stage, it is of considerable interest and importance. Vukobratović and his colleagues (Vukobratović, Frank and Juričić, 1970), for example, have defined a *stationary gait* as one in which there is constancy of the following events: average forward velocity, *stride* length or duration (a cycle that encompasses the successive touchdowns or liftoffs of one reference limb and all the others), phasing between limbs, and *duty factor* (stance duration expressed as a proportion of stride duration).

Computer models can be tested in which a simulated man or animal's progression can be examined under different sets of conditions. For example, a 'person' could be made to walk at a constant velocity and suddenly fall into a simulated hole. Then the time can be determined that is required for him to recover his original forward speed. If the gait pattern is stable it will reappear within some number of strides after a perturbation.

The gaits in which the speed demands are greatest and in which, almost paradoxically, the diversity of footfall patterns seems to be the greatest, are the high speed gallops. Galloping is an *asymmetrical* gait, since the successive footfalls of a hind- and forelimb on one side of the body are unevenly spaced in time (Muybridge, 1957) unlike those during the *symmetrical* gaits of walking and trotting. There is an abrupt transition between touchdowns of different limbs from trotting to galloping because the hindlimbs, which have been alternating, begin to strike almost simultaneously (Fig. 5.5). A true *bound* (Fig. 5.5d), in which the two hind- and then the two forelimbs strike simultaneously, is normal in only a few animals, such as the Siberian souslik or the weasel (Gambaryan, 1974). A smoother movement can result when the various limbs descend in a slightly staggered fashion and can more evenly support the body's weight from moment to moment. The more common forms of rapid galloping are therefore the *half-bound* (Fig. 5.5e), in which the hindlimbs strike simultaneously, but the forelimbs alternate; the *rotatory gallop* (Fig. 5.5f), in which the hindlimbs strike slightly out of phase and then the forelimbs descend in an opposite order; and the *transverse gallop* (Fig. 5.5g), in which the alternating forelimbs descend in the same order as the alternating hindlimbs.

Environmental contributions to interlimb co-ordination

It would be easier to understand the neuronal circuitry that underlies interlimb co-ordination if some relatively invariant coupling patterns could be obtained when animals move under different conditions of stepping. For at least moderate speeds of walking, trotting, and perhaps galloping (for at least one set of limbs) there appears to be a relatively constant interlimb interval between extension in the hindlimb and flexion of the ipsilateral forelimb (Miller and van der Burg, 1973). The coupling is not obligatory, however, since it differs between overground locomotion on open terrain and treadmill locomotion (Wetzel, Atwater, Wait and Stuart, 1975). Nor do interlimb timings or, especially, positioning of the moving limbs and body remain the same when animals perform under aversive testing conditions (avoidance of electric shock or a noisy air jet) as when they perform solely for a food reward (Lockard *et al.*, 1975). As might be suspected, animals tend to crouch when their behaviour is punished, even when escape is readily available by moving forward.

By general characteristics, however, movements can be quite similar across situations. A cat swimming in the water moves its limbs in a manner that is analogous to walking (Miller, van der Burg and van der Meché, 1975), and the amphibious newt 'swims' (undulates from side to side as it moves) whether in the water or on ground. The most universal rule of quadrupedal locomotion seems to be that when only one girdle, shoulder or pelvic, is examined, the two limbs usually alternate their step cycles. Most probably this pattern is readily shifted to being slightly out of phase or even completely in phase, as in the half-bound. To this extent an oscillatory model of interlimb co-ordination, in which the most stable positions are completely in phase or 50% out of phase, is appropriate.

Hindlimb-forelimb contributions to interlimb co-ordination

The fore- and hindquarters have specialized functions. In birds and humans, the forelimbs are structured for flight and the wholesale exploitation of the environment, respectively. Nonetheless, movements of the different limbs must permit balance and continued progression by the whole body even in these specialized animals, and the co-ordinated division of labour between front and back portions of the body is evident in all vertebrates. The simple requirement is to move the centre of gravity without undue acceleration in either vertical or lateral directions. In slow-moving ungulates the spine can be quite rigid, but in fast-moving carnivores the powerful thrust of the hindlimbs is accompanied by marked flexions and extensions of the spine, that impart additional speed. The forelimbs act primarily to prop and balance as the hindlimbs are successively gathered under the body for their thrusting actions. There is an intimate relationship between forelimb movements and up-down movements of the head and neck, again distinguishing the roles of the two sets of limbs. The quantification of roles by the different sets of limbs has been slow because it is difficult to measure forelimb events from any fixed reference point. The scapula, unlike the pelvic bones and the hip pivot point, slides along the chest wall as the animal moves along, allowing the length of the forelimb to vary as needed.

The movement of one limb is subservient to conditions that are imposed by the other limbs. Our understanding of how priorities of interlimb stepping are established has been increased greatly by some rather simple but revealing experiments that were conducted in Russia with the intact dog on a treadmill (Orlovsky and Shik, 1965; Shik and Orlovsky, 1967). It was first noted that when all four limbs run on the belt, braking one of them has little effect on the movements of the remaining three, so forward progression is safeguarded.

129

If, however, one or more limbs are raised with a cord and so prevented from touching the belt, interlimb programming is unmasked. There are strong, symmetrical connections between contralateral limbs. If one forelimb runs on the belt, a suspended forelimb will perform alternating 'steps' in the air. If the single, running limb is restrained during the stance its cycle is markedly perturbed, but if the other forelimb also runs on the belt, there is much less perturbation.

Connections between hind- and forelimbs are asymmetrical, with the fore-limb influence on the hindlimbs being more profound. With the hindlimbs suspended, the rhythm of the forelimbs is unchanged from that seen in four-legged running, but if the forelimbs are suspended the hindlimb rhythm changes. That the hindlimbs are not totally without ascending influence was suggested by experiments in which the overall forelimb input was 'reduced' by raising one of them while the hindlimbs stepped. In this situation, there were subtle changes in the running forelimb, whose cycle was synchronized, at times, with the hindlimb rhythms. The tightest relationships appear to be between contralateral limbs, the next tight from forelimb to hindlimb, and the weakest from hindlimb to forelimb. The same ordering of priorities has been found by counting the number of strides after the belt begins to move and until steady-state rhythms of different sets of limbs (i.e., gait stability) are established. First the contralateral limbs begin to alternate; then the hindlimb-forelimb movements are co-ordinated.

5.2 A tripartite model of the neuronal control programme for locomotion

Review of the locomotor movements themselves suggested that some routines of programming are common to most vertebrate species, at least terrestrial quadrupeds, and across individual situations of stepping, but that routines are readily rearranged to conform to moment by moment exigencies. A reasonable way to supply versatile programming economically would be for the nervous system to provide modules—sets of functional neural pathways—that could be deployed in various alternative combinations to suit different occasions. The output pool of motoneurons and muscles would be activated in functional ensembles whose specific constituents and timings might vary within wider or more narrow constraints. Such a scheme has the support of a good deal of experimental evidence, at least in the cat. For initial purposes of simplifying, only three compartments are proposed. First, a relatively autono-mous spinal programme, that resides in interneurons, can provide for rhythmic stepping. Second, the spinal system receives activating signals from supra-spinal sources. Third, there is segmental afferent input that interacts with the spinal programme by means of feedback either directly or as channelled through ascending neural pathways. These components are shown diagrammatically in Fig. 5.6.

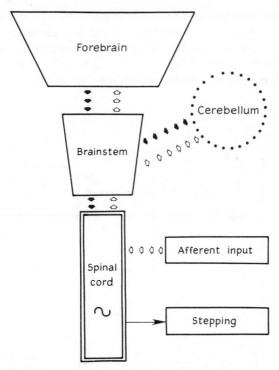

Fig. 5.6 The main features of a tripartite view of the neural control system for locomotion. The *spinal portion* interacts with both *afferent input* (open arrows) and *descending signals* (closed arrows). The cerebellum is outlined by intermittent closed circles to emphasize the uncertainty about how integral is its locomotor role (see text and Wetzel and Stuart, 1975). The most essential descending signals are probably those from the brainstem. At least part of the cyclic nature of stepping derives from afferent signals produced by the moving limbs themselves. This afferent information is utilized by the spinal cord together with pathways that involve the cerebellum.

5.2.1 Spinal aspects of the control programme

It was known as long ago as the latter part of the nineteenth century that the lumbar cord and, to lesser extent, the cervical cord could support stepping or stepping-like movements by the limbs even in mammals. From work by Graham Brown that eventually was continued by Shurrager and Dykman in the 1950s (1951) and Grillner and his colleagues in the 1970s (see below) the hindlimb stepping was determined to be relatively effective. Further, local afferent (peripheral) input was found to be non-essential for programming the main order of limb movements during the step cycle. A cat deprived surgically of afferentation from one limb retained sequential, rhythmic EMGs in that limb during controlled locomotion that were much like the EMGs of normal stepping (Grillner and Zangger, 1974). Presumably 'normality' was brought about by activity in central neurons together with afferentation from the remaining three intact and running limbs. It was also learned that a relatively effective ambulation was possible after the entire spinal cord of a monkey had been deafferented (Taub and Berman, 1968). All of this evidence argued for an intrinsic locomotor programme that stores great detail about the stepping

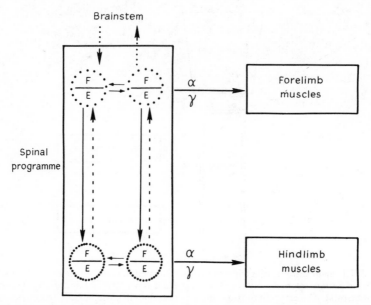

Fig. 5.7 The spinal control programme for stepping, as expressed by signals in γ- and α-motoneurons that innervate the limb muscles. Four interneuronal stepping generators are represented by circles, subdivided into half centres for flexors (F) and extensors (E) after the model of Graham Brown (see his 1916 review for comprehensive listing of reports). It is recognized, however, that a ring model may be more appropriate (see text). Between cervical and lumbar enlargements and between right and left generators are shown interconnecting pathways that probably include the long and short propriospinal fibres, respectively. The hindlimb-forelimb influence is weaker (dashed-line arrow) than that from fore- to hindlimb. Uncertainty about the arrangement and autonomy of forelimb generators, together with possible low brainstem connections that may be essential for forelimb stepping, is indicated by dotted lines.

movements which will be required by the animal even before, paradoxically, those movements occur. The logical major divisions of this programme are circuits for the step cycle of a single limb and pathways to serve interlimb co-ordination (see Fig. 5.7).

Lumbar stepping generators as a single-limb stepping model

While in the dogfish there appear to be a number of individual segmental stepping generators that can provide for the wavelike swimming motion of this creature (Grillner, 1974), the information about stepping generation in mammals is largely limited to the lumbar spinal cord of the cat. Two single-limb generators have been postulated, one for each hindlimb. No evidence has been found for inherently rhythmic pacemaker cells, although they could be present, as is suspected to be a component in the stepping generation for some invertebrates (see discussion by Pearson, Fourtner and Wong, 1973). Rather, the two models have been based on intercellular connections that could provide rhythmicity: a ring circuit, in which muscles are activated successively (perhaps by excitatory and inhibitory interconnections, Székely, 1965), or an oscillatory arrangement. The oscillator may be thought of as a special case of ring, in which there is simply alternating activation of flexors and extensors by means of mutual inhibitory interconnections (for review, see Gurfinkel and Shik, 1973).

The earliest oscillator model that dealt with mammalian stepping generators in any detail was that of Graham Brown (1916 review), who postulated paired *half-centres* of motoneurons or interneurons (the distinction was not clear). Each half-centre would serve either flexors and extensors, and they would be paired by reciprocal inhibitory pathways so as to provide rhythmic alternation of the two functions. Graham Brown's suggestions were made from observing stepping-like movements in a variety of largely cat preparations. After some years, his concepts were revived by Lundberg and his colleagues (Engberg and Lundberg, 1969; Lundberg, 1969) and given credence by careful work to discover whether or not appropriate interneurons exist in the lumbar cord that could drive rhythmic activity in the limbs (Jankowska *et al.*, 1967a, b). In the inert anesthetized preparation, they indeed found an appropriate set of neurons, but under rather special circumstances. Locomotor-type activity of these cells was 'unmasked' only under the influence of systemically administered DOPA, a noradrenergic precursor.

As described below a number of pharmacological investigations on DOPA and its locomotor effects were launched in the late 1950s and early 1960s. Some, such as that by Dahlstrom and Fuxe (1965) sought the supraspinal origin of the noradrenergic cells, since no such cell bodies were found in the spinal cord itself (Carlsson *et al.*, 1964).

In the anesthetized cat, activity in flexor motoneurons can be elicited at relatively short latency by electrical stimulation of high threshold muscle and other limb afferents (the *FRA* as defined by Holmquist and Lundberg, 1961; see, however, criticism of this terminology by Matthews, 1972). Under DOPA the Jankowska *et al.* (1967a, b) studies revealed that the pathway is shut down and a longer latency FRA response appears in the same flexor motoneurons and their contralateral extensor motoneurons.

The finding of alternative routes to the same motoneurons is compatible with a 'state' theory of locomotion, in which different sets of central neurons are active during locomotion than are active during quiescence or, perhaps, movements other than locomotor ones. Even though the final output is to the same muscles, their order of strength of activation could be programmed quite differently by the different inputs.

Although much of the evidence for an oscillatory system was circumstantial in the inert preparation, the Swedish group did show that after DOPA
there was late excitatory action in one pool of interneurons from stimulation
of the ipsilateral FRA and in another pool from stimulation of the contralateral
FRA. Both pools were concentrated in the lateral portion of Rexed's lamina
VII. Moreover, the two pools were joined by inhibitory interconnections, and
similar arrangements of pools were found for both the large α-motoneurons
and the small γ-motoneurons that supply the muscle fibres of the spindle organ
(Bergmans and Grillner, 1969). It was suggested that oscillation of activity
between the two 'half-centres' might be accomplished by refractoriness that
would accumulate in each pool in turn.

Two lines of study have continued that complement the Swedish work, in
which feasibility was determined of locating stepping generators in the lumbar
cord. One line showed that the hindlimbs of an adult cat that was acutely
spinalized could walk, in fact, on a treadmill when either DOPA or Clonidine
(a noradrenergic receptor site activator) was administered (Grillner, 1973).
The adult spinalized cat will not normally locomote, either during an acute,
short-term experiment, or if it is allowed to recover from the immediate effects
of surgery and spinal shock (Shurrager and Dykman, 1951; Grillner, 1973).

A second line of work concerned the further identification and description
of the interneurons by a combination of electrophysiological and pharmacological techniques. During controlled locomotion experiments, Orlovsky and
Feldman (1972b) recorded from lumbosacral interneurons whose species were
not identified but which had clear periods of activity or inactivity in different
portions of the single hindlimb step cycle. Unlike α-motoneurons, the firing
rates of the lamina VII and IX component of this sample increased with the
intensity of brainstem stimulation and speed on the belt, so that force could
conceivably be regulated by the graded activity of individual cells. In other
experiments with controlled locomotion, a particular species of interneuron
has been implicated, the characteristics of which have been of interest for a long
time, the so-called Ia inhibitory interneuron (Feldman and Orlovsky, 1974).
The primary Ia afferents from muscle spindles activate that same muscle or its
synergists by means of a monosynaptic connection to the α-motoneuron,
but the Ia afferents also send out a collateral that is connected to the antagonistic
muscle via an interposed (Ia inhibitory) neuron. The inhibitory cells are in a
relatively convenient location to study and have been found to be phasically
active, together with their corresponding motoneurons, even when the latter
were detached from their muscles by de-efferentation of the hindlimbs.

In summary, some spinal elements that may be involved in programming
rhythmic movements of a single limb are being identified. It should not detract
from this valuable work to keep in mind the wide separation that still exists
between what is known about spinal generation of stepping and the elaborately
co-ordinated muscle behaviours which occur throughout the step cycle.

134

Pathways for forelimb stepping and hindlimb-forelimb interactions

Both early and recent findings that the forelimbs of cats or dogs do not loco-mote well in a spinal preparation have shown probable differences between hindlimb and forelimb stepping generators. Such differences are to be expected, of course, in view of the different roles and morphologies of the forelimbs and hindlimbs. Expected also would be flexible linkages between generators for the four limbs in order to permit the mutable footfall patterns that are seen in all vertebrate species. While pathways in the brainstem that are related to movements of the head and neck may provide a significant contribution to stepping machinery for the forelimbs (see work by Gernandt and Gilman, 1961), most study to date has been directed to the propriospinal pathways that reside within the spinal cord itself. It has proved difficult to untangle ascending and descending components, but propriospinal neurons are logical candidates to serve interlimb co-ordination because they link the cervical and lumbar enlargements both ipsilaterally and bilaterally. Of the remaining pathways in the spinal cord, those to the cerebellum are perhaps of most inter-est to stepping and will be discussed in connection with their presumed role in processing afferent information.

Short propriospinal pathways at both the brachial and lumbar levels in the cat are known to interconnect freely, on the basis of electrophysiological experiments that were performed in the early 1940s (Lloyd, 1941; 1942). While the short crossing fibres evidently exist at all levels of the cord (Lloyd, 1941, 1942), as is reminiscent of the functional connections between the long-itudinal array of segments in the dogfish (Grillner, 1974), it is usually assumed that in the cat the important functional crossings are restricted to the enlarge-ments. This assumption is based largely on observations that functional re-covery of stepping by the hindlimbs after various hemisections of the cord cannot be accomplished by fibres between enlargements and is prevented by destroying fibres that cross at the lumbar enlargement (Jane, Evans and Fisher, 1964). Details of the connections made by short propriospinal neurons are not available, because only the brachial system has been studied anatomically (Sterling and Kuypers, 1968). It arises in the ventromedial part of the inter-mediate gray and distributes bilaterally via the ventral funiculus to moto-neurons. One would expect to find massive inputs from one side of the cord to the other, since contralateral limbs usually alternate their steps precisely during slow and moderately fast locomotion. The connections probably provide for ready access between short propriospinal cells, stepping generator cells, and motoneurons.

The connection to interneurons has been demonstrated for long ascending and descending propriospinal neurons, that project mostly on the same side and primarily to the ventromedial part of the intermediate gray from which, it will be remembered, the short propriospinal fibres arise (Giovanelli Barilari and Kuypers, 1969). There is also, however, anatomical (Giovanelli Barilari and Kuypers, 1969) and electrophysiological evidence for both the descending (Jankowska *et al.*, 1974) and ascending (Bergmans, Miller and Reitsma, 1973) systems that implies there are strong motoneuronal connections, many of which may be monosynaptic. A model of interlimb co-ordination might then have to include the possibility of interactions between a direct interlimb influence and the local stepping generator's influence on motoneurons.

The overall characteristics of the propriospinal system that are emerging from anatomical, electrophysiological, and behavioural information are remarkably congruent. This congruence is highlighted by evidence on asymmetry between fore- and hindlimb circuitry. The anatomical distribution of descending pathways is more bilateral than that of the ascending paths (Giovanelli Barilari and Kuypers, 1969), implying more pervasive descending control. The intraspinal influence of descending systems is more powerful than that of ascending systems, as has been documented by a variety of findings. While reflex effects in inert preparations are readily elicited by electrical stimulation of descending systems, those that proceed by way of ascending routes require special procedures such as 'preconditioning' stimulation in the brainstem (Gernandt and Megirian, 1961) or drug potentiation with L-DOPA (Bergmans *et al.*, 1973). As with the behavioural evidence, then, there is the strong implication that forelimb machinery is less dependent on information from the hindlimbs than vice versa.

5.2.2 Supraspinal contributions to locomotor control

It is certain that visual and other pre-emptive inputs that involve the cerebral cortex and probably act via motivation-related structures of the limbic system have immediate access to the spinal stepping programme to initiate, modify, or terminate locomotion. The speed with which a covey of quail takes flight or a jackrabbit swerves across a highway is self-evident. Fortunately, however, it has proved possible to dispense with consideration of the so-called 'higher centres' at this stage in our study of the neural control of stepping, because quite full expression of the locomotor movements can be observed in animals that retain only the brainstem and lower tissue. The degree to which the cerebellum is required in locomotion by such a reduced preparation may, however, be substantial.

Descending brainstem pathways that operate on spinal programming

Levels of transection of the brainstem that produce a particular, definable level of locomotor capability differ somewhat across species, with somewhat more rostral tissue apparently required in animals with more developed brains (for more detailed discussions see Laughton, 1924; Hinsey, Ranson and McNattin, 1930). The cat is probably fairly representative of an intermediate mammal. In every animal studied (monkey, cat, rabbit, guinea pig, and rat) the performance that can be elicited immediately or within 24 h after surgery (defining an *acute* preparation) will improve over days and weeks if the animal

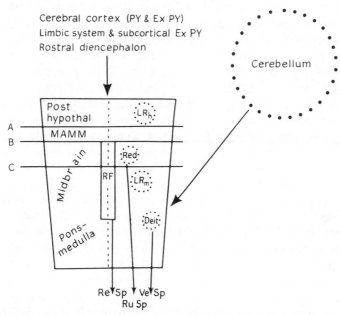

Fig. 5.8 Major descending influences upon the spinal stepping programme. A variety of 'higher centres' must contribute to stepping changes that are brought about by voluntary decisions and 'motivation.' None are essential for rudimentary stepping. After transection at *A* (premammillary) a cat will locomote spontaneously shortly after surgery (acute). Only a small portion of the posterior hypothalamus need be intact. With a *B* transection an acute cat will not locomote unless its limbs are set in motion on a moving treadmill belt by electrical stimulation of a 'locomotor region' under the inferior colliculus (LR_m) as determined by Shik, Severin and Orlovsky (1966). After section *C* the animal will not show this controlled locomotion, even though the LR_m is intact. A second locomotor region has been located in the posterior hypothalamus (LR_h). The 2 locomotor regions appear to be interconnected in complex fashion with each other, the pyramidal tract (PY), the reticular formation of the brainstem (RF) and perhaps descending extra pyramidal (ExPy) motor pathways (see text). The pre-eminent descending commands to the spinal stepping programme probably are those in the reticulospinal tract (ReSp). The RF and ReSp systems are shown for simplification as separate. The final ReSp outflow, together with at least 2 other outflows important for stepping, the vestibulospinal (VeSp) from Deiters nucleus (Deit) and the rubrospinal (RuSp) from the Red nucleus are all heavily influenced by descending, rhythmic signals from the cerebellum.

is allowed to survive (*chronic* preparation). Figure 5.8 shows the brainstem structures that are probably most important for stepping. The lowest transection of the neuraxis that will permit spontaneous stepping soon after surgery (acutely) is the premammillary cut, A, as shown in Fig. 5.8. More or less natural internal command signals may be assumed for this cat, although they recur

137

cyclically and, of course, the signals are not volitional or logically related to the environment. Progression is aimless in the sense that the animal will try to move forward even if its nose bumps against a wall.

A second cut, which is critical by comparison with the first one, is the post-mammillary transection, B, (Fig. 5.8). This cat will not locomote spontaneously in the acute state but retains the essential machinery, since it will step if (1) a *locomotor region* under the inferior colliculus is stimulated electrically (the controlled locomotion preparation of Shik, Severin and Orlovsky, 1966) or (2) it is allowed to recover for one or two weeks (Bard and Macht, 1958). These experiments suggest that cross-sectional cuts interrupt or inactivate, perhaps not in a very orderly way, different routes to the lower brainstem and possibly the spinal generators as well. The existence of alternative descending pathways has been confirmed by a series of experiments with controlled locomotion. The 'locomotor region' is not one with sharply defined functional or anatomical boundaries, nor is its stimulation effective unless there is intact tissue surroundings and rostral to it. The C transection in Fig. 5.8 abolishes controlled locomotion. The locomotor region itself has distinct dorsal and ventral portions, and the complex can be reached effectively either by direct stimulation, by activation of a similar region in the posterior hypothalamus (if this tissue remains intact), or (probably) by way of the pyramidal tract. The pyramidal tract, that most direct and often monosynaptic route from the cerebral cortex to spinal motoneurons, has a surprisingly complex role in stepping. At least part of its influence seems to be exerted indirectly by way of the locomotor region, since stimulation of the pyramids at the pontine level will initiate locomotion only if the medullary (more caudal) pyramids are severed and if the brainstem transection is at a level high enough to permit controlled locomotion (B cut, Fig. 5.8, data of Shik, Orlovsky and Severin, 1968).

The reticular formation of the pons and the emergent reticulospinal tract are contiguous to the 'locomotor region', but there are also other attractive candidates for pathways that might transmit command signals downward. In addition to the pyramidal tract, two conspicuous extrapyramidal pathways (that are interconnected with higher motor control structures as well as the cerebellum) have been suggested. All of them may include more than one component, but they are treated for present purposes as if unitary. In the inert, acute cat preparation, activation of the vestibulospinal tract by electrical stimulation furnishes excitation to extensor motoneurons, while the influence of the other three tracts is excitatory to flexor motoneurons. Their combined role during controlled stepping movements again (as with the spinal L-DOPA studies) encourages using the concept of a locomotor state, as opposed to a quiescent state, since they could exert their influence only in particular phases of the step. In some of these experiments activity from the various tracts was recorded during controlled locomotion that had been initiated by stimulation of the locomotor region (Orlovsky, 1970a, b; 1972a, b). Each tract was particularly active during only a portion of the step cycle. In other experiments, each tract could be stimulated separately to see how the two stimulation influences (locomotor region plus a tract) interacted (Orlovsky, 1972c). As might be anticipated, during controlled locomotion electrical stimulation of a particular tract was able to augment flexor activity during the swing phase or extensor activity during the stance, whichever was the excitatory effect that had been seen in the quiescent state.

The cells in the tracts evidently exert their effect on the spinal stepping programme in two ways: by an increased firing frequency beginning at the onset of controlled locomotion and by a phase-modulated discharge that conforms to the movements of the step cycle. It is tempting to speculate that the first activity concerns the switching-on of the stepping programme (the 'state') and the second provides for specialized excitatory reinforcement that matches the stepping movements themselves. The latter role would depend at least in part on peripheral feedback. As to the locomotor region of the brainstem, whatever pattern of discharge may occur in the intact animal, evidently the input need not be rhythmic. A steady flow of electrical pulses at a low frequency (20–50 p.p.s.) to even one side of the locomotor region can effectively elicit controlled locomotion.

Controlled locomotion is not prevented by severing the pyramidal tract, or by destroying the red nucleus (origin of the rubrospinal tract) or Deiters, lateral vestibular nucleus (origin of a component of the vestibulospinal tract), as established by the Russian work (Shik, Severin and Orlovsky, 1967). In this sense, all three pathways are optional. When present during controlled locomotion, however, their activity is co-operative and probably much like that which occurs during normal stepping by the intact animal. There are subtle differences in the percentages of cells that fire in different tracts, their firing frequency, and timing profiles. There is also indirect evidence that the reticulospinal pathway is the pre-eminent source for switching on the spinal stepping programme. Its cell bodies arise in the pons and medulla and mirror the excitability changes in cells of the locomotor region (Shik *et al.*,1967; Orlovsky, 1969). More reticulospinal cells work by sustained discharge than do cells in the other tracts. Moreover, there are noradrenergic reticulospinal cell bodies in this general complex (Dahlstrom and Fuxe, 1965) which would be suitable to initiate spinal stepping. The finding that L-DOPA or Clonidine could initiate stepping in spinal cats was significant because all known monaminergic cell bodies reside in sites above the spinal cord.

Although there is circumstantial evidence that reticulospinal neurons act as a final common pathway to the spinal stepping generators, it is too soon to say with assurance that the supraspinal input has a simple switching function alone. First, the descending inputs are phase-modulated. Second, serotonergic pathways may be involved as well as noradrenergic ones, as shown in work with rabbits by Viala and Buser (1969a, b; 1971). Finally, depending on the species, the brainstem structures may have quite a powerful role in interlimb co-ordination. The low spinal or high decerebrate cat can walk or trot (hind-limbs alternating) or gallop (hindlimbs nearly in phase). The decerebrate rabbit, however, shows only bilaterally in-phase rhythmic bursts in flexor and ex-tensor muscles, as if it were hopping. The spinal rabbit shows no rhythmic patterning unless L-DOPA or 5-HTP (a 5-HT serotonergic precursor) are given (after Nialamide, a monoamine oxidase inhibitor that potentiates the effect of DOPA), in which case an alternating rhythm appears that characterizes symmetrical gaits of walking and trotting.

Efferent cerebellar activity and integrated stepping

Examination of the role of the cerebellum in the stepping programme, or as an adjunct to it, emphasizes the difficulty in distinguishing the control of force from the control of rhythmicity. The classically recognized symptoms of cerebellar loss in humans and animals (see Dow and Moruzzi, 1958) include deficits in rate, force, and direction of movement. More recent studies of movements during locomotion have been made in partially or completely decerebellectomized dogs, and in high decerebrate cats under controlled locomotion before and after removal of the cerebellum (Denny-Brown, 1966; Orlovsky, Severin and Shik, 1966b). The surgery produced changes in swing–stance durations, there was less effective compensation to braking a limb, several measures were more variable, and there was general sluggishness. It was claimed that the four-legged stepping patterns occurred, nevertheless. Unfortunately, the dog whose cerebellum was completely removed evidently performed so poorly that it would be almost impossible to say whether the losses were greater for force or rhythm.

Together with study of 'normal' controlled locomotion and the descending tracts, counterpart experiments have been performed in which activity was recorded after the cerebellum had been removed (Orlovsky, 1970b; Orlovsky, 1972a, b; Orlovsky and Pavlova, 1972). The important new findings drew logically from a wealth of previous information about the cerebellum. Parti-cular routes from the cerebellum to each tract (rubro-, vestibulo-, and reticulo-spinal) and from the spinal cord to the cerebellum, as well as the topographical nature of the cellular arrangement in the cerebellum have been specified with some confidence from a variety of anatomical and electrophysiological in-vestigations and will not be summarized here. The most striking new finding was that decerebellation almost abolished the phase-modulated discharge in each tract. About the same number of cells worked during locomotion, but their frequency was much lower, in the case of the reticulospinal and rubro-spinal tracts. Rates almost doubled in vestibulospinal tracts, due to disin-hibition (removal of inhibitory input) of Purkinje cell and fastigial influence.

The powerful modulatory effect of the cerebellum that was demonstrated by these experiments is even more surprising when it is considered that there are direct collaterals (apparently non-modulatory) to the brainstem nuclei

140

from various ascending tracts. It might be thought that since these collaterals monitor afferent activity rather directly, they would profoundly influence descending activity. This is not the case.

The roles of the cerebellum in stepping that are emerging are, first, to reinforce by generalized excitation the influence of the descending tracts, so as to make the movements more forceful. Second, the cerebellum provides a powerful link in the route from limb afferents (and presumably incoming information from many sources, even the cerebral cortex, see Eccles, Ito and Szentágothai, 1967) to stepping generators in the spinal cord, and from there to motor outflow. This influence is rhythmic. By examining its nominally efferent functions, we have therefore anticipated the afferent role of the cerebellum, as discussed in the following section.

5.2.3 Afferent usage in locomotor control

The Russian studies in which priorities of single and interlimb co-ordination were established by braking or otherwise perturbing the stepping were significant. They showed that a given afferent input is always used against a background of what the rest of the limb, the other limbs, and the body are doing. As was true for discussion of the cerebellum, a long-term issue about the broad functions of afferentation has been whether peripheral inputs contribute more to locomotor rhythms or forces. More recently, there has been concern about the extent to which afferentation is involved in every single ongoing stepping movement, as opposed to being primarily a feedback system by which perturbations are sensed and dealt with centrally by reflex action (for full discussion see Grillner, 1975; Wetzel and Stuart, 1975). The more relevant afferent components are diagrammed in Fig. 5.9.

Signals from the single limb: afferent to afferent co-ordination

It was pointed out above, in discussion of the spinal programme, that afferent input is not required for establishing the gross rhythms of stepping. In all experiments in which one or another limb have been deafferented, however, whether in amphibians or mammals, the movements of the impaired limb appear to be weak and their excursions poorly controlled. Force production is therefore impaired selectively to some extent. In the intact limb rhythm and force are co-ordinated. It will be shown that the limb afferents usually provide their inputs during one phase of the step cycle. If a flexor muscle is providing tension, its afferents are more likely to be active during the swing; if an extensor muscle, its afferents discharge during the stance.

Skin afferents have immediate and pre-emptive input to the locomotor programme, as anyone can attest who has stepped on a sharp stone. It is also quite possible that movements of the skin across the bones of the limbs provide continuous input during normal stepping. Yet reports of many years standing have remained unchallenged in their implication that stepping movements are retained even after complete removal of the skin (e.g., Sherrington, 1910).

Joint afferents, and especially the muscle spindle and tendon organ afferents, have received more experimental attention with respect to stepping than have skin afferents. An important question has been whether their periods of activity

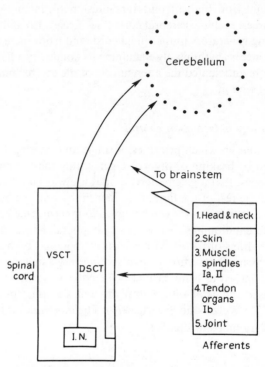

Fig. 5.9 Afferent aspects of the neural control programme for stepping. Different species of receptors provide input to the spinal cord and brainstem. It should be emphasized that even the initial terminations of sensory neurons are poorly understood (see text). The figure highlights the dorsal (DSCT) and ventral (VSCT) spinocerebellar tracts, which are shown as being more concerned with topographical input from the hindlimb or interneuronal (IN) activity, respectively. See discussion in text.

coincide with times of active muscle contraction or with times of passive muscle stretch, when external forces (tension of antagonistic muscles or gravity) move a joint. What little is known about joint afferent activity during loco-motion suggests that the joint's status (its angular excursion in degrees) and the muscle status (the amount of tension of the various muscles acting about a joint, even when the joint angle is locked) are both capable of modifying the joint afferent discharge. Normally, then, joint afferents are probably co-active to an unknown extent with muscle afferents (for fuller discussion see Wetzel and Stuart, 1975).

More is known about afferents from the muscle spindles, with the important data gained from controlled locomotion experiments in which ankle muscles (largely extensors), their afferent neurons, and their efferent neurons were studied (both studies by Severin *et al.*, 1967). Spindle secondary (group II) afferents are probably rung in together with spindle primary (group I) afferents (Goslow *et al.*, 1973b), but this important point has not been tested as yet. The spindle afferents (unidentified as to Ia or group II origin) from extensor muscles were active during the stance phase of controlled locomotion, at a time when the extensors were contracting. The finding was of great importance to theories of the much-studied alternative routes through which Ia afferents can be fired (for full review, see Matthews, 1972). Ia endings respond both to stretch of the extrafusal muscle fibres (served by large α-motoneurons) with which they are in parallel arrangement, and also to contraction of intrafusal fibres in the spindle itself (served by small γ-motoneurons). The γ input can cause Ia afferents to fire even while their extrafusal fibres are contracting. If there were no γ input to the spindle during the stance, discharge by the Ia afferents would pause, while tension is removed from them by contraction of the whole muscle. Indeed, novacaine block of γ input to medial gastrocnemius vastly reduced the ankle extensor's spindle discharge during the stance phase of the controlled step (Severin, 1970).

During controlled locomotion, the extensor spindle afferents were not highly responsive to passive stretch, particularly at faster rates of stepping. They fired only sporadically as the ankle joint was flexed during the F (flexion) epoch of the step cycle. Even during slow walking the amount of stretch during the F epoch is quite large for extensors, and so their relative silence at this time might result from complex visco-elastic effects (Goslow *et al.*, 1973b). The signals conveyed to the spinal cord by extensor spindle afferents were in a train of high-rate discharges with a sharp onset and offset but no peak firing rate, as if duration was the more important message. Although the point was not well established, it is probable that strength of the incoming signal was reported by individual fibres by means of an increased discharge rate as the vigour (force and speed) of locomotion increased from slow walking to more rapid movement. Less has been learned about spindle afferents from flexors than from extensors, but their active periods coincide with both contraction and passive stretch of these muscles.

Golgi tendon organs are found at the musculotendonous junctions and are supplied with a large group I afferent, called the Ib fibre. While Ia afferents signal length, tendon organs transduce force, being extremely sensitive to small changes in active muscle tension (for general review of transducing properties see Stuart, Mosher and Gerlach, 1972). The activity of an individual receptor will monitor tension of those muscle fibres (from several motor units) with which it is aligned in series arrangement. It will not necessarily show an activity profile that conforms to the tension profile of the whole muscle but rather the response mirrors the profile of force actually coupled to the receptor itself. Extensor tendon organs have been found to fire during the stance phase of controlled locomotion, however, and there are preliminary calculations that the ensemble, collective response of all the tendon organs of a muscle will reflect accurately the profile of net force developed at the tendon (Reinking, Stephens and Stuart, 1975).

In summary, the limb afferents from extensor muscles are found to be, if not completely co-active, at least with large overlapping and cascading periods of co-activity. The reflexly controlled response to spindle and tendon organ input is probably not either length or force separately, but the resultant stiffness (force per length change). The extensor Ia, Ib, and spindle group II fibres all fire during the stance phase and probably show their peak activity when force is maximal, at the E^2/E^3 junction. Flexor EMG and afferent activity are more complex (Goslow *et al.*, 1973b; both studies by Severin *et al.*, 1967) but particularly important at lift-off and touch-down, when these muscles are co-active with extensors. In view of these varied but composite inputs, we must assume that a good deal of information from the limb normally arrives simultaneously at the spinal cord.

The above picture of continuous net peripheral inputs from the limb as a whole is not incompatible with findings that the stepping movements in the cycle respond immediately to perturbations. The only proviso would be that the central locomotor programme be able to switch or re-route traffic quickly. It has been suggested that the cycle responds more to some inputs than to others, with a predominant influence from the hip. Movements of the hip have been reported to affect stepping much more than do movements at other joints. If one leg of a low spinal cat was stopped during the stance phase of stepping on a treadmill, the extensor EMG continued until the femur was pulled backward, but knee and ankle angles were not apparently critical (Grillner, 1975). In addition, the same stimulus of a tap to the dorsum of the foot or lower leg (a cue not unlike that provided by a swinging branch that brushes the leg in the natural environment) will show *reflex reversal*. That is, the input enhances extensor activity if delivered during the stance but enhances flexor activity if delivered during the swing (Forssberg, Grillner and Rossignol, 1975).

It may be concluded from the profuse information that is sent inward during a normal step cycle, and from the appropriate reflex responses to perturbation, that co-ordination of stepping by the central nervous system is concerned on a moment-to-moment basis with accurate descriptions of the external environment. Since so much information is processed simultaneously, however, the question does arise about how the smooth timings of efferent events are brought about by interactions between afferent and central neural pathways. The answer to this question has been sought by two kinds of experiments; those in which (1) the limb afferent pathways are traced inward, or (2) recordings are made of interneuronal activity directly.

144

Afferents and their initial spinal terminations

The kinematic data described earlier showed that the sequence of events in one limb is quite autonomous and rather impervious to perturbations of other limbs. Probably, then, the first priority of segmental-central interaction is a linkage between afferent pathways and the individual stepping generator for the homonymous limb. It has proved very difficult, however, to trace the presumably tight linkage between limb afferents and their homonymous motoneurons, let alone the interconnections involved in the stepping generator. Only one monosynaptic afferent-efferent pathway has been well documented, the excitatory spindle Ia-to-α-motoneuron arc, with all the other routes assumed to involve one or more interneurons. Intracellular records can be taken quite efficiently and easily from the large α-motoneurons, but small interposed cells are not easily penetrated by even sophisticated modern recording techniques.

A more serious problem has been to identify the particular afferent species from which the experimenter is recording. A good deal of information has been collected about polysynaptic pathways that link different muscles and joints (Eccles, Eccles and Lundberg, 1957a, b, c; R. M. Eccles and Lundberg, 1958; 1959a, b; Hongo, Jankowska and Lundberg, 1969). Interpreting most of this information depends, however, upon a now unacceptable assumption (reviewed by Matthews, 1972, and McIntyre, 1974). It can no longer be held that graded electrical stimulation of a whole muscle nerve (which progressively recruits sensory axons in the order large cross-sectional area to small) is a procedure that selectively singles out successive populations of single afferent species (i.e., spindle Ia's before Ib's, before spindle group II's, etc.). One of the new techniques is afferent-triggered averaging in which the discharge of an intact (i.e., in-continuity) functionally isolated single afferent or its cell body in the dorsal root ganglion is summated together with synaptic potentials of motoneurons. This approach has already revealed that spindle group II afferents have monosynaptic connections with homonymous motoneurons (Kirkwood and Sears, 1975). Afferent-triggered averaging can also be used for the study of disynaptic and possible trisynaptic pathways as well (Reinking *et al.*, 1975). Many more data must be generated, however, before the usefulness of this way of defining the reflex connections and function of different afferent species can be properly established.

Second order afferent influences upon intraspinal cells and tracts

In addition to attempts to trace the route inward from first order afferent fibres, another strategy has been to bypass the problem of central connections and record directly (extracellularly) from interneurons during locomotion. Clues may then be gained as to how afferent signals are actually used by the single limb's generator or more generally by the rest of the neural machinery for locomotion. During controlled locomotion it has been possible to record from interneurons whose activity is clearly synchronized with stepping by a single hindlimb, even if the cells (which lie in lamina VII and IX) cannot be identified unambiguously (Orlovsky and Feldman, 1972b). The interneurons characteristically fired during one part of the step; swing, stance or other subportion of the cycle.

145

Two other sets of interneurons are more easily identified, the dorsal and ventral spinocerebellar tracts (DSCT and VSCT). The cerebellum has been firmly established as a station to which locomotor signals are routed, on the basis of controlled locomotion experiments on these two tracts (see 5 studies by Arshavsky *et al.*, reported in 1972). Further, the cerebellum is supplied with at least two distinct kinds of information. One kind consists of relatively discrete messages from individual muscles or small muscle groups, that signal the phase of the cycle. These messages, the DSCT transmission, occur during the period that is appropriate for whichever muscles constitute the receptive field of individual DSCT cells. As was true for muscle and tendon organ afferents, DSCT rates increase to accompany increased vigour of locomotion. The phase modulation of DSCT cells is lost after deafferentation, implying that their afferent source is rather direct.

VSCT cell behaviour is quite different from that of DSCT cells and is seemingly indirect with respect to afferentation. The reasons for this view are, first, that the peak firing rate of an 'average' VSCT cell is reached considerably earlier in the stance than is true for the 'average' DSCT cell. When allowance is made for transmission time from the periphery, the DSCT cell presumably is more capable of reflecting accurately the profile of overall incoming extensor receptor activity. There are more extensors than flexors at the ankle, and they are stronger, so an 'average' cell is most likely to show peak activity near the E^2/E^3 junction, when extensor afferentation is strongest. The average VSCT cell, although its discharge is also phase-modulated, fires so early in the stance that it may not be able to transmit realistic information about that cycle's burst of peripheral input. Its profile may even include information from a previous step.

The second reason to believe that VSCT cells only indirectly are related to afferent input is that the phase-modulation of VSCT cells survives deafferentation. VSCT cells would appear, in summary, to be giving out information more from other interneuronal sources than from afferent sources (see, for example, Lundberg, 1971).

5.3 Behavioural implications of afferent-descending-spinal loops

As with other cells in afferent or interneuronal pathways, neurons in the cerebellar cortex or in the deep cerebellar nuclei show typical locomotor behaviours of: (1) modulation during the single limb's step cycle; (2) a peak firing rate; and (3) an increasing rate with increased speed and force of stepping (Orlovsky, 1972d, e). These three aspects of the cerebellar signals are then returned to the spinal cord by way of reticulospinal, vestibulospinal, and rubrospinal pathways that were described above. Nor should inputs from other brainstem or even higher structures be forgotten, in closing these widely distributed neuronal loops to and from the periphery.

It would appear that massive numbers of cells throughout the nervous system

146

are involved during locomotion. Recall that even now only a restricted sample has been characterized by recording techniques, the cells that fire in relation to movements of a single hindlimb. The co-ordination of various stepping movements, both within and between limbs, must be carried out by subordination of a multitude of pathways. The large number of cells are probably essential to maintain a sufficient level of motoneuronal excitability in order to activate sequentially every muscle of the four limbs, together with muscles of the head, neck, and spine.

Sharp turns, quick stops, and other swift adjustments of all four limbs that may be needed by the animal may also depend upon having many neurons already active. It is known that the spinal stepping generators of the high decerebrate cat respond to a summed input from descending and afferent pathways. Sacral ventral root filaments can be driven rhythmically by combining electrical stimulation of the brainstem locomotor region with stimulation of the sacral dorsal roots, when the strength of brainstem stimulation is too low to elicit the cyclic activity (Budakova, 1971). This finding implies that in the intact animal the intrinsic locomotor programme also responds to net inputs that reflect the most important goals of the animal, whether the directives be motivational ones from the brain (to attack, stalk, or climb) or pain signals from the feet. Only by the strict ordering and switching in and out of afferent, central, and efferent neurons, which are themselves organized in functional subsets, can the whole animal maintain forward progression. In this collective sense, then, it is reasonable to postulate a generalized locomotor state in which most of the nervous system is involved ultimately with the timing and amplitude of individual muscle tensions.

References

Ariano, M.A., Armstrong, R.B. and Edgerton, V.R. (1973) Hindlimb muscle fiber populations of five mammals. *J. Histochem. Cytochem.* **21**, 51–55.

Arshavsky, Yu.I., Berkinblit, M.B., Fukson, O.I., Gelfand, I.M. and Orlovsky, G.N. (1972a) Recordings of neurones of the dorsal spinocerebellar tract during evoked locomotion. *Brain Res.* **43**, 272–275.

Arshavsky, Yu.I., Berkinblit, M.B., Fukson, O.I., Gelfand, I.M. and Orlovsky, G.N. (1972b) Origin of modulation in neurones of the ventral spinocerebellar tract during locomotion. *Brain Res.* **43**, 276–279.

Arshavsky, Yu.I., Berkinblit, M.B., Gelfand, I.M., Orlovsky, G.N. and Fukson, O.I. (1972a) Activity of the neurones of the dorsal spinocerebellar tract during locomotion. *Biophys.* **17**, 506–514. (Translation from the Russian journal *Biofizika*.)

Arshavsky, Yu.I., Berkinblit, M.B., Gelfand, I.M., Orlovsky, G.N. and Fukson, O.I. (1972b) Activity of the neurones of the ventral spino-cerebellar tract during locomotion. *Biophys.* **17**, 926–941.

Arshavsky, Yu.I., Berkinblit, M.B., Gelfand, I.M., Orlovsky, G.N. and Fukson, O.I. (1972c) Activity of the neurones of the ventral spinocerebellar tract during locomotion of cats with deafferentated hind limbs. *Biophys.* **17**, 1169–1176.

Arshavsky, Yu.I., Kots, Y.M., Orlovsky, G.N., Rodionov, I.M. and Shik, M.L. (1965) Investigation of the biomechanics of running by the dog. *Biophys.* **10**, 737–746.

Bard, P. and Macht, M.B. (1958) The behavior of chronically decerebrate cats. In: *Neurological Basis of Behavior*, Wolstenholme, G.E.W. and O'Connor, C.M., Eds., pp. 55–71. Boston: Little, Brown.

Mechanics and energetics of animal locomotion

Bergmans, J. and Grillner, S. (1969) Reciprocal control of spontaneous activity and reflex effects in static and dynamic γ-motoneurones revealed by an injection of DOPA. *Acta physiol. scand.* **77**, 106–124.

Bergmans, J., Miller, S. and Reitsma, D.J. (1973) Influence of L-DOPA on transmission in long ascending propriospinal pathways in the cat. *Brain Res.* **62**, 155–167.

Budakova, N.N. (1971) Stepping movements evoked by a rhythmic stimulation of a dorsal root in mesencephalic cat. *Sechenov Physiol. J. U.S.S.R.* **57**, 1632–1640. (In Russian.)

Burke, R.E. (1975) A comment on the existence of motor unit 'types.' In: *The Nervous System: 25 Years of Progress, Vol. 1, Basic Neurosciences*, Brady, R., Ed., New York: Raven.

Burke, R.E. and Edgerton, V.R. (1975) Motor unit properties and selective involvement in movement. In: *Exercise & Sports Science Reviews, Vol. III*, Wilmer, J.H. & Keogh, J.F., Eds., pp. 33–81. New York: Academic Press.

Carlsson, A.T., Falk, B., Fuxe, K. and Hillarp, N.A. (1964) Cellular localization of monamines in the spinal cord. *Acta physiol. scand.* **60**, 112–119.

Cavagna, G.A. (1970) Elastic bounce of the body. *J. appl. Physiol.* **29**, 279–282.

Dahlström, A. and Fuxe, K. (1965) Evidence for the existence of monamine neurons in the central nervous system. II. Experimentally induced changes in the intraneuronal amine levels of bulbospinal neuron systems. *Acta physiol. scand.* **64**, *Suppl. 247*, 1–36.

Denny-Brown, D. (1966) *The Cerebral Control of Movement*. Springfield, Ill.: Thomas.

Dow, R.S. and Moruzzi, G. (1958) *The Physiology and Pathology of the Cerebellum*. Minneapolis: Univ. of Minnesota Press.

Eberhart, H.D., Inman, V.T. and Bressler, B. (1964) The principle elements of human locomotion. In: *Human Limbs and Their Substitutes*, Klopsty, P.E. and Wilson, P.D., Eds., pp. 327–349. New York: McGraw-Hill.

Eccles, J.C., Eccles, R.M. and Lundberg, A. (1957a) Synaptic actions on motoneurones in relation to the two components of the group I muscle afferent volley. *J. Physiol., Lond.* **136**, 527–546.

Eccles, J.C., Eccles, R.M. and Lundberg, A. (1957b) The convergence of monosynaptic excitatory afferents on to many different species of α-motoneurones. *J. Physiol., Lond.* **137**, 22–50.

Eccles, J.C., Eccles, R.M. and Lundberg, A. (1957c) Synaptic actions on motoneurones caused by impulses in Golgi tendon organ afferents. *J. Physiol., Lond.* **138**, 227–252.

Eccles, J.C., Ito, M. and Szentágothai, J. (1967) *The Cerebellum as a Neuronal Machine*. New York: Springer-Verlag.

Eccles, R.M. and Lundberg, A. (1958) The synaptic linkage of 'direct' inhibition. *Acta physiol. scand.* **43**, 204–215.

Eccles, R.M. and Lundberg, A. (1959a) Supraspinal control of interneurones mediating spinal reflexes. *J. Physiol., Lond.* **147**, 565–584.

Eccles, R.M. and Lundberg, A. (1959b) Synaptic actions in motoneurones by afferents which may evoke the flexion reflex. *Arch. ital. Biol.* **97**, 199–221.

Engberg, I. and Lundberg, A. (1969) An electromyographic analysis of muscular activity in the hindlimb of the cat during unrestrained locomotion. *Acta physiol. scand.* **75**, 614–630.

Feldman, A.G. and Orlovsky, G.N. (1974) Activity of interneurons mediating reciprocal Ia inhibition during locomotion. *Brain Res.* **84**, 181–194.

Forssberg, H., Grillner, S. and Rossignol, S. (1975) Phase-dependent reflex reversal during walking in chronic spinal cats. *Brain Res.* **85**, 103–107.

Frank, A.A. (1968) Automatic control systems for legged locomotion machines. Unpublished doctoral dissertation, University of Southern California.

Frank, A.A. and McGhee, R.B. (1969) Some considerations relating to the design of autopilots for legged vehicles. *J. Terramech.* **6**, 23–35.

Freund, H.J., Büdinger, H.J. and Dietz, V. (1975) Activity of single motor units from human forearm muscles during voluntary isometric contractions. *J. Neurophysiol.* **38**, 933–946.

Gambaryan, P.P. (1974) *How Mammals Run*. Jerusalem: Keter.

Gambaryan, P.P., Orlovsky, G.N., Protopopova, T.G., Severin, F.V. and Shik, M.L.

148

(1971) The activity of muscles during different forms of locomotion of cats and the adaptive function of the (hindlimb) musculature in the family Felidae. *Proc. Inst. Zool. Acad. Sci., U.S.S.R.* **48**, 220–239. (In Russian.)

Gernandt, B.E. and Gilman, S. (1961) Differential supraspinal control of spinal centers. *Exp. Neurol.* **3**, 307–324.

Gernandt, B.E. and Megirian, D. (1961) Ascending propriospinal mechanisms. *J. Neurophysiol.* **24**, 364–376.

Giovanelli Barilari, M. and Kuypers, H.G.J.M. (1969) Propriospinal fibers interconnecting the spinal enlargements in the cat. *Brain Res.* **14**, 321–330.

Goslow, G.E., Jr., Reinking, R.M. and Stuart, D.G. (1973a) The cat step cycle: hind limb joint angles and muscle lengths during unrestrained locomotion. *J. Morph.* **141**, 1–41.

Goslow, G.E., Jr., Reinking, R.M. and Stuart, D.G. (1973b) Physiological extent, range and rate of muscle stretch for soleus, medial gastrocnemius and tibialis anterior in the cat. *Pflügers Archiv* **341**, 77–86.

Goslow, G.E., Jr., Stauffer, E.K., Nemeth, W.C. and Stuart, D.G. (1972) Digit flexor muscles in the cat: Their action and motor units. *J. Morph.* **137**, 335–352.

Graham Brown, T. (1916) Die reflex functionen des zentralnerven mit besonderer berucksichtigung, der rhythmischen fatigkeiten bein sargeiter. *Ergebn. Physiol.* **15**, 480–790.

Grillner, S. (1972) The role of muscle stiffness in meeting the changing postural and loco-motor requirements for force development by the ankle extensors. *Acta physiol. scand.* **86**, 92–108.

Grillner, S. (1973) Locomotion in the spinal cat. In: *Control of Posture and Locomotion*. Stein, R.B., Pearson, K.G., Smith, R.S. and Redford, J.B., Eds., pp. 515–535. New York: Plenum.

Grillner, S. (1974) On the generation of locomotion in the spinal dogfish. *Exp. Brain Res.* **20**, 459–470.

Grillner, S. (1975) Locomotion in vertebrates: central mechanisms and reflex interaction. *Physiol. Rev.* **55**, 247–304.

Grillner, S. and Zangger, P. (1974) Locomotor movements generated by the deafferented spinal cord. *Acta physiol. scand.* **91**, 38A–39A.

Gurfinkel, V.S. and Shik, M.L. (1973) The control of posture and locomotion. In: *Motor Control*, Gydikov, A.A., Tankov, N.T. and Kosarov, D.S., Eds., pp. 217–234. New York: Plenum.

Henneman, E. (1968) Peripheral mechanisms involved in the control of muscle. In: *Medical Physiology, Vol. II*, Mountcastle, V.B., Ed., pp. 1697–1716. St. Louis: Moshey.

Herman, R., Cook, T., Cozzens, B. and Freedman, W. (1973) Control of postural reactions in man: the initiation of gait. In: *Control of Posture and Locomotion*, Stein, R.B., Pearson, K.G., Smith, R.S. and Redford, J.B., Eds., pp. 363–388. New York: Plenum.

Hildebrand, M. (1959) Motions of the running cheetah and horse. *J. Mammal.* **40**, 481–495.

Hinsey, J.C., Ranson, S.W. and McNattin, R.F. (1930) The role of the hypothalamus and mesencephalon in locomotion. *Arch. Neurol. Psychiat., Chicago* **23**, 1–43.

Holmquist, B. and Lundberg, A. (1961) Differential supraspinal control of synaptic actions elicited by volleys in the flexor afferents in α-motoneurones. *Acta physiol. scand.* **54**, *Suppl. 186*, 1–51.

Hongo, T., Jankowska, E. and Lundberg, A. (1969) The rubrospinal tract. II. Facilitation of interneuronal transmission in reflex paths to motoneurones. *Exp. Brain Res.* **7**, 365–391.

Jane, J.A., Evans, J.P. and Fisher, L.E. (1964) An investigation concerning the restitution of motor function following injury to the spinal cord. *J. Neurosurg.* **21**, 167–171.

Jankowska, E., Jukes, M.G.M., Lund, S. and Lundberg, A. (1967a) The effect of DOPA on the spinal cord. 5. Reciprocal organization of pathways transmitting excitatory action to alpha motoneurones of flexors and extensors. *Acta physiol. scand.* **70**, 369–388.

Jankowska, E., Jukes, M.G.M., Lund, S. and Lundberg, A. (1967b) The effect of DOPA on the spinal cord. 6. Half-centre organization of interneurones transmitting effects from flexor reflex afferents. *Acta physiol. scand.* **70**, 389–402.

Jankowska, E., Lundberg, A., Roberts, W.J. and Stuart, D. (1974) A long propriospinal system with direct effect on motoneurones and on interneurones in the cat lumbosacral cord. *Exp. Brain Res.* **21,** 169–194.

Kirkwood, P.A. and Sears, T.A. (1975) Monosynaptic excitation of motoneurones from muscle spindle secondary endings of intercostal and triceps surae muscles in the cat. *J. Physiol.* **245,** 64P–66P.

Laughton, N.B. (1924) Studies on the nervous regulation of progression in mammals. *Am. J. Physiol.* **70,** 358–384.

Lloyd, D.P.C. (1941) Activity in neurons of the bulbospinal correlation system. *J. Neurophysiol.* **4,** 115–134.

Lloyd, D.P.C. (1942) Mediation of descending long spinal reflex activity. *J. Neurophysiol.* **5,** 435–458.

Lockard, D.E., Traher, L.M. and Wetzel, M.C. (1976) Reinforcement influences upon topography of treadmill locomotion by cats. *Physiol. Behav.* **16,** 141–146.

Lundberg, A. (1969) Reflex Control of Stepping. Nansen Memorial Lecture to Norwegian Academy of Sciences & Letters. Oslo: Universitetsforlaget.

Lundberg, A. (1971) Function of the ventral spinocerebellar tract: A new hypothesis. *Exp. Brain Res.* **12,** 317–330.

Manter, J.T. (1938) The dynamics of quadrupedal walking. *J. exp. Biol.* **15,** 522–540.

Matthews, P.B.C. (1972) *Mammalian Muscle Receptors and Their Central Actions.* London: Arnold.

Messinger, D.S. (1974) Ankle extensor activity in the walking cat. *Am. Zool.* **14,** 1267.

Miller, S. and van der Burg, J. (1973) The function of long propriospinal pathways in the coordination of quadrupedal stepping in the cat. In: *Control of Posture and Locomotion,* Stein, R.B., Pearson, K.G., Smith, R.S. and Redford, J.B., Eds., pp. 561–577. New York: Plenum.

Miller, S., van der Burg, J. and van der Meché, F.G.A. (1975) Coordination of movements of the hindlimbs and forelimbs in different forms of locomotion in normal and decerebrate cats. *Brain Res.* **91,** 217–237.

Milner-Brown, H.S., Stein, R.B. and Yemm, R. (1973) The orderly recruitment of human motor units during voluntary isometric contractions. *J. Physiol.* **230,** 359–370.

Muybridge, E. (1957) *Animals in Motion.* New York: Dover (from *Animal Locomotion,* first published in 1887).

McGhee, R.B. (1967) Finite state control of quadruped locomotion. *Simulation* **9,** 135–140.

McGhee, R.B. (1968) Some finite state aspects of legged locomotion. *Mathemat. Biosci.* **2,** 67–84.

McGhee, R.B. and Frank, A.A. (1968) On the stability properties of quadruped creeping gaits. *Mathemat. Biosci.* **3,** 331–351.

McIntyre, A.K. (1974) Central actions of impulses in muscle afferent fibres. In: *Handbook of Sensory Physiology* III, 2. *Muscle Receptors,* Hunt, C.C., Ed., pp. 235–288. Berlin: Springer-Verlag.

Orlovsky, G.N. (1969) Electrical activity in brainstem and descending paths in guided locomotion. *Sechenov Physiol. J U.S.S.R.,* **55,** 437–444. (In Russian.)

Orlovsky, G.N. (1970a) Work of the reticulo-spinal neurones during locomotion. *Biophys.* **15,** 761–771.

Orlovsky, G.N. (1970b) Influence of the cerebellum on the reticulospinal neurones during locomotion. *Biophys.* **15,** 928–936.

Orlovsky, G.N. (1972a) Activity of vestibulospinal neurons during locomotion. *Brain Res.* **46,** 85–98.

Orlovsky, G.N. (1972b) Activity of rubrospinal neurons during locomotion. *Brain Res.* **46,** 99–112.

Orlovsky, G.N. (1972c) The effect of different descending systems on flexor and extensor activity during locomotion. *Brain Res.* **40,** 359–372.

Orlovsky, G.N. (1972d) Work of the Purkinje cells during locomotion. *Biophys.* **17,** 935–941.

Orlovsky, G.N. (1972e) Work of the neurones of the cerebellar nuclei during locomotion. *Biophys.* **17,** 1177–1185.

Orlovsky, G.N. and Feldman, A.G. (1972a) Role of afferent activity in the generation of stepping movements. *Neurophysiol.* **4**, 304–310. (Translated from the Russian journal, *Neirofiziologiya*.)

Orlovsky, G.N. and Feldman, A.G. (1972b) Classification of lumbosacral neurons by their discharge pattern during evoked locomotion. *Neurophysiol.* **4**, 311–317.

Orlovsky, G.N. and Pavlova, G.A. (1972) Effect of removal of the cerebellum on vestibular responses of neurons in various descending tracts in cats. *Neurophysiol.* **4**, 235–240.

Orlovsky, G.N. and Shik, M.L. (1965) Standard elements of cyclic movement. *Biophys.* **10**, 935–944.

Orlovsky, G.N., Severin, F.V. and Shik, M.L. (1966a) Effect of speed and load on co-ordination of movements during running of the dog. *Biophys.* **11**, 414–417.

Orlovsky, G.N., Severin, F.V. and Shik, M.L. (1966b) Effect of damage to the cerebellum on the coordination of movement in the dog on running. *Biophys.* **11**, 578–588.

Pearson, K.G., Fourtner, C.R. and Wong, R.K. (1973) Nervous control of walking in the cockroach. In: *Control of Posture and Movement*, Stein, R.B., Pearson, K.G., Smith, R.S. and Redford, J.B., Eds., pp. 495–514. New York: Plenum.

Philippson, M. (1905) L'autonomie et la centralisation dans le systéme nerveux des animaux. *Trav. Lab. Physiol. Solvay* **7**, 1–208.

Prochazka, V.J., Tate, K., Westerman, R.A. and Ziccone, S.P. (1974) Remote monitoring of muscle length and EMG in unrestrained cats. *Electroencephal. clin. Neurophysiol.* **37**, 649–653.

Reinking, R.M., Stauffer, E.K., Stuart, D.G., Taylor, A. and Watt, D.G.D. (1975) The inhibitory effects of muscle spindle primary afferents investigated by afferent triggered averaging methods. *J. Physiol., Lond.* **248**, 20–22P.

Reinking, R.M., Stephens, J.A. and Stuart, D.G. (1975) The tendon organs of cat medial gastrocnemius: significance of motor unit size and type for the activation of Ib afferents. *J. Physiol., Lond.* **250**, 491–512.

Severin, F.V., Orlovsky, G.N. and Shik, M.L. (1967) Work of the muscle receptors during controlled locomotion. *Biophys.* **12**, 575–586.

Severin, F.V., Shik, M.L. and Orlovsky, G.N. (1967) Work of the muscles and single motor neurones during controlled locomotion. *Biophys.* **12**, 762–772.

Sherrington, C.S. (1910) Flexion-reflex of the limb, crossed extension-reflex, and reflex stepping and standing. *J. Physiol., Lond.* **40**, 28–121.

Shik, M.L. and Orlovsky, G.N. (1965) Coordination of the limbs during running of the dog. *Biophys.* **10**, 1148–1159.

Shik, M.L., Orlovsky, G.N. and Severin, F.V. (1968) Locomotion of the mesencephalic cat elicited by stimulation of the pyramids. *Biophys.* **13**, 143–152.

Shik, M.L., Severin, F.V. and Orlovsky, G.N. (1966) Control of walking and running by means of electrical stimulation of the mid-brain. *Biophys.* **11**, 756–765.

Shik, M.L., Severin, F.V. and Orlovsky, G.N. (1967) Structures of the brain stem responsible for evoked locomotion. *Sechenov Physiol. J. U.S.S.R.* **53**, 1125–1132. (In Russian.)

Shurrager, P.S. and Dykman, R.A. (1951) Walking spinal carnivores. *J. comp. physiol. Psychol.* **44**, 252–262.

Stephens, J.A., Reinking, R.M. and Stuart, D.G. (1975) The motor units of cat medial gastrocnemius: Electrical and mechanical properties as a function of muscle length. *J. Morph.* **146**, 495–512.

Sterling, P. and Kuypers, H.G.J.M. (1968) Anatomical organization of the brachial spinal cord of the cat. III. The propriospinal connections. *Brain. Res.* **7**, 419–443.

Stuart, D.G., Mosher, C.G. and Gerlach, R.L. (1972) Properties and central connections of Golgi tendon organs with special reference to locomotion. In: *Research in Muscle Development and the Muscle Spindle*, Banker, B.Q., Pryzbylsky, R.J., van der Meulen, J.P. and Victor, M., Eds., pp. 437–464. Amsterdam: Exerpta Medica.

Székely, G. (1965) Logical network for controlling limb movements in urodela. *Acta physiol. acad. scient. hung.* **27**, 285–289.

Taub, E. and Berman, A.J. (1968) Movement and learning in the absence of sensory feedback. In: *The Neuropsychlogy of Spatially Oriented Behavior*, Freedman, S.J., Ed., pp. 173–192. Homewood, Ill.: Dorsey.

Mechanics and energetics of animal locomotion

Tokuriki, M. (1973a) Electromyographic and joint-mechanical studies in quadrupedal locomotion. I. Walk. *Jap. J. Vet. Sci.* **35**, 433–446.

Tokuriki, M. (1973b) Electromyographic and joint-mechanical studies in quadrupedal locomotion. II. Trot. *Jap. J. Vet. Sci.* **35**, 525–533.

Tokuriki, M. (1974) Electromyographic and joint-mechanical studies in quadrupedal locomotion. III. Gallop. *Jap. J. Vet. Sci.* **36**, 121–132.

Viala, D. and Buser, P. (1969a) The effects of DOPA and 5-HTP on rhythmic efferent discharges in hind limb nerves in the rabbit. *Brain Res.* **12**, 437–443.

Viala, A. and Buser, P. (1969b) Activités locomotrices rhythmicues stéréotypées chez les lapin sous anesthésie légére. *Ex. Brain Res.* **8**, 346–363.

Viala, D. and Buser, P. (1971) Modalités d'obtention de rhythmes locomoteurs chez le lapin spinal par traitements pharmacologiques (DOPA, 5-HTP, d'amphétamine). *Brain Res.* **35**, 151–165.

Vukobratović, M., Frank, A.A. and Juričić, D. (1970) On the stability of biped locomotion. *IEEE Trans. Bio-Medical Engineering, BME-17*, 25–35.

Wetzel, M.C., Atwater, A.E. and Stuart, D.G. (1975) Movements of the hindlimb during locomotion of the cat. In: *Neural Control of Locomotion*, Herman, R., Grillner, S., Stein, P.S.G. and Stuart, D.G., Eds., pp. 99–135. New York: Plenum.

Wetzel, M.C., Atwater, A.E., Wait, J.V. and Stuart, D.G. (1975) Neural implications of different profiles between treadmill and overground locomotion timings in cats. *J. Neurophysiol.* **38**, 492–501.

Wetzel, M.C., Gerlach, R.L., Stern, L.Z. and Hannapel, L.K. (1973) Behavior and histochemistry of functionally isolated cat ankle extensors. *Exp. Neurol.* **39**, 223–233.

Wetzel, M.C. and Stuart, D.G. (1976) Ensemble characteristics of cat locomotion and its neural control. *Progress in Neurobiology* **7**, 1–98.

Wilson, D.M. (1964) The origin of the flight-motor command in grasshoppers. In: *Neural Theory and Modeling*, Reiss, R., Ed., pp. 331–345. Palo Alto: Stanford Univ. Press.

Yager, J.G. (1972) The electromyogram as a predictor of muscle mechanical response in locomotion. Unpublished doctoral dissertation. University of Tennessee.

6 Energy cost of locomotion

G. Goldspink

6.1 Introduction

This chapter is concerned almost exclusively with the cost of locomotion in vertebrates. The reason for this is that except for flying insects there is a distinct lack of data for the invertebrates and what data is available seems very fragmentary. Before discussing the details of the energy cost of swimming, flying and terrestrial locomotion it is necessary to mention two concepts that are basic to metabolic studies involving whole animals.

One of these concepts is that of cost of transport. This is the amount of energy it costs to move one kilogram body weight through one metre irrespective of the time it takes to do this. The net cost of transport is obtained by subtracting the normal resting metabolic rate from that during locomotion. In the past it has been the usual practice to express the energy used as the number of ml O_2 kg^{-1} m^{-1} or calories kg^{-1} m^{-1}. However, when standard international units are used this is expressed as the number of mole O_2 kg^{-1} m^{-1}, m^3 O_2 kg^{-1} m^{-1} or Joules kg^{-1} m^{-1}.

Another basic concept is that of power (metabolic power). This differs from cost of transport in that it expresses the rate of energy expenditure and not the total amount used. Power, whether it be deduced from metabolic measurements or from principles of mechanics, is work done per unit time and in certain types of locomotion for example bird flight, power is just as important a parameter as cost of transport. It must be borne in mind that if one is estimating the power requirements for a certain type of locomotion using metabolic studies, these measurements will include 'internal work' as well as 'external work'. In the older units metabolic rate of metabolic power was usually expressed as ml O_2 g^{-1} h^{-1} or cal g^{-1} h^{-1} or even horse power per pound. In SI units the metabolic rate may be expressed as m^3 O_2 kg^{-1} s^{-1} or as (metabolic) power in Watts kg^{-1}. This latter method has the advantage that the results may be compared directly with the power requirements for locomotion as calculated from mechanics, aerodynamics or hydrodynamics. Hence these sort of values can be used to check whether the mechanical explanation for a given type of locomotion is reasonable.

In theory there are three possible methods of measuring the energy cost of locomotion. Energy that is not converted into useful work is degraded to heat. Therefore, direct calorimetric methods involving heat production measurements on animals during locomotion are a possibility. However, there are grave technical difficulties in applying the method to anything but small animals. The most convenient and most widely used method is indirect calorimetry involving the measurement of the exchange of the respiratory gases and calculating the energy equivalent. Another indirect method is to measure the change in the fuel content of the animal. In short-term exercise quite large errors are incurred with this method unless the changes in the levels of certain intermediates of metabolism are also measured. However, in certain circumstances, for example, bird migration, measurements of the depletion of fuel levels can yield very useful information.

As stated, the most widely used method is that of measuring the respiratory gas exchange. The rate at which oxygen is used or the rate at which carbon dioxide is expired or both, may be measured reasonably conveniently with virtually all kinds of animals. There are several problems associated with oxygen consumption measurements. One of the main ones being that the tissues of most animals, particularly the muscle tissue, can respire anaerobically for considerable periods of time. Therefore measurements usually have to be made under steady state conditions or some sort of estimate, for example measuring blood lactate, has to be made of how much energy is being derived from the anaerobic pathways.

6.2 Energy cost of swimming

Most of the measurements of the cost of swimming have been made on fish. This, of course, is concerned with subsurface swimming. The energy requirements for swimming in water rather than on it, are usually determined by measuring the amount of oxygen consumed or the amount of carbon dioxide expired whilst the fish is swimming in an enclosed system. There are two basic designs of fish respirometer in common use. These were originally designed by Blazka *et al.* (1960) and by Brett (1964) and are shown diagrammatically in Fig. 6.1. The design requirements for fish respirometers are that the water flow through the fish chamber should be laminar, or as near to laminar as possible. This is achieved by delivering the water through a trumpet-shaped pipe and/or by the use of plastic meshes. There should also be some means of maintaining a constant temperature as the metabolic measurements are very much affected by temperature. This provision is particularly important if the system involves a pump which may dissipate quite large quantities of heat. Changes in the oxygen level in the water may be monitored using an oxygen electrode or by titration using the Winkler method. Carbon dioxide may be

154

measured either by using an electrode or by titration. The oxygen content of the water should not be allowed to drop too far and normally measurements are made within the range of 100% to 95% saturation. This means there has to be provision for re-oxygenation or for changing the water between each set of measurements. Swimming velocity is very much dependent on the size of the fish and for this reason the results are often expressed in relative terms as body lengths per second.

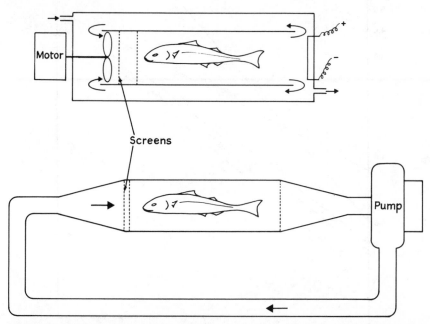

Fig. 6.1 Two designs of fish respirometer that are suitable for metabolic studies of swimming *Top*, designed by Blazka *et al.* (1960). *Bottom* designed by Brett (1964).

The metabolic cost of swimming has been found to be dependent on several parameters including swimming speed, environmental temperature and body size. As far as swimming speed is concerned the oxygen consumption has been found to increase in an exponential way. Often the results are presented as a semi-log plot and in this case a straight line is obtained. Therefore in order to avoid confusion, it is important to take careful note of the axes. In practice, it is difficult to obtain reproducible results particularly at the lowest and the highest swimming speeds. At the lowest swimming speeds the fish tend to move about erratically and at the highest swimming speeds the fish fatigue so quickly that there is insufficient time for a measurement to be made. Also at higher speed the percentage of energy derived from the anaerobic pathways is considerable so that the oxygen consumption on its own is no longer a reliable measure of (metabolic) power (Fig. 6.3).

Some of the most reliable results for the power required for swimming have been made by Brett and Sutherland (1965) on the pumpkinseed fish. These fish are by nature less excitable than other species and therefore they were easier to handle in the respirometer. However, the same sort of data has been obtained for salmon (Brett, 1964), haddock (Tytler, 1969), trout (Webb,

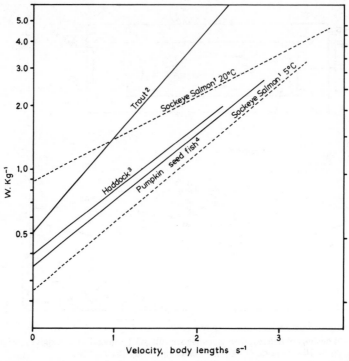

Fig. 6.2 Logarithmic plot of power (oxygen consumption) plotted against relative swimming speed in body lengths per second for a variety of species. Data from (1) Brett (1964), (2) Webb (1971), (3) Tytler (1969), (4) Brett and Sutherland (1964).

1971), and several other species of fish. These data are given in Fig. 6.2 in which the oxygen consumption (metabolic power) is plotted on a logarithmic axis against relative swimming speed which is plotted in a linear manner. In all cases, the oxygen consumption shows a logarithmic increase with increase in swimming speed; however, if these measurements were to be extended to the higher speeds there seems to be no doubt that a point would be reached at which the fish can no longer extract any more oxygen from its environment. At these higher speeds the amount of energy that can be derived from oxidative phosphorylation and from glycolysis (Fig. 6.3) is not adequate to meet the very high power requirements and therefore the fish fatigues very rapidly.

156

Environmental temperature also influences the cost of swimming (Brett, 1964; Smit *et al.*, 1971). As might be expected it increases the resting (standard) metabolic rate; however, the slope of the O_2/velocity plot is less steep at the higher temperatures. This means that the net cost of transport is less at the higher temperatures. The reason for this may be that there is a decreased internal resistance, that is to say, the internal tissue fluids will be less viscous at the higher temperatures. However, it may also be due to improved biochemical efficiency.

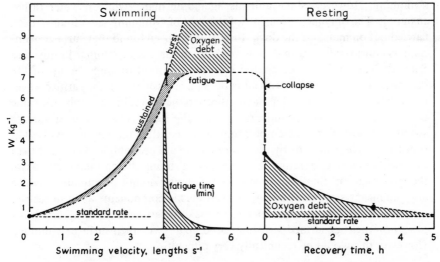

Fig. 6.3 Energy requirements for fish swimming plotted in a linear manner. Also shown is the amount of energy derived from anaerobic path both during exercise and after cessation of exercise. Taken from Brett (1964).

As far as the cost of transport is concerned, this increases up to about two body lengths s^{-1} or so, depending on the species, however, after this the power curve starts to rise steeply, therefore transport becomes very costly. It has been argued (Weihs, 1974) on theoretical grounds that fish should be able to move more economically by alternatively accelerating and gliding. It seems that some fish do employ this mode of swimming instead of maintaining a steady cruising speed. In theory at least, the accelerating and gliding method should decrease the net cost of transport of the fish. This situation is rather similar to the situation of intermittent flight in birds. However, in the case of fish, accurate metabolic measurements have yet to be carried out.

Not only is the cost of transport dependent on swimming speed, but it is also very much dependent on the size of the fish. Using Brett's data, Schmidt-Nielsen (1971) has shown that if the net cost of transport is plotted against

157

body size using logarithmic co-ordinates, a straight line relationship is obtained (Fig. 6.7). As will be seen from this figure, the smaller the fish the higher its cost of transport. It is in fact a general principle that the smaller the animal the higher the cost of transport and that this applies to flying and terrestrial locomotion as well as to swimming.

Interestingly, the shape of the fish does not seem to be particularly important in determining the cost of transport as the pinfish lies on the same line as the trout and the goldfish and even the eel which uses anguilliform propulsion falls more or less on the same line.

So far, the discussion has been concerned with the energy cost of subsurface swimming. However, some animals, including man, swim on water rather than in it. The only measurements of this sort of swimming seem to have been carried out on man and the duck. Briefly, it has been found that surface swimming is considerably more expensive than subsurface swimming. Prange and Schmidt-Nielsen (1970) compared surface swimmers to ships. A lot of the energy that is required for propelling the ship is dissipated in the surface waves that are formed when the ship moves through the water. Presumably, the same holds true for surface swimmers such as the duck. The action in surface swimming more closely resembles running rather than the actions involved in subsurface. These actions by their very nature are probably less efficient than the oscillating or undulating movements of fish and diving mammals. Also, the movements of the limbs in surface swimming do not presumably involve any significant storage of energy in their elastic components as in terrestrial locomotion. It would be of interest to make measurements on diving birds to see how their subsurface performance compares with their surface swimming. However, this would be technically very difficult.

6.3 Energy cost of flying

There are several kinds of flight; hovering flight, forward flapping flight, intermittent (swooping) flight and gliding (soaring) flight. Only the small birds and certain insects seem to use hovering flight to any great extent. Certain birds of prey apparently hover, however they usually do this in a situation where there is some lift from wind or updraughts. In this latter case they are really gliding on the updraughts and the hovering action is used merely to maintain station. Flapping forward flight is the main kind of power flight and is displayed by insects, birds and bats. Many of the very large birds such as vultures, storks and albatrosses, however, only rarely indulge in flapping flight. Intermittent (undulating or swooping) flight is often observed in small birds. In this type of flight the bird ascends by flapping flight and then descends, usually with its wings folded against its body. For small birds it seems that there are certain behavioural and energetic advantages in undulating flight. Gliding or soaring flight, on the other hand, seems to be confined to the large or medium-sized birds. This type of flight obviously represents an energy saving as the bird is in effect using the energy available from the environment. It seems that for energetic and aerodynamic reasons that it is only the larger birds that are able to exploit this method of flying. It is important to make the distinction between these different forms of flight because, as will be shown, the energy requirements differ very considerably.

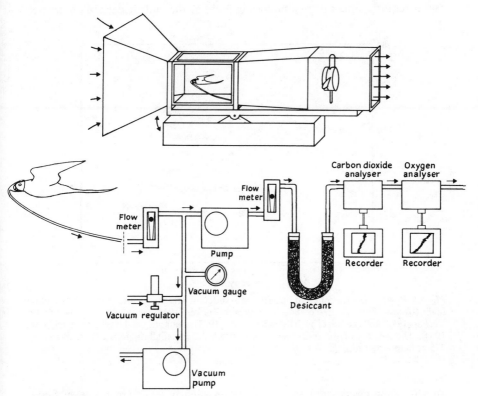

Fig. 6.4 Method of measuring oxygen consumption of birds during flight. From Tucker (1966, 1968, 1969).

One of the main problems involved in measuring the energy cost of bird flight is that birds fly very quickly and they fly in three dimensions. With the exception of some special cases such as hovering in the humming bird, the only really successful method of measuring energy cost during flight has been that used by Tucker (1966, 1968, 1969). Birds are flown in a wind tunnel in which the wind velocity can be varied and in which angle of the airflow can be adjusted by tilting the tunnel. The exchange of respiratory gases is then measured by fitting a clear plastic mask over the head of the bird. A pump is used to draw air through the mask and to collect the expired gases (Fig. 6.4). Using Tucker's method the levels of oxygen and carbon dioxide are measured by

159

passing the expired gases through Beckmann O_2 and CO_2 analysers and the flow of the airstream is very carefully monitored using very accurate flow gauges.

Using this system, Tucker, Schmidt-Nielsen and co-workers at Duke University have made some very interesting metabolic measurements for the different types of flight. Tucker for instance has shown that the metabolic power requirements for the undulating flight of the budgerigar are no greater

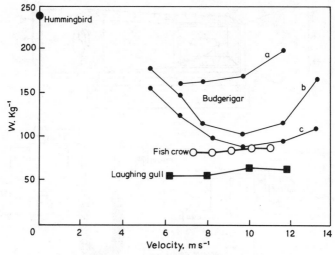

Fig. 6.5 Power input for different birds flying at different velocities. Budgerigar (a) ascending flight (b) level flight (c) descending flight. From Tucker (1968): laughing gull, level flight, Tucker (1972) and the fish crow; (Bernstein *et al.*, 1973): hummingbird hovering (Lasiewski, 1963). The weights of the birds were: budgerigar, 3.5 g; laughing gull, 277 g; fish crow, 275 g; hummingbird, 3 g.

and, in some circumstances, may be even slightly less than for level flight. When flying level, the metabolic power curve has a U-shape with the minimum power requirement occurring at about 10 m s^{-1} (Fig. 6.5). Descending flight also exhibits a U-shaped curve with the optimum occurring at about 10 m s^{-1}. Ascending flight, however, shows an exponential curve with the lowest power cost occurring at speeds less than about 8.0 m s^{-1}. At 10 m s^{-1} level flight has a requirement of about 119 W kg^{-1}. Ascending flight costs 164 W kg^{-1} and descending flight about 86 W kg^{-1}, giving a mean of 125 W kg^{-1} which is very little more than for level flight. At speeds slower than 7 m s^{-1} the power requirements for ascending flight actually become less than those for level flight, thus if the bird ascends at about 7 m s^{-1} and descends at about 10 m s^{-1} the power requirements are appreciably less than they are for level flight at slow speeds. Therefore if the bird is going to fly slowly it might as well engage in undulating flight. As the budgerigar is a ground feeder the comparatively low

160

power requirements for ascending and descending flight is obviously important as it presumably is for aerial feeders such as swallows and swifts.

As far as cost of transport is concerned, budgerigars travel most economically at 10–12 m s^{-1}. This is in fact almost as fast as they can fly. At this speed the budgerigar consumes about 0.03 % of its body weight in fat per kilometre. Tucker's results also show that the most economical speed for the laughing gull was 12.5 m s^{-1} which is also about as fast as the bird can fly and at this speed this bird consumes only 0.015 % of its body weight in fat per kilometre. Flying, like the other forms of locomotion, does in fact show the same type of relationship between the net cost of transport and body size. Again, if the cost of transport for flying is plotted against body size, using logarithmic co-ordinates, then a straight line relationship is obtained (Fig. 6.7). Other factors such as wing span and wing shape have also been shown to have a significant effect on the cost of flapping flight. Bernstein *et al.* (1973) compared the power requirement for flight in the fish crow which has a relatively short wing span with that for the laughing gull. They found that although the two species have about the same body weight, the power requirement for the fish crow was about 50 % greater than for the laughing gull.

Another advantage that large birds have, is that many of them exploit gliding as a means of energy saving. Metabolic measurements of the metabolic requirements for gliding have been carried out by Baudinette and Schmidt-Nielsen (1974). These authors found that the metabolic rate of the herring gull during gliding was only about twice the resting level whereas during flapping flight (Tucker, 1972) it is about seven times the resting level. A fascinating study has been carried out by Pennycuick (1972) in which he demonstrated how some East African vultures commute over quite long distances by soaring on thermals and by gliding from thermal to thermal. In this sort of situation the net cost of transport for gliding is probably very low as compared with that for flapping flight.

In contrast to gliding flight, the power requirements for hovering flight are high. Lasiewski (1963) measured the oxygen consumption of the humming bird in a closed system and attained values equivalent to 240 watts kg^{-1} body weight. As the flight muscles constitute 25 % of the body weight (Greenewalt, 1962) this means that they are producing 0.3 watts per gram (assuming an efficiency of 30 % for the conversion of lipids into work) which must be near the upper limit of sustainable power by any type of muscle.

6.4 Energy cost of terrestrial locomotion

There are several modes of terrestrial locomotion found in the vertebrates. These include hopping, crawling, bipedal and quadrupedal locomotion. In terrestrial locomotion energy is required to overcome the effects of gravity and to overcome air resistance (drag). The effect of gravity is also to cause friction within the animals' own tissues, particularly in the joints. The percentage of energy that is needed to overcome these different forces can be estimated to some extent by carrying out metabolic measurements on different animals under different conditions.

The methods of measuring the metabolic rate of terrestrial animals are more varied and technically less demanding than the methods used for fish swimming or bird flight. The methods that have been used include open systems, closed systems, constant pressure and constant volume systems. Usually, the animal is fitted with a mask although in the case of small animals, the animal and exercise apparatus may be enclosed in an airtight chamber. For large animals the expired gases may be collected in a lightweight bag (Douglas Bag). The volume of gases collected is measured by deflating the bag through an accurate gas volume meter after an aliquot has been subjected to gas analysis. One advantage of the Douglas bag is that it can be used with unrestrained animals and it is particularly useful for field studies on the larger animals, including man. Another method is to use a gasometer bell (spirometer) apparatus. The bell of the gasometer floats on a liquid seal. A soda lime cannister is included in the circuit to absorb the carbon dioxide so that the amount of oxygen used can be determined from the drop in the height of the bell. This is a very accurate and convenient method; however it has some disadvantages, including the fact that it cannot be used on unrestrained animals. Also in common with the Douglas bag method it has the disadvantage that breathing is often more laboured than it would be normally because a certain amount of pressure is required to open and close the valves of the different parts of the circuit. In animal experiments this can lead to stress and hence erratic behaviour and it is for this reason that an open circuit system is often preferred in which the air is drawn through the respiratory mask at a steady rate. This latter method is essentially the same as that used in measurements on bird flight metabolism. In most cases, the animals are run on a treadmill so that the speed of locomotion can be accurately controlled.

For most terrestrial animals, a straight line relationship is obtained between the metabolic power measurements and the speed of locomotion (Fig. 6.6). However, there is one notable exception – animals that hop. Dawson and Taylor (1973) carried out measurements on the kangaroo and found that as the speed of the animal increases, the oxygen consumption actually decreases slightly. The reason for this state of affairs is that at the highest speeds the distance covered with each hop is increased although the energy expenditure per hop is the same as at low speeds. As shown by Alexander and Vernon (1975) the total energy per hop is kept constant as speed increases because the increased kinetic energy is matched by increased elastic savings. Similar results have been obtained by Dawson (1976) for the oxygen consumption of the Australian hopping mouse; in this case the oxygen consumption does not actually go down although it is certainly less than would be expected for a quadrupedal animal of the same size. The mechanics of hopping are dealt with in detail in Chapter 7.

Most of the discussion is concerned with animals that use legs; however what about animals such as snakes which crawl? It will be noted from Fig. 6.6 that the snake has quite low energy requirements for transport as compared with other terrestrial animals of similar size. In fact the net cost of transport for the snake is only about one half of that required for a lizard of similar weight and about 10 times less than a small quadrupedal mammal, such as the mouse, Chodrow and Taylor (1973). Although the frictional resistance between the animal and the ground is obviously quite high in snake locomotion, the internal

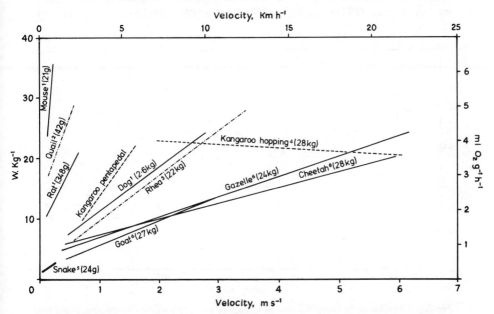

Fig. 6.6 Power input required for terrestrial locomotion. Crawling —, quadrupedal running ——, bipedal running —·—·—, hopping, ————. Data taken from (1) Taylor *et al.* (1970), (2) Fedak *et al.* (1974), (3) Taylor *et al.* (1971), (4) Dawson and Taylor (1973), (5) Chodrow and Taylor (1973), (6) Taylor *et al.* (1974).

resistance is presumably low as this type of locomotion does not involve as much joint articulation as quadrupedal or bipedal locomotion.

Another interesting observation is that quadrupedal animals of the same size have about the same energy requirements irrespective of their shape and limb configuration. Taylor *et al.* (1974) found that the increase in oxygen consumption with increase in running speed was essentially the same for the gazelle, the goat and the cheetah, although the thickness and weight of their limbs differ considerably. This is an important observation as it presumably means that very little energy is needed for swinging the legs forwards and backwards. The animal in order to accelerate and decelerate its limbs has to expend energy. However, this is apparently not very significant. Most of the

163

energy is apparently going into tensioning the tendons that are involved in the energy storage mechanism and in doing other internal work and this is essentially the same in all quadrupeds of the same size. There are some exceptions to this, for example the lion requires more energy for locomotion than would be expected. The lion has very thick limbs and the energy required to accelerate and decelerate them seems to be much more appreciable (Taylor, personal communication). Hence, movement for the lion is relatively more expensive.

Bipedal running has been studied and compared with quadrupedal running by Taylor *et al.* (1971), Taylor and Rowntree (1972) and Fedak *et al.* (1974).

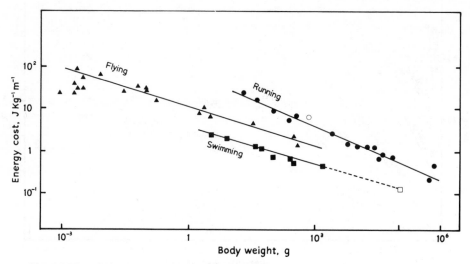

Fig. 6.7 The relation between the cost of transport and body size for swimming, flying and running as a double logarithmic plot. Data taken from Schmidt–Nielsen (1971). o, Man; □, Dolphin.

Briefly, what has been found is that for large mammals or birds, bipedal running is more expensive than quadrupedal running. However, for smaller mammals and birds the reverse situation is true. It has been shown by Fedak *et al.* (1974) that the regression lines for the cost of transport versus body size for bipedal and quadrupedal running cross at a body weight of just less than one kilogram. The reason for this difference in the energy requirements for bipedal and quadrupedal locomotion is not immediately obvious. For large animals it can be argued that the quadruped has more flexibility in its gait, i.e. it can walk, trot or gallop, whereas the biped can only walk or run. It is particularly difficult to see why small bipedal animals are more efficient than small quadrupedal animals. The situation is further confused by the studies of Taylor and Rowntree (1972) on primates that can walk either bipedally or

164

quadrupedally. Rather surprisingly they found that there was no significant difference in the metabolic power required for these two different modes of locomotion. However, quadrupedal locomotion in the chimpanzee and capu-chin monkey is relatively costly; it costs about 50 % more than in other animals of similar size. Locomotion on the ground for the primates does therefore seem to be rather expensive. Man, of course, is a bipedal primate; however, his cost of transport is also somewhat higher than for other bipedal animals of about the same size.

6.5 Comparison of the energy requirements for swimming, flying and terrestrial locomotion

If one compares the values given in Figs. 6.2, 6.5 and 6.6, it will be apparent that bird flight demands much more power (energy spent per unit time) than either terrestrial locomotion or fish swimming. Birds have to overcome drag both frictional and induced, however, their main problem is that the wings have to produce lift as well as thrust, hence, in flapping flight, much of the energy goes into actually keeping the bird airborne. Fish do not have quite the same problem of maintaining vertical position because they are in a denser medium and there is in fact no problem at all for fish of neutral buoyancy. Their problem is one of overcoming drag caused by moving through a more viscous medium. Nevertheless the power requirements and cost of transport are considerably less for fish swimming than for running and flying. The low oxygen content of water as compared with air means that fish are not capable of sustaining high power production. Indeed, even though the power require-ments for swimming are relatively low, fish cannot meet these requirements except at low speeds, without resorting to anaerobic metabolism. In terrestrial animals, the drag due to air resistance is almost negligible except at very high running speeds; however, terrestrial animals are, like birds, subjected to the effects of gravity. For an animal that moves on the ground the effect of gravity is to vertically compress the animal. This causes friction in the joints and moving parts of the limbs. Although the power requirements for terrestrial locomotion are less than for bird flight, birds do as a rule move much more rapidly than terrestrial animals. Therefore they can actually cover a given distance with less energy expenditure than terrestrial animals.

As far as cost of transport is concerned, fish have an optimum velocity at which the gross cost is at a minimum. Some birds such as the budgerigar also have an optimum speed, although with other birds the cost of transport decreases with increased flying speed. Most terrestrial animals have a straight line relationship between the power required and the speed of locomotion. Therefore their cost of transport is almost independent of speed. These are generalizations and small discrepancies and deviations from these general rules can be found when very careful measurements are made on individual animals. Nevertheless it is striking that such generalizations can be made for a particular type of locomotion and that they can be made for such a wide range of animal sizes. This presumably means that the mechanical principles for a particular mode of locomotion are basically the same whether that animal be a mouse or a gazelle. Their mechanical principles are discussed in detail in Chapters 7, 9 and 10.

Acknowledgements

The author wishes to acknowledge the very helpful comments on the manuscript of this chapter that he received from both Professor K. Schmidt-Nielsen and Professor R. McNeill Alexander.

References

Alexander, R.McN. and Vernon, A. (1975) The mechanics of hopping by kangaroos (Macropodidas). *J. Zool., Lond.* **177,** 265–303.

Baudinette, R.V. and Schmidt-Nielsen, K. (1974) Energy cost of gliding flight in herring gulls. *Nature* **248,** 83–84.

Bernstein, M.H., Thomas, S.P. and Schmidt-Nielsen, K. (1973) Power input during flight of the fish crow, *Corvus ossi fragus. J. exp. Biol.* **50,** 401–410.

Blazka, P., Volf, M. and Cepela, M. (1960) A new type of respirometer for the determination of the metabolism of fish in an active state. *Physiologia Bohemoslovenica* **9,** 553–560.

Brett, J.R. (1964) The respiratory metabolism and swimming performance of young sockeye salmon. *J. Fish. Res. Board Can.* **21,** 1183–1226.

Brett, J.R. and Sutherland, D.B. (1965) Respiratory metabolism of pumpkinseed (*Lepomis gibbosus*) in relation to swimming speed. *J. Fish. Res. Board Can.* **22,** 405–409.

Chodrow, R.E. and Taylor, C.R. (1973) Energetic cost of limbless locomotion in snakes. *Fed. Proc.* **32** Als. 1128.

Dawson, T.J. and Taylor, C.R. (1973) Energetic cost of locomotion in kangaroos. *Nature* **246,** 313–314.

Dawson, T.J. (1976) Energetic cost of locomotion in Australian hopping mice. *Nature* **259,** 305–306.

Fedak, M.A., Pinshow, B. and Schmidt-Nielsen, K. (1974) Energy cost of bipedal running. *Am. J. Physiol.* **227,** 1036–1944.

Greenewalt, C.H. (1962) Dimensional relationships for flying animals. *Smithson. misc. Collns.* **144,** No. 2.

Lasiewski, R.C. (1963) Oxygen consumption of torpid, resting, active and flying humming birds. *Physiol. Zool.* **36,** 122–140.

Pennycuick, C. (1972) Soaring, behaviour and performance of some East African birds observed from a motor-glider. *Ibis* **114,** 178–218.

Prange, H.D. and Schmidt-Nielsen, K. (1970) The metabolic cost of swimming in ducks. *Exp. Biol.* **53,** 763–777.

Schmidt-Nielsen, K. (1971) Locomotion: energy cost of swimming, flying and running. *Science* **177,** 222–226.

Smit. H., Amelink-Koutstaal, J.M., Vijverberg, J. and von Vaupel-Klein, J.C. (1971) Oxygen consumption and efficiency of swimming goldfish. *Comp. Biochem. Physiol.* **39A,** 1–28.

Taylor, C.R., Omiel, R., Fedak, M. and Schmidt-Nielsen, K. (1971) Energetic cost of running and heat balance in a large bird, the rhea. *Am. J. Physiol.* **221,** 597–601.

Taylor, C.R. and Rowntree, V.J. (1972) Running on two or on four legs: which consumes more energy? *Science* **179,** 186–187.

Taylor, C.R., Shkolnik, A., Dmiel, R., Baharav, D. and Borut, A. (1974) Running in cheetahs, gazelles and goats: energy cost and limb configuration. *Am. J. Physiol.* **227,** 848–850.

166

Taylor, C.R., Schmidt-Nielsen, K. and Raab, J.L. (1970) Scaling of energetic cost of running to body size in mammals. *Am. J. Physiol.* **219**, 1104–1108.

Tucker, V.A. (1966) Oxygen consumption of a flying bird. *Science* **154**, 150–151.

Tucker, V.A. (1968) Respiratory exchange and evaporative water loss in the flying budgerigar. *J. Exp. Biol.* **48**, 67–87.

Tucker, V.A. (1969) The energetics of bird flight. *Sci. Am.* **220**, 70–78.

Tucker, V.A. (1972) Metabolism during flight in the laughing gull, *Larus atricilla*. *Am. J. Physiol.* **222**, 237–245.

Tytler, P. (1969) Relationship between oxygen consumption and swimming speed in the haddock *Melanogrammus aeglefinus*. *Nature* **221**, 274–275.

Webb, P.W. (1971) The swimming energetics of trout oxygen consumption and swimming efficiency. *J. exp. Biol.* **55**, 521–540.

Weihs, D. (1974) Energetic advantages of burst swimming of fish. *J. theor. Biol.* **48**, 215–229.

7 Terrestrial locomotion

R. McN. Alexander

7.1 Introduction

This chapter is mainly about mammals, because we know much more about how they walk and run than we do about the terrestrial locomotion of any other group of animals. There are short sections about birds, lower vertebrates and arthropods. Invertebrates such as worms and snails which burrow or crawl without legs are discussed in Chapter 8.

7.2 The gaits of mammals

7.2.1 Gait patterns

Walking and running are very different activities. A horse looks and sounds quite different when it trots than when it walks or gallops. The patterns of these and other gaits have been described in detail by many authors including Hildebrand (1965), Sukhanov (1974) and Gambaryan (1974) so only a brief summary seems necessary here.

Steady locomotion involves regularly repeated cycles of leg movement. In the course of a cycle each foot is placed once on the ground and lifted once. The time taken to perform a cycle is called the cycle time. Any gait can be described in outline by giving two quantities for each foot, the duty factor and the relative phase (McGhee, 1968). The duty factor of each foot is the fraction of the cycle time for which that foot is on the ground. The relative phase in-dicates the stage in the cycle at which that foot is placed on the ground. The cycle is reckoned to start when one arbitrarily chosen foot is set down, so the relative phase of that foot is zero and those of the others are fractions of a cycle.

There are three common bipedal gaits, the walk and the run which are used by men and the bipedal hop which is used by kangaroos and jerboas. In walking and running the feet are moved alternately, at equal intervals: if the relative phase of one foot is taken as 0 that of the other is 0.5. In walking the duty factor of each foot is greater than 0.5 so there are double support phases when both feet are on the ground. In running the duty factors are less than 0.5 so there are floating phases when neither foot is on the ground. In bipedal hopping the feet move synchronously (relative phase is zero for both) and there are floating phases while they are moved forward.

Table 7.1 gives typical duty factors and relative phases for the principal gaits, and Fig. 7.1 shows three gaits of the horse. In the quadrupedal walk each pair of legs moves in the manner of the bipedal walk but there is a phase difference of a quarter cycle between the fore and hind legs. All the other gaits have duty factors less than 0.5 so there are periods when neither fore foot, nor neither hind foot, is on the ground. In trotting and racking each pair of legs

Table 7.1 Characteristics of some quadrupedal gaits. The duty factor is generally about equal for all the feet so only one value is given for each gait. *L*, *R* signify left, right, but canters and gallops with left and right transposed would be given the same names. The values given are typical, and subject to variation

		Walk		*Trot*		*Rack*	
Duty factor		>0.5		0.3–0.5		0.3–0.5	
		L	R	L	R	L	R
Relative phase	fore	0	0.5	0	0.5	0	0.5
	hind	0.75	0.25	0.5	0	0	0.5

		Canter		*Gallop (a)*		*Gallop (b)*	
Duty factor		0.3–0.5		<0.4		<0.4	
		L	R	L	R	L	R
Relative phase	fore	0	0.8	0	0.8	0	0.8
	hind	0.8	0.5	0.6	0.5	0.5	0.6

performs a bipedal run. The pairs are half a cycle out of step with each other in trotting, but in step in racking. In walking, trotting and racking the right and left legs of a pair are half a cycle out of phase with each other but this is not so in cantering and galloping. In a canter one fore and one hind foot are set down together but in a gallop the two fore feet are set down immediately after each other, followed by the two hind feet. Horses and other large mammals gallop with the feet of each pair slightly out of phase with each other. Rodents tend to use the form of gallop known as the bound, in which the feet of each pair are set down simultaneously (Gambaryan, 1974). The distinctions between the gaits are less sharp than this paragraph may suggest.

The quadrupedal walk is the typical slow gait of mammals and the gallop is the fastest gait. Most mammals trot at intermediate speeds but camels and giraffes rack. These mammals have long legs, relative to the distance from hip to shoulder, and the fore and hind legs of each side might be apt to

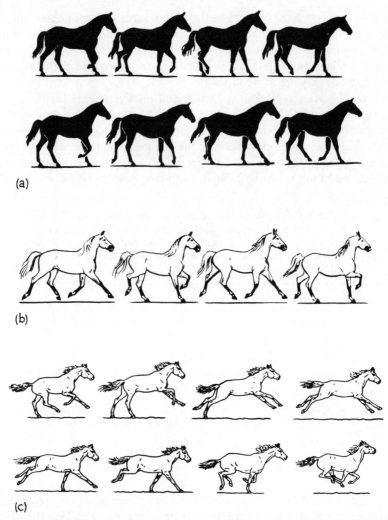

(a)

(b)

(c)

Fig. 7.1 Outlines traced from cinematograph films of horses (a) walking, (b) trotting and (c) galloping (from Gambaryan, 1974).

get in each other's way if they were half a cycle out of phase, as in the trot. Horses can be trained to rack. The gnu (*Connochaetes taurinus*) generally changes directly from a walk to a canter, without trotting, as it increases speed (Pennycuick, 1975).

7.2.2 Stability

A three-legged stool is stable but a two-legged stool is not. Similarly a quadruped can only be stable when it has three or four feet on the ground. The condition

170

for stability is that a vertical line through the centre of mass of the body must pass through the triangle (or quadrilateral) of which the corners are the points of contact of the three (or four) feet with the ground. If the quadruped is to remain stable as it walks, it must move only one foot at a time so the average duty factor of the feet must be at least 0.75. Further, it must move the feet in appropriate sequence. If the duty factor is between 0.75 and 0.83, only one sequence will give constant stability: it is the sequence LF, RH, RF, LH shown in Table 7.1 and Fig. 7.2. Any other sequence would make the vertical through the centre of mass pass outside the triangle of support, at some stage in the cycle of leg movements. This sequence seems to be used for walking

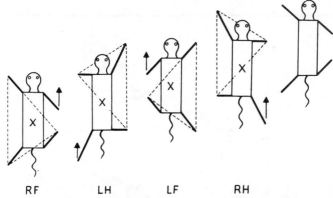

RF   LH   LF   RH

Fig. 7.2 Diagrams of a quadruped walking in such a way as to be stable at all times. *X* marks the centre of mass and the broken lines show the triangle of support. Each diagram represents the instant when the foot marked with an arrow (and indicated by initials below) is lifted. In each case the centre of mass will move forward before the next change of supporting feet. *RF*, right forefoot; *LH*, left hindfoot, etc. (from Alexander 1975a).

by all quadrupeds, even in fast walks involving stages with only two feet on the ground. Two other sequences in addition to this one would permit constant stability when the duty factor exceeded 0.83, but no animal is known to use them. They are LF, RF, RH, LH and LF, LH, RH, RF (McGhee and Frank, 1968).

Walking with duty factors high enough for constant stability is only performed at very low speeds. All faster gaits involve periods of instability.

7.2.3 Speed, stride length and gait

A stride is a complete cycle of leg movements and the stride length, λ, is the distance travelled in a stride. The step length, $2b$, is the distance travelled by the trunk while a particular foot is on the ground. The ratio $2b/\lambda$ is approximately equal to the duty factor (not precisely equal because the velocity fluctuates in the course of a step). Stride frequency, n, is the number of strides taken in unit time. Let the mean velocity (averaged over a complete stride) be $\bar{u}$. Then $\bar{u} = n\lambda$: to increase its speed an animal must increase its stride frequency or its stride length or both.

So that animals of different size can be compared it is convenient to deal with dimensionless numbers (ratios) rather than with quantities such as λ, b, n and $\bar{u}$. Let h be the height of the hip joint from the ground, when the animal is standing, and let g be the acceleration of free fall. They are used in the definitions of the following dimensionless numbers:

$$\hat{\lambda} = \lambda/h \qquad \hat{b} = b/h$$
$$\hat{n} = n(h/g)^{1/2} \qquad \hat{u} = \bar{u}/(gh)^{1/2}$$

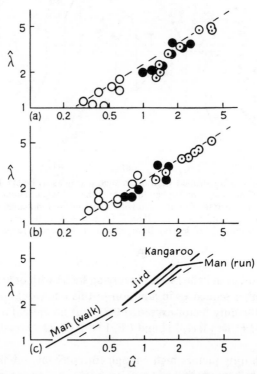

Fig. 7.3 Graphs plotted on logarithmic coordinates showing the relationship between stride length, speed and gait. The dimensionless numbers $\hat{\lambda}$ (representing stride length) and $\hat{u}$ (representing speed) are defined in the text. The broken line on each graph shows $\hat{\lambda} = 2.3\hat{u}^{0.6}$. (a) Data from Muybridge's (1957) photographs of horses, mostly with riders (photographs of horses drawing carts have not been used). (b) Data from Muybridge's photographs of other mammals and an ostrich. (c) Data for men walking (Zarrugh, Todd and Ralston, 1974) and running (Cavagna, Margaria and Arcelli, 1965), for a kangaroo hopping (Alexander and Vernon, 1975a) and for a jird (*Meriones unguiculatus*; unpublished observations by Mr. S. N. G. Frodsham and the author). o, walk; ●, run, trot or rack; ⊙, canter or gallop (from Alexander, 1977).

172

The circumflex accent is used throughout this chapter, to distinguish dimensionless numbers from their dimensional equivalents.

In Fig. 7.3, dimensionless stride length $\hat{\lambda}$ is plotted against dimensionless velocity $\hat{u}$. It appears that for a remarkably wide range of mammals $\hat{\lambda} \simeq 2.3\hat{u}^{0.6}$. This implies $\hat{n} \simeq 0.44\hat{u}^{0.4}$. Mammals increase both stride length and stride frequency as they increase their speed, and these relationships appear to hold for mammals of all sizes.

Figure 7.4 shows the range of values of $\hat{u}$ at which various mammals use each gait. Most mammals probably change from a walk to a trot or run at some

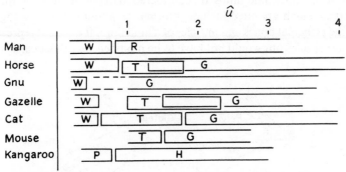

Fig. 7.4 A diagram showing the ranges of speed (expressed as the dimensionless number $\hat{u}$) at which mammals use their various gaits. (Data from Alexander and Vernon, 1975a; Cavagna, 1969; Goslow, Reinking and Stuart, 1973; Heglund, Taylor and McMahon, 1974; Muybridge, 1957; and Pennycuick, 1975). *G*, gallop or canter; *H*, hop; *P*, pentapedal gait; *T*, trot; *W*, walk.

value of $\hat{u}$ in the region of 0.8. The kangaroo changes from the pentapedal gait to the hop when $\hat{u} \simeq 0.8$. (The pentapedal gait involves all four legs and the tail; Bausschulte, 1972.) Most mammals probably change from a trot to a gallop at some value of $\hat{u}$ between about 1.3 and 2.0.

Information on the maximum speeds of mammals is sparse. Sporting records show that athletes have reached speeds up to about 12 m s^{-1} ($\hat{u} \simeq 4.0$) when sprinting and that racehorses (with riders) and greyhounds have run short races at speeds up to about 19.5 m s^{-1} and 18.5 m s^{-1}, respectively ($\hat{u} \simeq 5.5$ and $\hat{u} \simeq 8$). It has been claimed that pronghorn (*Antilocapra*) and cheetah (*Acinonyx*) reach 27 and 25 m s^{-1}, respectively (McWhirter and McWhirter, 1971) but the maximum speeds of various African ungulates seem to be well below the highest values reported by casual observers (Alexander, Langman and Jayes, 1977).

Though mammals increase stride length as they increase their speed, they generally keep step length fairly constant. For many mammals, step length is about equal to hip height ($\hat{b} \simeq 0.5$), but $\hat{b}$ tends to be larger for small mammals than for large ones and may have any value between about 0.3 and 0.75 (Gambaryan, 1975; McMahon, 1975). If $\hat{b} = 0.5$ and $\hat{\lambda} = 2.3\hat{u}^{0.6}$, the duty factor (approximately $2b/\lambda$) is about 0.75 when $\hat{u} = 0.4$, and 0.5 when $\hat{u} = 0.8$. Constant stability requires a duty factor of 0.75 or more, and is only achieved in very slow walks with $\hat{u}$ less than about 0.4. Floating phases can only occur in running, trotting and racking if the duty factor is less than 0.5, and these gaits are generally adopted as $\hat{u}$ passes 0.8.

7.3 The legs of mammals

7.3.1 *Structure*

To save space, only hind limbs are considered in detail but most of the books and papers which are cited include information about fore limbs. Figure 7.5 shows the principal bones and muscles of a hind limb. It does not represent any particular species but would not have to be modified very drastically to repre-

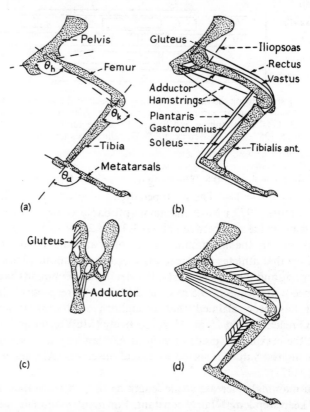

Fig. 7.5 Diagrams of the hind limb skeleton and muscles of a generalized mammal. (a), (b) and (d) are side views and (c) is an anterior view. (a) defines angles referred to in the text, (b) shows the lines of action of muscles and (c) and (d) show the arrangement of fibres within muscles. Further explanation is given in the text.

sent any species known to me. (For detailed information on various species see text books of human and veterinary anatomy, Bausschulte, 1972 etc.)

Figure 7.5a shows the general arrangement of the bones, but omits the details of the ankle and foot. The pelvic girdle is rigidly attached to the vertebral column. The hip joint is a ball and socket joint but the knee and ankle are essentially hinge joints.

Figure 7.5b shows the arrangement of the principal muscles. Many small muscles are omitted but the muscles which are shown represent about 80% of the total mass of hindlimb muscles in men, dogs and kangaroos (Alexander, 1974; Alexander and Vernon, 1975a, b and unpublished data). The gluteus, iliopsoas, vastus and hamstrings are not single muscles, but groups.

Figure 7.5c and d shows how the muscle fibres are arranged in some of the principal muscles (Alexander, 1974; Alexander and Vernon, 1975a, b). The hamstring muscles and adductor femoris are parallel fibred. The gluteus, rectus, vastus, soleus, gastrocnemius and plantaris are all pennate. The functional significance of this difference in structure will be discussed later.

The principal differences between mammal hind limbs are differences in proportion and in the structure of the foot. Marked differences occur in the ratio of the lengths of the femur, tibia and longest metatarsal. This ratio is commonly about $1:1:0.3$ in rodents, about $1:1:0.4$ in cats, dogs, pigs, etc., and about $1:1:0.7$ in ruminants and horses. Ruminants and horses have only one functional metatarsal in each foot (in ruminants it represents the metatarsals of two toes, fused together). Ruminants have only two functional toes on each foot and horses have only one. The additional metatarsals and toes which are present in other mammals are lost in ruminants and horses or reduced to vestiges. The fore limbs are modified in the same way as the hind limbs. These peculiarities make the distal parts of the limbs of ruminants and horses relatively long and light.

Reducing the number of toes makes it possible to lighten the foot for the following reason. The stresses set up in limb bones in locomotion are generally mainly due to components of force at right angles to the bones, which exert bending moments on the bones (Alexander, 1974; Alexander and Vernon, 1975a). If a bone or any beam is to withstand a bending moment M it must have a cross-sectional area of at least $kM^{0.67}$, where k depends only on the shape of the cross-section and the strength of the material (see for instance Alexander, 1968). If two equal beams are together to withstand the bending moment each needs a cross-sectional area $k(0.5M)^{0.67} = 0.63kM^{0.67}$, so their total cross-sectional area must be $1.26kM^{0.67}$. Hence a single metatarsal can be lighter than two or any greater number.

The long, light distal parts of the limbs of ruminants and horses are generally regarded as adaptations for speed although a pronghorn cannot gallop much faster than a cheetah (which lacks these adaptations) nor can a horse gallop much faster than a greyhound. The significance of these adaptations is discussed further in Section 7.5.1.

175

The adaptations of primates for specialized forms of terrestrial locomotion are discussed in Jenkins (1974).

7.3.2 Movements

The movements of limbs are generally studied by cinematography, and sometimes by X-ray cinematography. Valuable advice on cinematography and on analysis of films is given by Grieve *et al.* (1975). It is often convenient to film an animal running on a treadmill so that its speed can be controlled and it can be kept in front of the camera for as many strides as may be required. However, locomotion on a treadmill may be slightly abnormal because of the need to control velocity accurately to match the speed of the belt (Wetzel *et al.*, 1975).

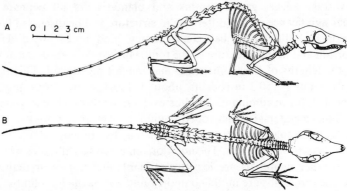

Fig. 7.6 The skeleton of the tree shrew *Tupaia glis* in walking posture, based on X-ray cinematographs of living specimens (from Jenkins, 1974).

Mammals can be divided by their limb movements into two main groups, which have been called the cursorial and non-cursorial mammals (Jenkins, 1971). The cursorial group includes the horse (Fig. 7.1) and other large mammals, and some smaller mammals such as domestic cats. They stand with their legs fairly straight. The non-cursorial group includes the tree shrew *Tupaia* (Fig. 7.6) and other small to medium-sized animals including rodents, small carnivores such as *Mustela* and the opossum *Didelphis*. They stand and move on much more strongly flexed legs. This chapter is mainly about cursorial mammals.

7.3.3 Muscle action

The lengths of muscles change as an animal moves. The distances from their origins to their insertions could be measured directly from X-ray cinemato-

176

graph films. They can also be calculated from conventional (light) cinemato-graph films, and this has been done in studies of various mammals (Goslow, Reinking and Stuart, 1973; Alexander, 1974; Alexander and Vernon, 1975a). It is necessary to know the moment arms of the muscles about the joints, and these can often be measured most conveniently from X-radiographs. Figure 7.7 shows how some muscles in the hind leg of the cat change their length in

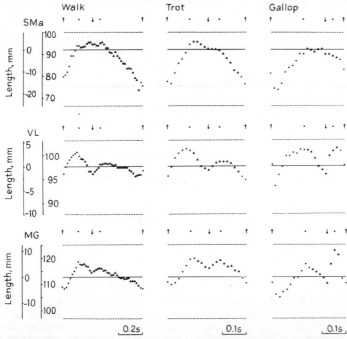

Fig. 7.7 Graphs of length against time for three muscles in the hind legs of cats in walking, trotting and galloping. The muscles are *SMa*, semimembranosus anterior; *VL*, vastus lateralis and *MG*, gastrocnemius medialis. Their lengths were calculated from measurements on film. Dots indicate that the muscle is electrically inactive and triangles that it is electrically active. The continuous horizontal line shows the resting length and the broken horizontal lines show the maximum and minimum *in situ* lengths. The outer scale gives length relative to resting length and the inner scale gives absolute length. Upward arrows show when the foot was lifted and downward ones when it was set down (from Goslow *et al.*, 1973).

various gaits. The semimembranosus anterior is a hamstring muscle which inserts on the femur. It lengthens while the foot is off the ground and being brought forward, and shortens while the foot is on the ground. The iliopsoas (not illustrated) does the reverse. The gastrocnemius and vastus, however, extend and shorten twice in each stride, once while the foot is off the ground and once while it is on the ground.

Changes in the overall length of a muscle may be due to lengthening or shortening of the muscle fibres, to elastic extension or recoil of the tendons, or to both. These possibilities will be discussed further when it has been shown how muscle forces can be calculated.

Action potentials occur in muscles, when they are active. The technique of recording these electrical events is known as electromyography and is a valuable means of finding out which muscles are active and when, in particular activities (Basmajian, 1967; Grieve *et al.*, 1975). Engberg (1964) and Engberg and Lundberg (1969) used it in studies of cat locomotion. Their principal findings are summarized in Fig. 7.8. The semimembranosus anterior, vastus lateralis and

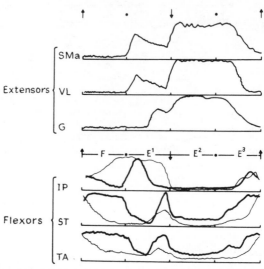

Fig. 7.8 Diagrammatic graphs showing how the intensity of electrical activity in cat hind limb muscles changes in the course of a stride. The stride has been divided into four stages which have been allotted equal lengths of the abscissa though their relative durations change with changes of speed. The graphs for extensor muscles apply equally to walking and trotting: electrical activity is more intense in trotting but its changes follow the same pattern as in walking. The thin lines for flexor muscles apply to walking and the thick lines to trotting. Upward arrows show when the foot is lifted and downward ones when it is set down. *SMa*, semimembranosus anterior; *VL*, vastus lateralis; *G*, gastrocnemius; *IP*, iliopsoas; *ST*, semitendinosus; *TA*, tibialis anterior (from Goslow, Reinking and Stuart, 1973).

gastrocnemius have been chosen to exemplify the extensors of the hip, knee and ankle, respectively. They become electrically active just before the foot is placed on the ground and remain so for most of the time it is on the ground. The pattern is essentially the same for walking and trotting and (at least for the gastrocnemius) galloping, but the amplitude of the activity increases with speed. The flexor muscles of the hip, knee and ankle show most electrical activity when the foot is about to be raised, or is off the ground.

Electromyographic investigations have been made of many human activities (Basmajian, 1967) including walking (Paul, 1971; van der Straaten *et al.*,

1975) and jumping (Kamon, 1971). There has also been a study of the role of forelimb muscles in rat locomotion (Cohen and Gans, 1975).

7.4 Forces exerted in locomotion by mammals

7.4.1 Forces on the ground

The forces which animals exert on the ground as they walk or run can be recorded by means of a force platform. Figure 7.9 shows in principle how force platforms work. It shows a small platform set into the floor and mounted on springs. Any force which acts downwards on the platform presses it down, reducing the distance y. Any forward force reduces x and any backward force increases x. Thus if changes in x and y are recorded in some way the record

7

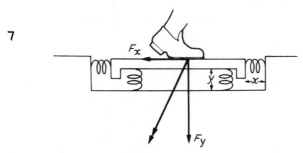

Fig. 7.9 A diagram illustrating the principle of the force platform (from Alexander, 1975b).

shows the longitudinal and vertical components of forces acting on the platform. Transverse components of force can also be determined, from transverse displacements of the platform.

Early force platforms were mounted on springs like the platform shown in Fig. 7.9 and their displacements were recorded by systems of levers. Such platforms may provide adequate records of forces which change slowly, but not of forces which change rapidly. The reason is this. A mass m mounted on a spring of stiffness S has a natural period of vibration which is $2\pi(m/S)^{1/2}$. Any disturbance will tend to start vibrations of this period. The vibrations can be eliminated by damping but even with critical damping the response time of an instrument cannot be reduced below (approximately) its natural period (see Alexander, 1968). Modern force platforms are very stiffly mounted and the forces are sensed by piezo-electric or other transducers. Björck (1958) used instrumented horse shoes instead of stationary force platforms in a study of the forces exerted by draught horses. Further information on force platforms is given by Grieve *et al.* (1975).

A selection of force platform records of locomotion is shown in Fig. 7.10. Other records of human walking and running can be found in many books and

papers including Carlsöö (1972), Kimura and Endo (1972) and Cavagna *et al.*
(1971). Records of walking by cats (Manter, 1938), dogs (Barclay, 1953;
Kimura and Endo, 1972) and horses (Björck, 1958) have also been published
as have records of jumping by men (Ramey, 1970) and dogs (Alexander, 1974).

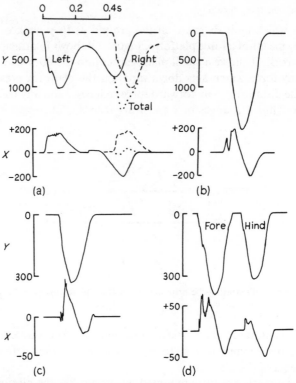

Fig. 7.10 Tracings of force platform records of (a) a 68 kg man walking; (b) the same running;
(c) a 10.5 kg wallaby (*Protemnodon rufogrisea*) hopping and (d) a 35 kg dog trotting. Vertical
(*Y*) and horizontal (*X*) components of force are shown, and the *X* component is shown positive when it acts forward on the ground. The forces are given in Newtons. The broken lines
in (a) show forces presumably exerted by the other foot (which did not land on the platform)
and the dotted lines show the total force exerted by the two feet together (unpublished records
by the author and Miss A. Vernon).

In walking, both of men (Fig. 7.10a) and of other mammals, the vertical
component of the force exerted by each foot has two maxima. Also, there is
a double support phase and the total force exerted on the ground is greater
then than at any other stage in the step (Fig. 7.10a, dotted line). In human running, dog trotting and wallaby hopping (Fig. 7.10b, c, d) the vertical component
simply rises to a maximum and falls again. In every case the foot exerts a
forward force on the ground in the first half of its period of contact and a

backward force in the second half. Transverse forces are relatively small in locomotion along a straight path.

Records of ground forces which occur in locomotion are generally reasonably smooth curves, except for brief disturbances which often occur immediately after impact of the foot with the ground. These impact disturbances occur if the foot is still moving with appreciable velocity when it hits the ground. Impact disturbances often have the form of a damped oscillation superimposed

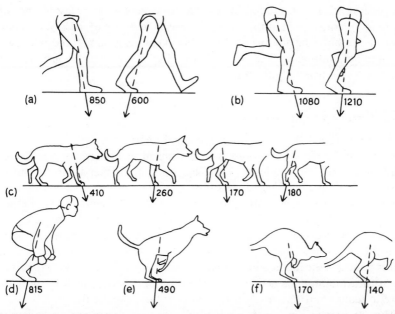

Fig. 7.11 Outlines traced from films of men and other animals on a force platform, showing the magnitude and direction of the force exerted by one foot. A 68 kg man is shown (a) walking, (b) running and (d) taking off for a standing jump; a 35 kg dog is shown (c) trotting and (e) taking off for a running jump and a 10.5 kg wallaby is shown (f) hopping (after Alexander, 1974 and 1976, and Alexander and Vernon, 1975a and b).

on an otherwise smooth curve, presumably because the impact sets the leg vibrating (see Alexander and Vernon, 1975a).

The use which can be made of force platform records is greatly increased, if cinematograph film is taken simultaneously. It is necessary to know which instant in the force record corresponds to each frame of the film. This can be done by inference: the beginning and end of a recorded force presumably correspond to the frames which show a foot being set down and lifted, and intermediate parts of the record can be matched to the film by interpolation. Alternatively, it may be useful to have an electronic device which makes marks simultaneously on the force record and on the margin of the film (Calow and Alexander, 1973).

Figure 7.11 shows outlines traced from films, with forces obtained from force platform records superimposed. In human walking and running, but not jumping, the line of action of this force passes close to the hip (except during impact disturbances). Similarly when a dog trots, but not when it jumps, the line of action of the force exerted by each hind foot always passes close to the hip and that of each fore foot passes close to the posterior dorsal corner of the scapula. The scapula rotates about an instantaneous centre in this region so the general rule is that ground forces tend to be kept more or less in line with the most proximal joint in the limb. However, the centre of mass of a wallaby is well anterior to the hip so a wallaby which kept the ground force in line with its hip would fall on its face. The wallaby (Fig. 7.11f) kept the ground force more or less in line with its centre of mass.

7.4.2 Forces exerted by muscles

The forces exerted in walking by one of the forelimb muscles of a horse have been measured directly, by means of a transducer attached surgically to a tendon (Barnes and Pinder, 1974). This technique depends on the muscle having a long tendon. The forces exerted by muscles can also be calculated from force platform records (Alexander, 1974; Alexander and Vernon, 1975a, b). Ambiguities arise because there are so many muscles in the leg and can only be resolved by making plausible assumptions. It generally seems plausible to assume that equal stresses act in closely co-operating muscles (for instance, in the rectus and vastus). This seems particularly likely to be near the truth for strenuous activities in which all appropriate muscles probably have to make contributions near the limits of their capability.

The assumptions involved in calculating muscle forces cannot at present be avoided. Electromyographs may be useful in formulating appropriate assumptions since they can show which muscles are active at a given instant and which are not. They may give some indication of changes in intensity of activity but it is not yet possible to determine by electromyography how much force a muscle is exerting (see Milner-Brown and Stein, 1975).

If the force exerted by a muscle and the dimensions of the muscle are known, the stress can be calculated (Chapter 1). Data from force platform records have been used to calculate stresses in limb muscles of men, dogs and kangaroos, in various activities (Alexander, 1974; Alexander and Vernon, 1975a, b). The extensors of the knee and ankle of man were found to exert stresses around 400 kN m^{-2} in running and jumping, but these values may be rather too high because the dimensions of muscles were obtained from a cadaver which probably had smaller muscles than the living subjects. Stresses up to about 300 kN m^{-2} occurred in the leg muscles of dogs and kangaroos, in strong jumps.

Figure 7.12 is based on a force platform record of a wallaby hopping, and on film taken simultaneously. The forces exerted by muscles were calculated by taking moments about joints and the lengths of muscles from joint angles. In the case of the hip muscles, moment has been plotted against joint

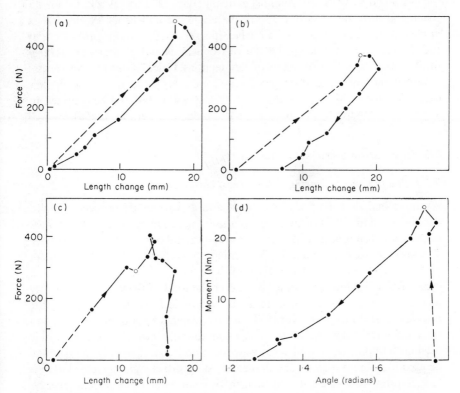

Fig. 7.12 Graphs calculated from a force platform record and film of a 10.5 kg wallaby hopping at 2.4 m s^{-1}. They show the forces exerted by (a) the gastrocnemius (b) the plantaris and (c) the extensor muscles of the knee, plotted against the changes in muscle length, during the period when the feet were in contact with the platform. (d) is a graph of the moment exerted about the hip by the extensor muscles of the hip, plotted against hip angle (θ_h, Fig. 7.5). The points represent measurements at intervals of 10 ms except for a single interval of 40 ms (indicated by a broken line) spanning the period when vibrations following impact made it difficult to obtain reliable values. Open symbols show the instant at which potential and external kinetic energy were minimal (from Alexander and Vernon 1975a).

angle rather than muscle force against muscle length because several muscles with different moment arms are involved.

The extensor muscles of the hip shorten throughout the period of contact of the foot with the ground (Fig. 7.12d). Since they shorten while exerting a force they do work, and the amount of work is indicated by the area under

183

the graph. The extensor muscles of the knee, on the other hand, lengthen while exerting a force (Fig. 7.12c). They do not do work; rather, work is done on them. The same idea can be expressed by saying that the work done by them is negative (see Chapter 3). The area under the graph represents the negative work involved. The gastrocnemius and plantaris muscles lengthen as the force increases and shorten as it diminishes again (Fig. 7.12a, b). Work is done on them, and then they do work. Elastic materials stretch when a tensile force is applied to them and shorten again when it is removed, and it is probable that the changes in overall length of the gastrocnemius and plantaris are largely due to elastic extension and recoil of their tendons. Calculations using the cross-sectional areas of the tendons and Young's modulus for tendon collagen indicated that the maximum forces shown in Fig. 7.12a, b could be expected to stretch each tendon by about 11 mm (Alexander and Vernon, 1975a).

7.5 Energetics of mammal locomotion

7.5.1 Fluctuations of kinetic, potential and elastic energy

As an animal runs, its body as a whole and its various parts accelerate and decelerate, so that the kinetic energy of the animal changes. The centre of mass of the animal rises and falls, so that its gravitational potential energy changes. Tendons are stretched and recoil. Stretching stores elastic strain energy in them and the recoil releases it again. Each of these forms of energy goes through a cycle of changes in every stride. Whenever the total of these forms of energy increases, positive work must be done by muscles. When it decreases, the muscles must do negative work. These requirements set a minimum value to the metabolic energy requirement for locomotion, since metabolic energy is used both in positive and in negative work. Additional metabolic energy may be used if different muscles simultaneously do positive and negative work.

It is often convenient to distinguish two components of kinetic energy, the external kinetic energy due to movement of the body as a whole and the internal kinetic energy due to movement of the parts of the body relative to each other. Consider a system of particles. Each particle has its individual mass m_j and has components of velocity u_j, v_j, w_j in a system of Cartesian co-ordinates. The total mass of the system is $\sum m_j$ and the components of velocity of its centre of mass are $\bar{u}$, $\bar{v}$, and $\bar{w}$. It can be shown that

$$\underset{\text{total kinetic energy}}{\tfrac{1}{2}\sum [m_j(u_j^2 + v_j^2 + w_j^2)]} = \underset{\text{external KE}}{\tfrac{1}{2}(\sum m_j)(\bar{u}^2 + \bar{v}^2 + \bar{w}^2)} +$$

$$+ \underset{\text{internal KE}}{\tfrac{1}{2}\sum m_j\{(u_j - \bar{u})^2 + (v_j - \bar{v})^2 + (w_j - \bar{w})^2\}}. \quad (7.1)$$

It is possible to determine the kinetic and potential energy of an animal, at each stage during a stride, by careful measurements on film. The body must be treated as a collection of rigid segments (head, trunk, thigh, shank, etc.).

The mass of each segment must be known, and the position of its centre of mass. The height and velocity of the centre of mass of each segment must be determined from the film, for each stage in the stride. The potential and kinetic energy of each segment can then be calculated. In a rigorous analysis, it would be necessary to determine the moments of inertia of each segment about three axes through its centre of mass, and the components of its angular velocity about these three axes. These data would be used to calculate the kinetic energy associated with rotation of each segment, but the kinetic energies of rotation of segments will generally be small compared to their kinetic energies of translation.

It may be possible to simplify the procedure by making approximations (see Alexander and Vernon, 1975a) but it will usually still be laborious after any reasonable simplification. It will also generally be subject to quite severe errors due to the difficulty of locating precisely the centre of mass of the trunk in successive frames of the film.

An alternative technique uses force platform records to determine the fluctuations of potential and external kinetic energy (Cavagna, 1975). The technique has been used in studies of human locomotion (reviewed by Cavagna, 1969) and has also been applied to hopping by a wallaby (Alexander and Vernon 1975a).

Figure 7.13a shows how the potential and external kinetic energy of the human body fluctuate in normal walking. The potential energy is lowest and the kinetic energy highest, at the stage in the step when both feet are on the ground. Since the fluctuations of potential and kinetic energy are similar in amplitude and out of phase with each other, the fluctuations of their total are small. The muscles need only do positive work, x, at one stage in the step and y at another (and equal amounts of negative work at other times). Figure 7.13b shows that in fast walking the fluctuations of potential energy are smaller and the fluctuations of kinetic energy much larger, and much more work is required of the muscles in each stride. In running the fluctuations of kinetic and potential energy are in phase with each other and the amplitude of the fluctuations of their total is large. However, the work required of the muscles in each step is less than this amplitude, because of elastic storage of energy. The Achilles and patellar tendons exert their maximum forces (and must be maximally stretched and store maximum elastic energy) at the time when kinetic and potential energy are lowest.

The fluctuations of potential, kinetic and elastic energy which occur when a kangaroo hops, follow the same pattern as in human running (Fig. 7.14). Potential and external kinetic energy fluctuate in phase with each other. Fluctuations of internal kinetic energy are quite small. The Achilles tendon exerts large forces and is stretched, at the time when potential and kinetic energy are low. Figure 7.14c is derived from the same force platform record as Fig. 7.12 which shows that the maximum force in the Achilles tendon was 850 N. It has already been observed that this must have stretched the tendon by 11 mm. A force rising from zero to 850 N and stretching an elastic body 11 mm will (if Hooke's Law is obeyed) do work amounting to $\frac{1}{2} \times 850 \times 0.011 = 4.7$ J. The total amount of energy stored elastically in the Achilles tendons of the two legs must thus have been about 9.4 J, or 0.9 J kg body weight^{-1}. Elastic storage in the tendons must have reduced substantially the amount of work required of the muscles in each hop.

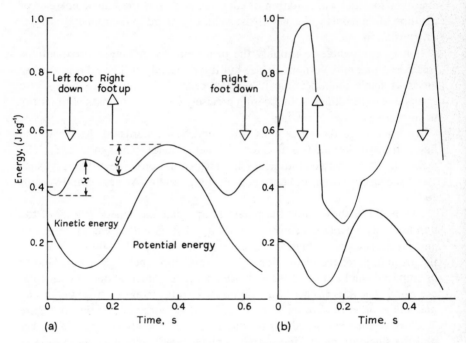

Fig. 7.13 Graphs showing the fluctuations of potential energy and external kinetic energy which occurred when a man walked at (a) 1.4 m s^{-1} and (b) 2.8 m s^{-1} (from data given by Cavagna and Margaria, 1966).

The fluctuations of external kinetic energy were much greater for a kangaroo hopping fast (Fig. 7.14a) than hopping slowly (Fig. 7.14b). However, the feet were on the ground for 45 % of the time at the lower speed, but only 28 % of the time at the higher. They had to exert much larger forces at the higher speed since the average value of the vertical component of the ground force must have equalled the weight of the body. (An animal of weight *mg* could be supported by an upward force *mg* acting continuously, or a force 2 *mg* acting for half the time, and so on.) Since the forces were larger, more elastic energy must have been stored in the tendons at the higher speed. This probably explains Dawson and Taylor's observation (1973) that the amount of metabolic energy used per hop by a kangaroo is almost the same for all speeds of hopping.

186

The long metacarpals and metatarsals of ruminants and horses (Section 7.3.1) probably increase the energy saving which can be made by elastic storage. The longer the metatarsals the more work is done by a given ground reaction

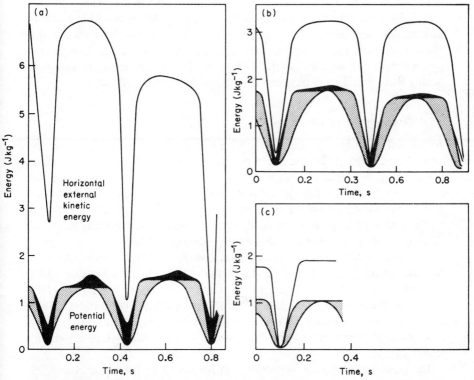

Fig. 7.14 Graphs showing the fluctuations of kinetic and potential energy which occurred when a young kangaroo (*Macropus rufus*) and a wallaby (*Protemnodon rufogrisea*) hopped. (a) and (b) show data calculated from a film of the kangaroo hopping on a treadmill at 6.2 and 2.7 m s⁻¹ respectively. (c) was calculated from a force platform record of the wallaby hopping at 2.4 m s⁻¹. Energy is expressed in J (kg body mass)⁻¹. The external kinetic energy has been divided into two components associated with the horizontal and vertical components of the velocity of the centre of mass. The vertical component is stippled. The internal kinetic energy (omitted from (c)) is shown black (from Alexander and Vernon, 1975a).

when it bends the ankle through a given angle. Energy must be saved by elastic storage in the check ligaments of horses (Camp and Smith, 1942) as well as in the tendons of muscles.

In fast locomotion the ground force exerts large moments about the ankle and stretches the Achilles tendon. The ground force and the gastrocnemius and plantaris muscles together exert large moments about the knee and stretch the patellar tendon. Energy is saved by elastic storage in these long tendons and less would be saved if the tendons were short. The corresponding muscles have to exert large forces and pennate structure has the advantages of enabling them to do this without being too bulky, and of giving them long tendons. The ground force generally exerts relatively small moments about the hip in steady locomotion and there is no obvious way in which much energy could be saved by elastic storage in the extensors of the hip. These muscles exert large moments about the hip in jumping (Fig. 7.11d, e) and possibly in acceleration, and it is probably in these activities that they have their major roles. These roles require performance of a lot of work in a single contraction, which only bulky muscles can do. There would be no apparent advantage in having these muscles pennate and they are in fact parallel-fibred.

7.5.2 Other energy requirements

An animal running in air must do work against the drag which the air exerts on its body. Pugh (1971) measured the oxygen consumption of men walking and running on a treadmill in a wind tunnel and calculated that drag accounts for only 7.5% of the metabolic cost of running at 6 m s^{-1} when there is no wind. It has been shown by simple calculation that the cost of overcoming drag is unlikely to be a large fraction of the cost of locomotion for any mammal running in still air, except at high speeds (Alexander, 1976).

Energy must also be used overcoming viscous forces in the tissues (including 'friction' in the joints). It can be shown that when a man of mass m walks, the work done in each step against friction in the supporting hip is about 0.03 μmg J, where μ is the coefficient of friction for the joint (Alexander, 1968). As the coefficient is about 0.003 the work per step is only about 0.07 J, which is very small compared to the fluctuations of (potential + kinetic) energy (Fig. 7.13).

The power required for breathing and for pumping the blood round the body are greater during activity than at rest, because oxygen has to be supplied faster to the muscles. It is not necessary to take separate account of this in the calculations which follow because the values which are used for the efficiency of muscular work were obtained from experiments with men on sloping treadmills. These men must have increased the power used for breathing and for pumping blood, as they increased their rate of climbing.

7.5.3 The direction of the ground force

The positive and negative work done changing the external kinetic energy of the body generally represent the major part of the energy cost of fast locomotion (Figs. 7.13b, 7.14). They arise because the forces exerted on the ground have horizontal components. The vertical component of the ground force is plainly required to counteract the weight of the animal. The horizontal component is less obviously useful, and it would seem at first sight that energy could be saved by eliminating it. It can be shown that this is not so because, if there were no horizontal component, energy would be wasted by muscles of one joint doing work against those of another (Alexander, 1976, 1977).

7.5.4 Theory of walking

This and the next two sections of the chapter present simple mechanical models of particular gaits. The costs of transport they predict will be compared with costs of transport calculated from measurements of oxygen consumption (Chapter 6).

Drag will be neglected so the models will show that each gait requires equal quantities of positive and negative work. It will therefore be convenient to express the cost of transport in terms of an efficiency η, defined in such a way

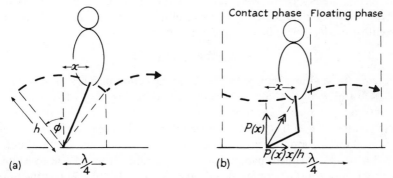

Fig. 7.15 Diagrams of the models of (a) walking and (b) running, which are explained in the text.

that $1/\eta$ units of metabolic energy are required to do one unit of positive work plus one unit of negative work.

The models describe bipeds, but can be applied to many quadrupedal gaits by treating the quadruped as two bipeds in tandem. It will be assumed that the ground reaction is kept in line with the hip joint. It will also be assumed initially that the legs have no mass, so that fluctuations of internal kinetic energy do not occur.

There is a range of possible walking styles, of which the extremes have been named the stiff walk and the compliant walk (Alexander, 1976). In the stiff walk the length of each leg is kept constant while its foot is on the ground, so that the hip moves forward in a series of arcs of circles (Fig. 7.15a). It is highest, and the potential energy of the body is greatest, as it passes over the supporting foot. However, the kinetic energy of the body is lowest at this time because the body has been decelerated by the horizontal component of the ground force. Since the length of the leg is kept constant, no elastic storage of energy can occur.

In the compliant walk the leg bends while the foot is on the ground, so that the potential energy of the body is lowest as the hip passes over the supporting foot. Potential and kinetic energy fluctuate in phase with each other, as in running, and elastic storage can occur.

It has been shown that in human walking, potential and kinetic energy fluctuate out of phase with each other (Fig. 7.13). Force platform records show that the same is true of walking by dogs (Kimura and Endo, 1972), cats (Manter, 1938) and horses (Björck, 1958). Cursorial mammals in general seem to use the stiff walk.

Consider a biped of mass m with legs of length h, executing the stiff walk (Fig. 7.15a). It takes strides of length λ lifting each foot as the other is set down, so that the double support phase is infinitesimally brief. (This assumption is unrealistic but convenient.) The mean velocity over a complete step is $\bar{u}$. The net cost of transport for walking, T_w, is the metabolic energy required (in excess of resting requirements) to move unit mass of animal a unit distance. From Alexander (1977) using $\hat{\lambda}$ and $\hat{u}$ as defined on p. 172

$$T_w \simeq (\hat{\lambda}^3 g/256\eta) \coth^2(\hat{\lambda}/4\hat{u}). \tag{7.2}$$

Figure 7.15a shows that in stiff walking the centre of mass must fall $h(1 - \cos\phi) \simeq \lambda^2/32h$ while the animal advances half a step (i.e. while x increases from 0 to $\lambda/4$, which takes time $\lambda/4\bar{u}$). If $\bar{u}$ were greater than $(gh)^{1/2}$ (i.e. if $\hat{u}$ were greater than 1) this would involve falling with acceleration greater than g. Hence the maximum possible speed for the stiff walk is just a little above the speed ($\hat{u} \simeq 0.8$) at which mammals normally change from a walk to a run or trot. Athletes achieve walking speeds over 3.7 m s^{-1} ($\hat{u} = 1.2$) but they use a peculiar hip action which reduces the vertical excursion of the centre of mass (Dyson, 1973).

7.5.5 Theory of running and trotting

The model presented in this section represents a biped running but it can also be applied to quadrupeds trotting and racking, which resemble two bipeds running in tandem. It is set out in detail by Alexander (1977) and presented only in outline here.

The model is shown in Fig. 7.15b. The step length ($2b$) is less than half the stride length (i.e. less than $\lambda/2$) so floating phases occur. At a given instant while a foot is on the ground the hip joint is a distance x in front of the foot and the vertical component of the ground reaction is $P(x)$. We make $P(x)$ proportional to $\cos(\pi x/2b)$ so as to be zero when the foot first touches the ground ($x = -b$), rise to a maximum as the hip passes over the foot ($x = 0$) and fall to zero again as the foot leaves the ground ($x = b$). So that the average ground reaction over a complete stride equals mg, we make $P(x) = (\pi mg\lambda/8b)\cos(\pi x/2b)$. We give the horizontal component of the ground reaction a value $P(x)x/h$ (where h is the height of the hip joint when the animal is standing) so that the ground reaction is more or less in line with the centre of mass. These assumptions are designed to imitate force platform records of men running and dogs trotting (Figs. 7.10b, d, 7.11).

To determine the positive and negative work which the muscles must do in each stride, we must calculate the fluctuations of kinetic, potential and elastic energy. The horizontal component of the ground reaction decelerates the animal while x is negative, and accelerates it while x is positive. It can be shown that the kinetic energy W_K gained and lost in each half stride must be about $0.09 \, mg \, b\lambda/h$.

During the floating phase the animal travels a distance $(\frac{1}{2}\lambda - 2b)$ (Fig. 7.15b). Its velocity during the floating phase is only slightly different from its main velocity $\bar{u}$ so the duration of the floating phase is about $(\frac{1}{2}\lambda - 2b)/\bar{u}$. For half this time it is rising and for half falling under the influence of gravity and it is easily shown that it must rise and fall through height $g(\lambda - 4b)^2/32\bar{u}^2$ gaining and losing potential energy $mg^2(\lambda - 4b)^2/32\bar{u}^2$. In addition, it falls and rises while the foot is on the ground and it can be shown that the total amount of potential energy W_P which is lost and regained in a half stride is $(mg^2 \lambda/32\bar{u}^2)$ $(\lambda - 2.9b)$.

The leg shortens and extends again in each step, by flexion and extension of its joints. These changes in its length are partly due to changes in length of muscles and partly to elastic stretching and recoil of tendons. Let the change in length *attributable to elastic stretching only* caused by a force P on the foot be PC: C is then the elastic compliance of the leg. The elastic strain energy stored in the tendons will be $\frac{1}{2}P^2 C$. The maximum value of the ground reaction in running occurs when $x = 0$ and is $(\pi mg\lambda/8b)$. Hence the elastic energy W_E gained and lost in each half stride is $\frac{1}{2}(\pi mg\lambda/8b)^2 C$.

Potential and kinetic energy are greatest in the floating phase but the elastic strain energy has a maximum in the contact phase. Hence the negative and positive work done by the muscles in each half stride are each equal to $(W_P + W_K - W_E)$. In the half stride, the animal of mass m travels a distance $\lambda/2$ so the cost of transport is given by

$$T_r = 2(W_P + W_K - W_E)/m\lambda\eta \simeq \frac{g}{16h\eta}\left[\frac{gh}{\bar{u}^2}(\lambda - 2.9b) + 2.9b - \frac{2.5 \, mg \, C\lambda h}{b^2}\right].$$

This equation can be simplified by using the dimensionless numbers $\hat{\lambda}$, $\hat{b}$ and $\hat{u}$ which have already been defined, together with an additional one $\hat{C}$ $(=mgC/h)$

$$T_r \simeq \frac{g}{16\eta}\left[\frac{(\hat{\lambda} - 2.9\hat{b})}{\hat{u}^2} + 2.9\hat{b} - \frac{2.5\hat{C}\hat{\lambda}}{\hat{b}^2}\right]. \tag{7.3}$$

Equation 7.3 can be modified to give the cost of transport for bipedal hopping by changing $\hat{\lambda}$ to $2\hat{\lambda}$ (since only one step occurs in each stride) and $\hat{C}$ to $\frac{1}{2}\hat{C}$ (since two legs have half the compliance of one)

$$T_h \simeq \frac{g}{16\eta}\left[\frac{(2\hat{\lambda} - 2.9\hat{b})}{\hat{u}^2} + 2.9\hat{b} - \frac{2.5\hat{C}\hat{\lambda}}{\hat{b}^2}\right]. \tag{7.4}$$

7.5.6 Theory taking account of limb mass

So far the mass of the legs and the internal kinetic energy changes due to leg movement have been ignored. Let each leg have moment of inertia J about the hip joint. While each foot is on the ground the limb has an angular velocity of about $\bar{u}/h$ about the hip, and the associated kinetic energy is $\bar{u}^2 J/2h^2$. In each stride the foot is on the ground for time $2b/\bar{u}$ and off the ground for $(\lambda - 2b)/\bar{u}$ so the limb must be brought forward in the recovery stroke with angular velocity $(\bar{u}/h)[2b/(\lambda - 2b)]$ and kinetic energy $(\bar{u}^2 J/2h^2)[4b^2/(\lambda - 2b)^2]$. Each leg gains and loses both these amounts of kinetic energy once in each stride. This adds an amount T' to the cost of transport where

$$T' \simeq \frac{\bar{u}^2 J}{h^2 m \lambda \eta} \left(1 + \frac{4b^2}{(\lambda - 2b)^2}\right).$$

Using the dimensionless numbers already defined and in addition $\hat{J} = J/mh^2$ this becomes

$$T' \simeq \frac{g \hat{u}^2 \hat{J}}{\hat{\lambda} \eta} \left(1 + \frac{4\hat{b}^2}{(\hat{\lambda} - 2\hat{b})^2}\right). \tag{7.5}$$

7.5.7 Comparison of theory with observation

Many measurements have been made of the oxygen consumption of mammals walking and running (Chapter 6). Metabolism involving 1 cm^3 oxygen releases about 20 J chemical energy. Measurements have been made of the efficiencies of positive and negative work performance by men and other animals (Chapter 3) so a value can be assigned to η. It is therefore possible to compare actual costs of transport with the predictions of Equations 7.2 to 7.5.

The quantities $\hat{b}$, $\hat{\lambda}$, $\hat{J}$ and $\hat{C}$ which appear in the equations are dimensionless and so will have the same values for geometrically similar animals of different size moving in geometrically similar ways. If a mouse was simply a scale model of an elephant and moved like a scaled-down elephant, all these quantities would have the same value for both. A mouse is not a scale model of an elephant but we can nevertheless hope to obtain graphs applicable to mammals of a wide range of sizes, by using typical values of $\hat{b}$, $\hat{\lambda}$, $\hat{J}$ and $\hat{C}$.

For most mammals $\hat{b} \simeq 0.5$ and $\hat{\lambda} \simeq 2.3\hat{u}^{0.6}$ (Section 7.2.3). $\hat{J}$ is about 0.035 for man and 0.02 for kangaroos and quail, so a value of 0.025 will be used. An experiment by Cavagna (1970) gave $\hat{C} = 0.03$ for man, and as no values are available for any other mammal this one will be used. It is thus possible to draw theoretical graphs of cost of transport against dimensionless speed, $\hat{u}$.

Figure 7.16 shows graphs constructed in this way. Two components of the cost of walking are shown separately in Fig. 7.16a. The cost for potential and external kinetic energy fluctuations, calculated from Equation 7.2, increases with increasing speed. The cost for internal kinetic energy changes, calculated from Equation 7.5, is relatively small at high speeds but larger at low speeds when the forward swing of the foot has to be accomplished very quickly if $\hat{b}$ is kept constant. The graph has not been extended to speeds lower than $\hat{u} = 0.4$ because $\hat{b}$ would probably be reduced at these very low speeds.

Four components of the cost of running or trotting are shown separately in Fig. 7.16b. The costs for potential and external kinetic energy and the saving

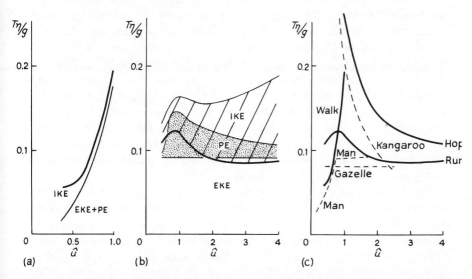

Fig. 7.16 Graphs of cost of transport (represented by $T\eta/g$) against velocity (represented by $\hat{u}$). (a) and (b) show theoretical costs for walking and running, respectively. In (b) the saving due to elastic storage (hatched) is subtracted from the total of the other components to obtain the net cost which is shown by the thick line. In (c) the continuous lines show theoretical net costs for walking, running or trotting and bipedal hopping. The broken lines show actual costs calculated from measurements of oxygen consumption by men walking and running (Passmore and Durnin, 1955), gazelles walking and trotting (Taylor *et al.*, 1974) and kangaroos hopping (Dawson and Taylor, 1973). It has been assumed that $\eta = 0.25$. EKE, IKE, external and internal kinetic energy: PE, potential energy.

due to elastic strain energy have been calculated from the appropriate terms in Equation 7.3. The cost for internal kinetic energy has been obtained from Equation 7.5. The total cost, shown by the thick line, does not change very much with speed because the increasing cost of internal kinetic energy changes is more or less balanced by increased elastic storage. When $\hat{u} < 0.8$, $\hat{\lambda} < 4b$ and the run becomes a compliant walk. The calculations used to extend the graph to these low speeds have been explained elsewhere (Alexander, 1977).

Figure 7.16c shows the total costs for walking and running, and also for bipedal hopping (from Equations 7.4 and 7.5). The theoretical costs for walking and running intersect at $\hat{u} = 0.8$, so that walking is the more economical of these gaits at lower speeds and running or trotting at higher speeds. This seems to explain why many mammals change from walking to running or trotting at about $\hat{u} = 0.8$. The theoretical cost for bipedal hopping is higher than for running at all speeds, but the disadvantage is quite small at high speeds.

It has, regrettably, not yet been possible to formulate a satisfactory model of galloping. A very crude model would represent a galloping quadruped as two hopping bipeds in tandem but it would not be satisfactory: it can be shown by considering the fluctuations in overall length of the trunk that the forequarters exert forces on the hindquarters and vice versa, so the quadruped is not equivalent to two independent bipeds (Alexander, 1977). These considerations suggest that the cost of galloping might be even greater than the cost of bipedal hopping. However, it seems possible that in galloping energy is saved by elastic storage in the tendons and ligaments of the back. The component of the cost of transport for internal kinetic energy becomes large at high speeds, so a mechanism for elastic storage of internal kinetic energy would be valuable. The Achilles and patellar tendons and their equivalents in the forelimb are well placed for potential and external kinetic energy to be transferred to them, but there is no apparent means of transferring to them internal kinetic energy associated with limb movements.

Figure 7.16c also shows some costs of transport calculated from measurements of oxygen consumption. The experimental curve for human walking agrees well with the theoretical one. The experimental curves for human running and for the gazelle are reasonably near the level of the theoretical curve for running and trotting but neither the gazelle curve nor similar graphs for other quadrupeds (goat and cheetah, Taylor *et al.*, 1974; dog, Taylor *et al.*, 1970), show the expected dip at walking speeds. The experimental curve for the kangaroo falls more steeply with increasing speed than the theoretical curve for bipedal hopping; the elastic compliance calculated from observations on man is probably too low for the kangaroo (see the discussion of compliance in Alexander and Vernon, 1975a).

The theoretical predictions are reasonably near the observed costs of transport for mammals around the size of dogs and men. They are far too low for mice and other small rodents. For instance, the value of $T\eta/g$ calculated from measurements of the oxygen consumption of $20g$ mice (Taylor *et al.*, 1970) is 1.4. So far, the discrepancy has not been explained (Alexander, 1977, but see McMahon, 1975).

The distal parts of the limbs of ruminants and horses are long and slender (Section 7.3.1), so that $\hat{J}$ must be lower than for most other mammals. This can be expected to reduce cost of transport. It should reduce the cost quite substantially at high speeds, at which internal energy changes account for a large proportion of the cost. However, Taylor *et al.* (1974) found little difference in cost of transport between goats and gazelles (which have low values of $\hat{J}$) and cheetahs (which have much larger ones).

194

7.6 Other animals

7.6.1 Birds

Some species of bird habitually run but others use the bipedal hop. Clark and Alexander (1975) studied running by quail (*Coturnix*) at moderate speeds (mainly between 0.4 and 0.9 m s^{-1}: $\hat{u}$ between about 0.5 and 1.1). They found that the gait was the one described in Section 7.5.4 as a compliant walk: the duty factor was greater than 0.5 but the total force exerted on the ground had a minimum in the double support phase. There was little evidence of elastic storage of energy. Muybridge (1957) illustrates an ostrich running at 2.9 m s^{-1} ($\hat{u} = 0.9$), with a short floating phase. Unpublished force platform records by Mr. D. J. Letten show that the gait of the domestic duck is a stiff walk. I know of no published studies of the mechanics of hopping.

The knees of birds are much closer to a transverse line through the centre of mass than are the hips, and neither pigeons nor quail move the hip joint much as they run (Cracraft, 1971; Clark and Alexander, 1975). They move their legs from the knee much as we move ours from the hip.

7.6.2 Lower vertebrates

Mammals and birds place their feet on the ground below the trunk but amphibians and reptiles place theirs more laterally (Figs. 7.17, 7.18). This makes it seem likely that the ground reaction has a large transverse component as indicated in Fig. 7.17, but this has been demonstrated only for the toad (Barclay, 1946). If the ground reaction acts in the direction indicated by the broken lines in the figure, it will exert large moments about the elbow but only small moments about the shoulder. If it acted vertically, the converse would be true.

The stable walk illustrated in Fig. 7.2 is used occasionally, for instance by tortoises (Sukhanov, 1974). The relative phases of the limbs are then more or

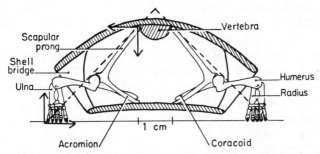

Fig. 7.17 A diagrammatic transverse section through a standing turtle showing the position of the fore limbs and forces which are supposed to act on the feet and vertebral column (from Walker, 1971).

195

less as shown for the walk in Table 7.1. However, it is usual for the limbs to be moved more nearly in the phase relationship of the trot, even at low speeds and with high duty factors. For instance Walker (1971) reproduces film of a turtle (*Chrysemys*) walking with relative phases as follows: *LF* 0, *RF* 0.5, *LH* 0.6, *RH* 0.1. The duty factor is 0.8 which is high enough for a stable walk but because of the phase relationships there are periods with only two feet on the ground.

Newts and most reptiles bend their bodies from side to side as they run (Fig. 7.18). This enables them to make longer steps than would otherwise be possible because a fore foot (for instance) is set down while its shoulder is turned forward and is not lifted until its shoulder is turned back (Roos, 1964). It is not clear how this affects the cost of transport.

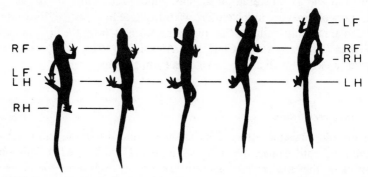

Fig. 7.18 Successive stages in the stride of a newt (*Triturus vulgaris*) (from Roos, 1964).

Note in Fig. 7.18 that the bends do not travel along the newt's body, as in a swimming eel (see Chapter 9). Rather, the newt repeatedly bends into an S-shape and straightens again. The bends do not form travelling waves, but standing waves like those of a vibrating string. The limb girdles are more or less at the nodes where no bending occurs, but where the body rotates through the largest angle in each cycle of movement. If the standing waves are to be used to increase the step length as much as possible the limbs must move with the relative phases characteristic of the trot.

Ordinary lizards make standing waves like the newt but *Chalcides*, which has very short legs, makes movements intermediate between standing and travelling waves (Daan and Belterman, 1968). Limbless lizards and snakes usually propel themselves by travelling waves which push against irregularities in the ground, driving them forwards. Various other techniques of locomotion are also used by limbless reptiles: they have been reviewed by Gans (1962) and Gray (1967). The progress of a snake is resisted by a frictional force μmg where μ is the coefficient of friction with the ground. The cost of transport should therefore be $\mu g/\eta$. Costs of transport calculated from measurements of oxygen consumption on a treadmill correspond to $\mu = 0.25$ (assuming $\eta = 0.25$) (Chodrow and Taylor, 1973).

Many species of lizard lift their fore feet off the ground at high speeds and run bipedally (Snyder, 1962). Speeds up to 7 m s^{-1} have been recorded on film. Species which run in this way depend on the tail for balance: amputation of a short length of tail can have a serious effect on ability to run bipedally. The duty factor falls as speed increases and may be as low as 0.35 (Daan and Belterman, 1968).

The mechanics of jumping by frogs was studied by Calow and Alexander (1973) who measured ground forces and calculated muscle stresses. To jump a distance λ a frog of mass m must give itself kinetic energy $\frac{1}{2}mg\lambda$ (see for instance Alexander, 1968) which is lost when it lands. The cost of transport should thus be $0.5g/\eta$, much higher than for running (Fig. 7.16). Jumping is not an economical technique of locomotion but it is probably an effective method of escape from predators because its direction is unpredictable.

7.6.3 Arthropods

Most of the arthropods which possess walking legs stand with their feet far apart so that the angle θ (Fig. 7.19) is small. This is probably necessary if they are to avoid being overturned by wind or (if they live in water) by water movements. The drag exerted by wind or water tends to overturn them while their weight tends to prevent overturning. Overturning will occur if the moment exerted by drag about the downwind feet exceeds the moment exerted by weight about the same feet. This will occur if the drag exceeds (weight) $\times \cot \theta$. For animals of the same shape in the same wind, drag is proportional to (body length)^2 (i.e. to area) and weight to (body length)^3 so small animals need a smaller value of θ, if they are to avoid being overturned. For typical insects $\theta \simeq 30°$ while for mammals such as dogs $\theta \simeq 80°$. The Pauropoda stand higher on their legs than most other small arthropods with $\theta \simeq 50°$ (Manton in Gray, 1968) but they live in exceptionally sheltered places, among leaf litter and in crevices in soil. Lobsters and many other crustaceans are as large as small mammals but nevertheless have low θ. They need low θ in spite of their size

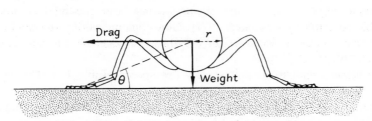

Fig. 7.19 A diagrammatic front view of an insect. Further explanation is given in the text (from Alexander, 1971).

because they live in water where even slow currents exert substantial drag, and much of their weight is counteracted by buoyancy. The plausibility of these explanations of the stance of arthropods has been tested by rough calculation (Alexander, 1971). Some other explanation is required for the rather similar stance of reptiles (Fig. 7.17).

Many studies have been made of the gaits of Orthoptera (Hughes and Mill, 1974; Burns, 1973) but few of other insects (Manton, 1973). A particularly detailed study has been made of the cockroach *Periplaneta* (Delcomyn, 1971). Its stride length is more or less constant (about 3 cm, $\hat{\lambda} \simeq 3$) over the range of speeds from 2 to 50 cm s^{-1}, but increases a little at higher speeds. Table 7.2 shows that in its normal gait, used at speeds of 10 cm s^{-1} and above, *Periplaneta* moves its legs in groups of three. Legs 1 and 3 of one side and leg 2 of the other

Table 7.2 Gaits of *Periplaneta*. L, R signify left, right and *1, 2, 3* signify the first, second and third legs. Data from Delcomyn (1971)

Speed		<3 cm s^{-1}		$10 \to 80$ cm s^{-1}	
Duty factor		0.8		$0.7 \to 0.5$	
		L	R	L	R
Relative phase	1:	0	0.5	0	0.5
	2:	0.6	0.1	0.5	0
	3:	0.2	0.7	0	0.5

move simultaneously. Duty factors do not fall below 0.5 even at the highest speeds so there are always at least three feet on the ground, forming a triangle of support. The insect is stable throughout the stride. This seems to be generally true of insects though duty factors slightly lower than 0.5 have been observed for *Campodea* at high speeds (Manton, 1973). It is a striking difference from mammals which usually maintain stability throughout the stride only at extremely low speeds ($u < 0.4$), and have duty factors below 0.5 when $u > 0.8$. The top speed of 80 cm s^{-1} observed in the study of *Periplaneta* corresponds to $\hat{u} \simeq 2.5$. It has been argued that constant stability is required by insects because of their small size: they would be blown off balance far more easily than mammals by gusts of wind because their ratio of surface area to mass is so much larger. A very rough calculation has been presented in support of this contention (Alexander, 1971). Note that increasing the number of legs from four to six decreases the duty factor needed for constant stability from 3/4 (0.75) to 3/6 (0.5).

Periplaneta changes its gait at very low speeds (Table 7.2). The legs in each group of three move slightly out of phase.

Manton (1973 and earlier papers) has studied the gaits of myriapods. Examples are shown in Table 7.3. In each case there is a more or less constant phase difference between each leg and its neighbour on the same side.

198

In millipedes, each leg moves slightly after the one behind so metachronal waves travel anteriorly along the body. In centipedes such as *Scolopendra*, each leg moves slightly before the one behind so the waves travel posteriorly. Millipedes move the legs of a pair simultaneously but centipedes move them alternately. Both millipedes and centipedes reduce the duty factor as they increase their speed but the ranges of duty factor which they use scarcely overlap. Typical millipedes are burrowing herbivores. They cannot run particularly fast but the very high duty factors which they use at low speeds ensure that most of the legs can push simultaneously as they force their way through the soil. *Scolopendra* and similar centipedes are carnivores which

Table 7.3 Gaits of typical millipedes and of the centipede *Scolopendra*. Data from Manton (1973). Other details as for Table 7.2

		Millipedes Slow		Fast		Scolopendra Slow		Fast	
Duty factor		0.8		0.4		0.5		0.15	
		L	R	L	R	L	R	L	R
Relative phase	1:	0	0	0	0	0	0.50	0	0.50
	2:	0.96	0.96	0.9	0.9	0.33	0.83	0.08	0.58
	3:	0.92	0.92	0.8	0.8	0.67	0.17	0.16	0.66
	4:	0.88	0.88	0.7	0.7	0	0.50	0.24	0.74
	5:	0.84	0.84	0.6	0.6	0.33	0.83	0.32	0.82
	6:	0.80	0.80	0.5	0.5	0.67	0.17	0.40	0.90
	7:	0.76	0.76	0.4	0.4	0	0.50	0.48	0.98
	etc.								

may crawl through existing crevices but do not generally burrow actively. Though their legs are quite short, they can crawl fast: a 38 mm *Cryptops* has been observed crawling at 0.26 m s^{-1}. The duty factor probably has to be low, for high speeds to be possible. If the velocity is u, the stride length λ, the stride frequency n, the step length $2b$ and the duty factor β, $u = n\lambda = 2nb/\beta$. Since the legs are fairly short b is quite small and n would have to be very large at high speeds, if β were not small. Even so, *Cryptops* at top speed uses a stepping frequency of 25 Hz.

Figure 7.20a shows *Scolopendra* running. Each foot moves slightly before the next foot posterior to it so when two adjacent feet are on the ground the more anterior one is at a later stage in the step: the two legs converge. *Scolopendra* has quite short legs but *Scutigera* (Fig. 7.20b) has long ones. If it ran in the manner of *Scolopendra* adjacent legs would cross over each other while their feet were on the ground. It does not run in this way: each foot moves after the next posterior (metachronal waves travel anteriorly) so that adjacent legs with feet on the ground diverge. *Scutigera* is the fastest known centipede. Specimens only 22 mm long can run at 0.42 m s^{-1} but because the legs are so long the stepping frequency required for this is only 13 Hz.

The gaits of many arthropods are irregular. Some foot sequences are more common than others but there is no fixed sequence even at a given speed. This has been stressed by Wilson (1967) in a study of spiders and particularly by MacMillan (1975) in a study of the lobster.

Crabs habitually walk sideways, with one side of the body rather than the anterior end leading. Ghost crabs (*Ocypode*) run sideways amazingly fast (Burrows and Hoyle, 1973). Specimens only 2–3 cm across the carapace have

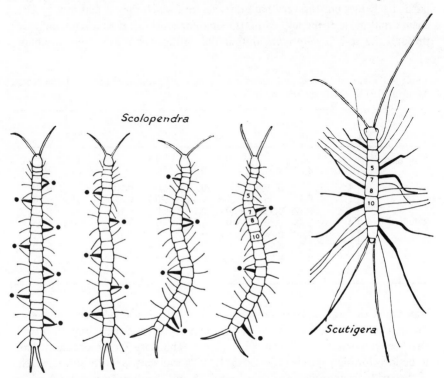

Fig. 7.20 Outlines traced from photographs of the centipedes *Scolopendra* (a–d) and *Scutigera* (e) running. Legs making contact with the ground are indicated by thick lines. (a) to (d) show running at progressively higher speeds (from Manton, 1965).

top speeds around 2 m s^{-1} ($\hat{u} \simeq 3$). Their stride frequencies at this speed are 15–20 Hz and their stride lengths around 12 cm ($\hat{\lambda} \simeq 3$). The gait is a trot which uses only two legs of each side and involves floating phases.

A great deal of information about the functional morphology of terrestrial locomotion in arthropods, as well as about gaits, can be found in reviews by Manton (1973) and Gray (1967).

Fleas, locusts, click beetles and some other insects make quite large jumps. These require large accelerations and high power output because the acceleration has to be achieved over a very short distance. The mechanics of jumping by several species has been studied in some detail (see Bennet-Clark, 1975), and it has been shown that high power is achieved by catapult mechanisms. Energy is stored prior to the jump in an elastic structure by a relatively slow muscular contraction, and released very rapidly at take-off by an elastic recoil.

References

Alexander, R.McN. (1968) *Animal Mechanics*, London: Sidgwick and Jackson.
Alexander, R.McN. (1971) *Size and Shape*, London: Edward Arnold.
Alexander, R.McN. (1974) Mechanics of jumping by a dog. *J. Zool.*, *(Lond.)* **173**, 549–573.
Alexander, R.McN. (1975a) *The Chordates*, Cambridge: Cambridge University Press.
Alexander, R.McN. (1975b) *Biomechanics*, London: Chapman and Hall.
Alexander, R.McN. (1976) Mechanics of bipedal locomotion. In: *Perspectives in Experimental Biology* **1**, 493–504, Spencer-Davies, P., Ed., Oxford: Pergamon.
Alexander, R.McN. (1977) Mechanics and scaling of terrestrial locomotion. In: *Scale Effects in Animal Locomotion*, Pedley, T.S., Ed., London: Academic Press.
Alexander, R.McN., Langman, V.A. and Jayes, A.S. (1977) Fast locomotion of some African ungulates. *J. Zool.*, *(Lond.)* (In press).
Alexander, R.McN. and Vernon, A. (1975a) The mechanics of hopping by kangaroos (Macropodidae). *J. Zool.*, *(Lond.)* **177**, 265–303.
Alexander, R.McN. and Vernon, A. (1975b) The dimensions of knee and ankle muscles and the forces they exert. *J. Human Movement Studies* **1**, 115–123.
Barclay, O.R. (1946) The mechanics of amphibian locomotion. *J. exp. Biol.* **23**, 177–203.
Barclay, O.R. (1953) Some aspects of the mechanics of mammalian locomotion. *J. exp. Biol.* **30**, 116–120.
Barnes, G.R.G. and Pinder, D.N. (1974) *In vivo* tension and bone strain measurement and correlation. *J. Biomech.* **7**, 35–42.
Basmajian, J.V. (1967) *Muscles Alive*, 2nd edn., Baltimore: Williams and Wilkins.
Bausschulte, C. (1972) Morphologische und biomechanische Grundlagen einer funktionellen Analyse der Muskeln der Hinterextremitat (Untersuchung an quadrupeden Affen und Kanguruhs). *Z. Anat. und EntwGesch.* **138**, 167–214.
Bennet-Clark, H.C. (1975) The energetics of the jump of the locust *Schistocerea gregaria*. *J. exp. Biol.* **63**, 53–84.
Björck, G. (1958) Studies on the drag force of horses. *Acta Agric. Scand.* suppl. 4.
Burns, M.D. (1973) The control of walking in Orthoptera. I. Leg movements in normal walking. *J. exp. Biol.* **58**, 45–58.
Burrows, M. and Hoyle, G. (1973) The mechanism of rapid running in the ghost crab, *Ocypode ceratophthalma*. *J. exp. Biol.* **58**, 327–349.
Calow, L.J. and Alexander, R.McN. (1973) A mechanical analysis of a hind leg of a frog. *J. Zool.*, *(Lond.)* **171**, 293–321.
Camp, C.L. and Smith, N. (1942) Phylogeny and function of the digital ligaments of the horse. *Univ. Calif. Mem. Zool.* **13**, 69–124.
Carlsöo, S. (1972) *How Man Moves*, London: Heinemann.
Cavagna, G.A. (1969) Travail mecanique dans la marche et la course. *J. Physiol.*, *(Paris)* **61**, suppl. 1, 3–42.
Cavagna, G.A. (1970) Elastic bounce of the body. *J. appl. Physiol.* **29**, 279–282.
Cavagna, G.A. (1975) Force platforms as ergometers. *J. appl. Physiol.* **39**, 174–179.
Cavagna, G.A. and Margaria, R. (1966) Mechanics of walking. *J. appl. Physiol.* **21**, 271–278.
Cavagna, G.A., Margaria, R. and Arcelli, E. (1965) A high speed motion picture analysis of the work performed in sprint running. *Res. Film* **5**, 309–319.
Chodrow, R.E. and Taylor, C.R. (1973) Energetic cost of limbless locomotion in snakes. *Fed. Proc.* **32**, 422 Abs.
Clark, J. and Alexander, R.McN. (1975) Mechanics of running by quail (*Coturnix*). *J. Zool.*, *(Lond.)* **176**, 87–113.
Cohen, H. and Gans, C. (1975) Muscle activity in rat locomotion: movement analysis and electromyography of the flexors and extensors of the elbow. *J. Morph.* **146**, 177–197.

Cracraft, J. (1971) The functional morphology of the hind limb of the domestic pigeon, *Columba livia. Bull. Am. Mus. nat. Hist.* **144** (3) 171–268.

Daan, S. and Belterman, T. (1968) Lateral bending in locomotion of some lower tetrapods. I and II. *Koninklijke Nederlandse Akademie van Wetenschappen. Proceedings C* **71**, 245–266.

Dawson, T.J. and Taylor, C.R. (1973) Energy cost of locomotion in kangaroos. *Nature* **246**, 313–314.

Delcomyn, F. (1971) The locomotion of the cockroach *Periplaneta americana. J. exp. Biol.* **54**, 443–452.

Dyson, G.H.G. (1973) *The Mechanics of Athletics*, 6th edn., London: University of London Press.

Engberg, I. (1964) Reflexes to foot muscles in the cat. *Acta physiol. scand.* **62**, suppl. 235.

Engberg, I. and Lundberg, A. (1969) An electromyographic analysis of muscular activity in the hind limb of the cat during unrestrained locomotion. *Acta physiol. scand.* **75**, 614–630.

Gambaryan, P.P. (1974) *How Mammals Run*, New York: Wiley.

Gans, C. (1962) Terrestrial locomotion without limbs. *Am. Zool.* **2**, 167–182.

Goslow, G.E., Reinking, R.M. and Stuart, D.G. (1973) The cat step cycle: hind limb joint angles and muscle lengths during unrestrained locomotion. *J. Morph.* **141**, 1–42.

Gray, J. (1968) *Animal Locomotion*, London: Wiedenfeld and Nicolson.

Grieve, D.W., Miller, D., Mitchelson, D., Paul, J. and Smith, A.J. (1975) *Techniques for the Analysis of Human Movement*, London: Lepus Books.

Heglund, N.C., Taylor, C.R. and McMahon, T.A. (1974) Scaling stride frequency and gait to animal size: mice to horses. *Science* **186**, 1112–1113.

Hildebrand, M. (1965) Symmetrical gaits of horses. *Science* **150**, 701–708.

Hughes, G.M. and Mill, P.J. (1974) Locomotion: terrestrial. In: *The Physiology of Insecta*, 2nd edn., Rockstein, M., Ed., **3**, Chapter 5, New York: Academic Press.

Jenkins, F.A. (1971) Limb posture and locomotion in the Virginia opossum (*Didelphis marsupialis*) and in other non-cursorial mammals. *J. Zool., (Lond.)* **165**, 303–315.

Jenkins, F.A. (ed.) (1974) *Primate Locomotion*, New York: Academic Press.

Kamon, E. (1971) Electromyographic kinesiology of jumping. *Arch. phys. Med. Rehab.* 152–157.

Kimura, T. and Endo, B. (1972) Comparison of force of foot between quadrupedal walking of dog and bipedal walking of man. *J. Fac. of Sci., Univ. Tokyo* (V) **4**, 119–130.

McGhee, R.B. (1968) Some finite state aspects of legged locomotion. *Mathematical Biosci.* **2**, 57–66.

McGhee, R.B. and Frank, A.A. (1968) On the stability properties of quadruped creeping gaits. *Mathematical Biosci.* **3**, 331–351.

McMahon, T.A. (1975) Using body size to understand the structural design of animals: quadrupedal locomotion. *J. appl. Physiol.* **39**, 619–627.

Macmillan, D.L. (1975) A physiological analysis of walking in the American lobster (*Homarus americanus*). *Phil. Trans. R. Soc., Ser. B* **270**, 1–59.

McWhirter, N. and McWhirter, R. (1971) *The Guinness Book of Records*, 18th edn. London: Guinness Superlatives.

Manter, J.T. (1938) The dynamics of quadrupedal walking. *J. exp. Biol.* **15**, 522–540.

Manton, S.M. (1965) The evolution of arthropodan locomotory mechanisms. Part 8. Functional requirements and body design in Chilopoda. *J. Linn. Soc. (Zool.)* **45**, 251–484.

Manton, S.M. (1973) The evolution of arthropodan locomotory mechanisms. Part 11. Habits, morphology and evolution of the Uniramia (Onychophora, Myriapoda, Hexapoda) and comparisons with the Arachnida, together with a functional review of uniramian musculature. *J. Linn. Soc., (Zool.)* **53**, 257–375.

Milner-Brown, H.S. and Stein, R.B. (1975) The relation between surface electromyogram and muscular force, *J. Physiol.* **246**, 549–569.

Muybridge, E. (1957) *Animals in Motion*, 2nd edn., New York: Dover.

Passmore, R. and Durnin, J.V.G. (1955) Human energy expenditure. *Physiol. Rev.* **35**, 801–840.

Paul, J.P. (1971) Comparison of emg signals from leg muscles with the corresponding force actions calculated from walkpath measurements. *Conference on Human Locomotor Engineering* 13–28. London: Institute of Mechanical Engineers.

Pennycuick, C.J. (1975) The running of gnu (*Connochaetes taurinus*) and other animals. *J. exp. Biol.* **63**, 775–800.

Pugh, L.G.C.E. (1971) The influence of wind resistance in running and walking and the mechanical efficiency of work against horizontal or vertical forces. *J. Physiol., (Lond.)* **213**, 255–276.

Ramey, M.R. (1970) Force relationships of the running long jump. *Med. Sci. Sports* **2**, 146–151.

Roos, P.J. (1964) Lateral bending in newt locomotion. *Koninklijke Nederlands Akademie van Wetenschappen. Proceedings C.* **67**, 223–232.

Smidt, G.L. (1973) Biomechanical analysis of knee flexion and extension. *J. Biomech.* **6**, 79–92.

Snyder, R.C. (1962) Adaptations for bipedal locomotion of lizards. *Am. Zool.* **2**, 191–203.

Sukhanov, V.B. (1974) *General System of Symmetrical Locomotion of Terrestrial Vertebrates and some Features of Movement of Lower Tetrapods*, New Delhi: Amerind Publishing Co.

Taylor, C.R., Schmidt-Nielsen, K. and Raab, J.L. (1970) Scaling of energetic cost of running to body size in mammals. *Am. J. Physiol.* **219**, 1104–1107.

Taylor, C.R., Shkolnik, A., Dmi'el, R., Baharav, D. and Borut, A. (1974) Running in cheetahs, gazelles and goats: energy cost and limb configuration. *Am. J. Physiol.* **227**, 848–850.

van der Straaten, J.H.M., Lohman, A.H.M. and Linge, B. van (1975) A combined electromyographic and photographic study of the muscular control of the knee during walking. *J. Human Movement Studies* **1**, 25–32.

Walker, W.F. (1971) A structural and functional analysis of walking in the turtle, *Chrysemys picta marginata. J. Morph.* **134**, 195–214.

Wetzel, M.C., Atwater, A.E., Wait, J.V. and Stuart, D.G. (1975) Neural implications of different profiles between treadmill and overground locomotion timings in cats. *J. Neurophysiol.* **38**, 492–501.

Wilson, D.M. (1967) Stepping patterns in tarantula spiders. *J. exp. Biol.* **47**, 133–151.

Zarrugh, M.Y., Todd, F.N. and Ralston, H.J. (1974) Optimization of energy expenditure during level walking. *Eur. J. appl. Physiol.* **33**, 293–306.

8   Crawling and burrowing

E. R. Trueman and H. D. Jones

8.1 Introduction

Arthropods and vertebrates rely principally on rigid skeletal elements acting as levers to transmit muscular forces to the substratum in order to effect locomotion. The rigid elements are subject to compressive forces and allow pairs of muscles to be mutually antagonized. However, many soft-bodied invertebrate animals lack such rigid elements and compressive forces are opposed by fluid-filled body cavities. It is important that the fluid in the cavities is of low viscosity and incompressible in order to transmit forces rapidly and with the minimum of loss. The fluid used in animals is always water or an aqueous solution and such a skeleton is called a fluid or hydrostatic skeleton. A few invertebrates utilize the comparatively weak forces produced by cilia as a means of locomotion.

This chapter will be principally concerned with locomotion of animals using hydrostatic skeletons and thus will necessarily be concerned with animals of great taxonomic diversity but exhibiting varying degrees of convergence in the organization or mode of application of their locomotory systems.

8.2 Ciliary gliding

Some small bottom-dwelling animals are able to use a ciliated ventral surface to glide over the substratum. The use of cilia for gliding is restricted to more or less flattened animals such as some Turbellaria (Clark, 1964) and some Nemertea or to animals which are relatively light such as some aquatic gastropod molluscs (Miller, 1974a, b; Jones, 1975). Ciliary locomotion in gastropods is related neither to the systematic position, nor, more surprisingly, to the size of the animal (Clark, 1964). Mucus plays an essential role in ciliary locomotion in providing a medium of increased viscosity in which the cilia beat and obtain a good purchase.

204

8.3 Muscular pedal crawling

8.3.1 Turbellaria

The utilization of muscular forces for crawling allows animals to move appreciably faster and/or to be of a larger size than animals using ciliary gliding. The transition to muscular locomotion may be seen in the triclads. Many aquatic triclads utilize faster muscular locomotion in the form of large retrograde waves, as an escape reaction (normal locomotion being ciliary), (Pearl, 1903; Trueman, 1975). Some terrestrial triclads utilize muscular forces for locomotion and because of the circular body in such forms, locomotion resembles the peristaltic locomotion of nemertines and earthworms.

Polyclads are more reliant on muscular locomotion than triclads, muscular waves passing backwards (retrograde) alternatively down either half of the body (ditaxic) (Clark, 1964). The skeletal antagonism in Turbellaria is provided by the deformable parenchyma and by the geodesic fibre system in the basement membrane (Clark, 1964; Trueman, 1975).

8.3.2 Muscular pedal waves of gastropods and chitons

The types of pedal waves. Most species of gastropod when viewed through a transparent surface whilst crawling exhibit waves of muscular activity passing along or even across the sole of the foot. Pedal waves have been classified according to their direction relative to movement and the number of sets of waves on the sole (Miller, 1974b). The first division is between *direct* and *retrograde* waves. Direct waves move along the foot in the same direction as the animal is moving. Retrograde waves move in the opposite direction to that of the animal.

The next division of both direct and retrograde waves is into *monotaxic* and *ditaxic*. Monotaxic waves pass along the sole in a single set over more or less of the whole width. Ditaxic waves pass along the sole in two alternating sets, one on either side of the mid-line. Examples of these four basic types are shown in Fig. 8.1 and the vast majority of gastropods have pedal waves of one of these four types. Other much rarer types of pedal muscular waves are described by Miller (1974b).

The number of pedal waves in any one set is usually small – two or three – at any one time. The notable exception to this is the terrestrial pulmonates where in *Limax maximus* there are up to nineteen waves on the sole at any time (Jones, 1975).

Ditaxic gastropods can turn by lengthening the step on one half of the sole – the waves remaining strictly alternate. Some ditaxic species are more versatile than this and can reverse the direction of locomotion on one half of the sole, thus turning in the animal's own length (Jones and Trueman, 1970). Mono-

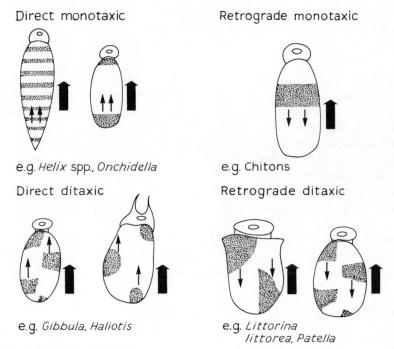

Direct monotaxic

e.g. *Helix* spp., *Onchidella*

Retrograde monotaxic

e.g. Chitons

Direct ditaxic

e.g. *Gibbula*, *Haliotis*

Retrograde ditaxic

e.g. *Littorina littorea*, *Patella*

Fig. 8.1 Diagrams of the four most common locomotory wave patterns in gastropods and chitons. Not to scale.

taxic species turn by deforming the anterior portion of the foot to one side or the other.

8.3.3 The mechanism of gastropod pedal waves

During the passage of a direct wave over the sole any region on the sole is lifted off the substratum, moved forwards (if the animal is moving forwards), longitudinally compressed then re-elongated and placed back onto the substratum, the pedal wave continuing forwards (Fig. 8.2a).

As a retrograde wave passes over the sole, the sole is lifted, elongated and moved forwards (again during forward movement of the animal) then stopped, re-compressed and placed back down on the substratum, the pedal wave continuing backwards (Fig. 8.2b). There is thus a clear difference in configuration between direct and retrograde waves.

During the passage of either a direct or a retrograde wave the sole is moved forwards relative to the substratum and relative to the body of the animal. Where the sole is stationary on the substratum it is applying a backthrust to the substratum in order to pull the body of the animal continuously forwards. Lissmann (1945 and 1946) and Jones (1975), and Jones and Trueman (1970) have provided experimental evidence of this for direct and retrograde waves respectively.

Jones and Trueman (1970) have proposed a model for the retrograde waves of *Patella* which relies almost exclusively on the copious dorso-ventral muscle fibres in the foot acting on the numerous blood spaces in the foot. The blood spaces are assumed to be incompressible and of constant volume though of differing shape. The mechanism is illustrated in Fig. 8.2b.

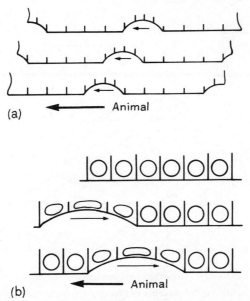

(a)   ◄────── Animal

(b)   ◄────── Animal

Fig. 8.2 (a) Diagrams to illustrate longitudinal compression of direct locomotory waves. The figures represent side-views of the sole during locomotion. (b) Diagrams illustrating elongation of the sole during the passage of retrograde locomotory waves, and the deformation of blood spaces above the sole (after Jones and Trueman, 1970).

Direct waves cannot function on this system as they are simultaneously lifted and compressed and thus the sole is not of constant volume during the passage of a wave. Jones (1973) has been able to demonstrate a possible locomotory mechanism by fixing slugs instantaneously. The sole of the foot of *Agriolimax* contains a highly vascular region crossed by two sets of oblique muscle fibres and some transverse muscle fibres. The oblique fibres of either set are parallel in a resting slug (Fig. 8.3a). At the commencement of locomotion the first wave appears towards the front of the sole (Lissmann, 1945; Jones, 1973) by contraction of some anterior oblique fibres thus forming a region of compression and uplifting and stretching the sole behind the wave to compensate (Fig. 8.3b). Further waves are propagated posteriorly as the first wave moves forwards until in a real slug some six or seven waves are passing along the sole. The body of the slug thus moves forwards continuously (refer to the small triangle on Fig. 8.3b–f) while any point on the sole is alternately moved forwards and stationary (refer to the two dots on Fig. 8.3b–f), the compensatory stretching allowing the slug to remain at more or less constant length.

The anterior oblique fibres contract to form the waves whereas the posterior oblique fibres contract when their ventral insertion is stationary adhering to

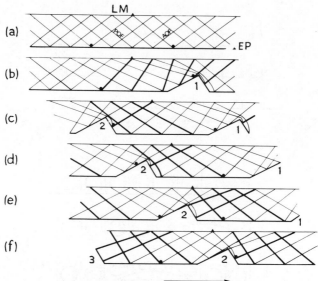

Direction of waves and animal

Fig. 8.3 Diagrams illustrating the sequence of events in the pedal musculature of *Agriolimax*. (a) At rest, the fibres of the anterior oblique muscles (AOF) are parallel as are the posterior oblique muscles (POF). LM, longitudinal muscle; EP, pedal epithelium. (b) First pedal wave (1) appears near front of sole by contraction of some AOF. This causes stretching of the sole behind the wave. (c–f) New waves appear posteriorly (2, 3) while the first wave moves forwards. Contracting muscles are drawn thick – note the changing length of fibres as waves pass. Refer to text for further elaboration.

the substratum, thus moving the body of the slug forwards and exerting a backthrust on the substratum. The two sets of muscles work against the blood pressure in the pedal haemocoel. The mechanical arrangement of these muscles is similar to that of pinnate muscles found in arthropods and vertebrates.

8.3.4 *Other types of molluscan crawling*

'Galloping' has been described in several *Helix* spp. (Jones, 1975) and occurs when the head of the animal is raised, protracted and put down again thus forming an arch about one third the length of the foot. The process is repeated until two or three such retrograde waves are formed and the animal progresses faster than normal. Normal direct pedal waves continue unabated. The mechanism involves alternate contraction of circular and longitudinal muscles of the body wall acting on the haemocoel spaces in the head-foot (Jones, 1975). 'Looping' occurs in *Otina otis* and in the Ellobiidae, when the anterior half of the foot is moved forwards and attached followed by the posterior half. This forms a large retrograde wave (Jones, 1975). *Truncatella* shows a particularly unusual form of looping where the snout and foot move alternately, somewhat in the manner of a leech (Morton, 1964).

'Leaping' is found in the Strombacea (Morton, 1964) and here the operculum is used as a lever to push the animal forwards, the foot acting as a muscular column and the plantar sole being much reduced.

8.3.5 Muscular creeping in other animals

Some anthozoan sea-anemones are able to creep by means of muscular waves passing across the pedal disc. Waves may be retrograde or direct depending on the species (Hyman, 1940).

Cristatella (Ectoprocta) forms linear colonies which are able to creep either way in the direction of the long axis on the flattened pedal surface. The mechanism is presumed to be muscular (Hyman, 1959).

8.4 Crawling of cylindrical animals

8.4.1 Peristalsis

Many soft-bodied vermiform animals use peristaltic contractions of the body wall to crawl over and burrow into substrata and the waves of contraction, like pedal muscular waves in gastropods may pass in the same direction as locomotion (direct) or in the opposite direction to locomotion (retrograde).

It is the muscles of the body wall that are used for locomotion and these are almost invariably arranged in two sets, an outer layer of circular muscle fibres and an inner layer of longitudinal fibres. The body wall can be considered to be a muscular tube enclosing a deformable incompressible fluid which forms the hydrostatic skeleton of these animals. The fluid may be aqueous as in earthworms or sipunculids or it may be deformable parenchyma or even gut contents as in nemertines and leeches.

Septate worms

The earthworm *Lumbricus*, has a body cavity (coelom) which is divided transversely into numerous more or less watertight compartments corresponding to each segment. Thus each segment is effectively of constant volume for the contained fluid may be prevented from passing to other regions of the body. Any segment can elongate or shorten by contraction of circular or longitudinal

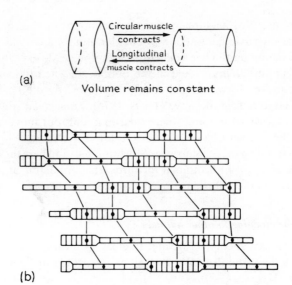

Fig. 8.4 (a) Diagram to illustrate reversible change of shape of a cylinder at constant volume by alternate longitudinal and circular muscle contraction. (b) Series of diagrams representing the locomotion of a septate animal, such as an earthworm, by retrograde peristaltic waves.

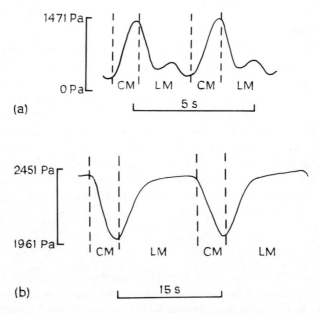

Fig. 8.5 Pressure recordings from the coelom of an earthworm during (a) surface crawling; (b) burrowing. CM, circular muscle contraction; LM, longitudinal muscle contraction. Pressure calibration in Pascals and recording read from left to right (after Seymour, 1969).

muscles respectively (Fig. 8.4a). The segmental length and circumference vary reciprocally and the longitudinal and circular muscles are mutually antagonistic, the coelomic fluid being the skeletal agent.

When an earthworm moves forwards the segments first elongate (Fig. 8.4b) and push forwards. The region immediately behind the elongation remains thickened and forms a point of attachment (*point d'appui*) to the ground and exerts backthrust on the ground (Gray, 1968). The protrusion of setae while the segment is thickened assists the fixation of the segment. The wave of elongation is followed by a wave of thickening by longitudinal muscle contraction which re-anchors the segments. The waves of contraction are necessarily retrograde.

The hydrostatic pressure in the coelom of any segment varies according to its state of elongation and whether the worm is crawling or burrowing. During crawling, the pressure is highest during elongation (Fig. 8.5a) these forces being necessary to push the worm forwards. During burrowing, peak pressure is attained during thickening (Fig. 8.5b) maximum forces being necessary to enlarge the burrow laterally after the elongating segments have penetrated a pre-existing crevice in the soil (Seymour, 1969).

Nemertines

Although nemertines are not segmental nor septate they crawl using retrograde waves. The parenchymatous tissue which forms the hydrostatic skeleton is cellular and thus any region of the body is of constant volume. Circular and longitudinal muscle fibres are again mutually antagonistic (Gray, 1968). The appearance of the waves is similar to those in some Platyhelminthes and there is a geodesic fibre system in the basement membrane which assists the control of the length and circumference ratio (Clark, 1964).

Non-septate worms

In a non-septate worm, e.g. *Polyphysia* (Elder, 1972; 1973), it is possible for the longitudinal and circular muscles to contract simultaneously in any segment, reducing both the length and circumference and thus the volume of the segment, but with consequent enlargement of another region of the body. Direct peristaltic waves may now effect locomotion (Fig. 8.6). Retrograde waves are also possible but the control of muscular contractions in a non-septate worm is more difficult as pressure changes in the body cavity are transmitted throughout the animal.

Arenicola contains septa in the head and tail segments but the trunk is non-septate. Either direct or retrograde waves can pass along the trunk. Seymour (1971) has noted that pressures in *Arenicola* are more erratic than in an earthworm due to continuous stretch-receptor activity in the whole trunk region, rather than in just one segment as in an earthworm.

211

The force that a worm may exert on the substratum should be equal to the internal pressure times the area of application. Seymour (1970) has simultaneously measured lateral forces, coelomic pressure and area of contact and found that the measured force exceeds calculated force (pressure times area) in *Arenicola* by 85% and in *Lumbricus* by 145%. This 'excess lifting force' is considered to be due to the intrinsic rigidity of the musculature. The excess lifting force is greater in *Lumbricus* due to the transverse muscular septa. The septa resist any tendency for the worm to flatten and lateral forces may be

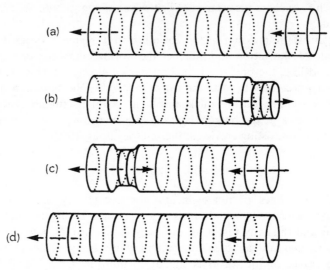

Fig. 8.6 Diagram illustrating direct peristaltic progression in the mid-trunk region of *Polyphysia*. Head is to the left. Note that during the passage of a wave the body is simultaneously shortened and narrowed. Arrows represent displacement of coelomic fluid (from Elder, 1973).

concentrated at two or three circumferential points, clearly advantageous to an animal burrowing by enlarging pre-existing crevices in soil. *Arenicola* cannot concentrate the lateral forces and burrows in homogeneous substrata, lateral forces being exerted round the entire circumference.

The geophilomorph centipede *Orya* is also notable for its ability to burrow into crevices in the soil and may develop haemocoelic pressures of up to 40 000 Pa (Trueman, 1975). Maximum coelomic pressures in *Lumbricus* are of the order of 37 000 Pa (Chapman, 1950; Seymour, 1971).

8.4.2 Looping

Many insect larvae are able to crawl along with the aid of non-peristaltic direct muscular waves (Hughes and Mill, 1974). The muscles involved are dorsal and ventral longitudinal muscles (Roberts, 1971), presumably working against general internal hydrostatic pressure and the exoskeleton, though the latter is not particularly rigid. It seems that any segment remains roughly of constant volume though complete inter-segmental septa are lacking.

Whereas in blow-fly larvae the wave of contraction is about half the body-length, in looping caterpillars the wavelength is equal to or greater than the body length, but the mechanism is otherwise similar. Attachment is provided for by the legs and prolegs.

Looping in *Malacobdella* (Clark, 1964) and leeches (Gray, 1968) is similar in principle, the mechanism of attachment being by suckers rather than limbs.

8.5 Burrowing in sand and mud

Many marine animals burrow in soft unstable substrata, ranging from sands containing some granules of diameter greater than 2 mm, through clean fine sand (particle diameter about 0.2 mm) to fine estuarine silts. Attention has been drawn (Webb, 1969; Trueman, 1975) to the different physical properties of marine soils and of their importance to infaunal animals. A dilatant sand becomes hard and more resistant to shear as increased forces are applied whilst thixotropic systems have the opposite property, showing reduction in resistance with increased rate of shear. The movement of animals through marine sand is commonly dependent on its physical characteristics, anchorage requiring a material of dilatant qualities whereas motion may be facilitated in a thixotropic system (Trueman and Ansell, 1969). Investigation of the forces required for penetration of different substrata by bivalve molluscs has been carried out by Nair and Ansell (1968).

In all burrowing animals the burrowing process requires the firm anchorage of part of the body whilst another region moves forward. These two events are carried out so as to produce a stepwise penetration of the substratum by soft-bodied animals, such as worms and molluscs, and with each step or digging cycle a series of actions takes place which are repeated in successive cycles. By contrast hard-bodied animals, such as crabs and heart urchins, use multiple appendages to effect a more continuous movement into the sand. In addition, some fish and *Amphioxus* (Webb, 1973) use undulatory swimming movements about an axial skeleton as a means of burial.

8.6 Burrowing in soft-bodied animals

8.6.1 Introduction

Clark (1964) suggested that all soft-bodied animals burrow by means of an essentially similar mechanism based on the formation of two types of anchors. The first is produced by dilation of the body above the distal extremity (penetration anchor) to prevent the animal from being pushed out of the substratum as it thrusts downward; the second is formed by the distension of the extremity (terminal anchor) and allows the body to be drawn into the burrow by contraction of longitudinal or retractor muscles (Fig. 8.7). These anchorages are applied alternately until burial is complete so that the resulting movement commonly takes place as a series of distinct steps (Fig. 8.8) each of which

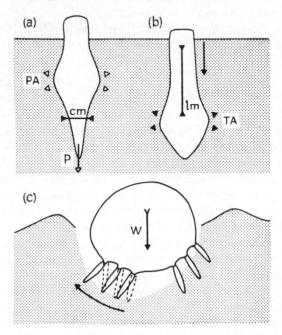

Fig. 8.7 Diagrams contrasting the burrowing process of a generalized soft-bodied animal (a and b) with that of an animal with hard exoskeleton (c). (a) Formation of penetration anchor (PA) which holds animal while distal region is elongated (P) by contraction of circular muscles (cm); (b) dilation producing terminal anchor (TA) allowing contraction of longitudinal or retractor muscles (lm) to pull the animal into the substratum (arrow) accompanied by high pressures in the fluid system; (c) a cavity is formed by the scraping action of some appendages while the body is anchored by others, the weight (W) may cause downward motion (from Trueman and Ansell, 1969).

involves both types of anchorage, extension of the burrow and the body being pulled forwards. The precise series of events in each cycle is characteristic of each different group of animals and is particularly clearly defined in polychaete worms, bivalve and gastropod molluscs (Trueman, 1975). The duration of active burrowing, termed the digging period, may be conveniently divided into two parts; (i) initial penetration, when cycles occur only sporadically; (ii) movement into the substrate when cycles follow in regular succession (Fig. 8.8).

214

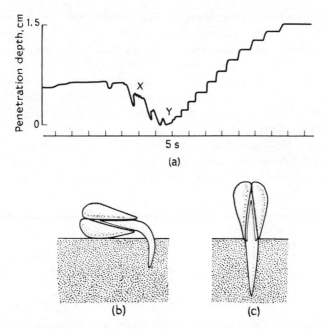

Fig. 8.8 (a) Recording of a complete digging period of the bivalve mollusc *Donax vittatus* obtained by attaching a thread from the posterior of the shell to an isotonic transducer. Probing of the foot occurs at X to make an initial penetration of the sand as in (b). At Y the foot is sufficiently buried to pull the shell erect (c) and the second part of the digging period follows. Each upstroke of the trace represents the shell being pulled into the sand (Trueman, 1975).

8.6.2 *Initial penetration of substratum*

When an animal lies on the surface of the substratum, anchorage can only be obtained by friction between its body and the sand. Those animals, which are circular in cross-section, have little contact and therefore adhesion with the substratum and only relatively weak penetrative scraping movements may be made. Eversion of part of the body, for instance the proboscis in the lugworm, *Arenicola*, or the base of the column in the anemone, *Peachia*, is commonly utilized to accomplish initial penetration (Trueman and Ansell, 1969). Other worms, e.g. *Nephtys* (Clark and Clark, 1960) swim rapidly towards the sand by means of undulatory movements until the anterior segments are buried while bivalve molluscs achieve penetration by a repeated stabbing motion with the foot (Trueman, 1975).

In *Arenicola marina* the papillate proboscis is everted with a centrifugal scraping movement so as to penetrate, thrust aside and drag back the sand. The proboscis is contained within the head coelom which is separated from the single trunk coelom by a septum. Pressure has been measured during eversion principally from the trunk coelom for insertion of a pressure-recording cannula

215

in the head coelom appears to prevent full proboscis eversion. Eversion of the proboscis in this species normally occurs at relatively low pressures (200–500 Pa) in the trunk coelom and recordings indicate that only low pressures occur until at least several segments have entered the sand (Fig. 8.9). In general the pressures observed in *Arenicola* during initial penetration are compatible with relatively poor anchorage on the sand surface and the ability of the worm to use small forces to achieve rapid penetration without pushing itself backwards. Other similar animals, e.g. *Cerebratulus* (Wilson, 1900), *Priapulus* (Hammond, 1970) and *Sipunculus* (Trueman and Foster-Smith, 1976) have longer proboscides than *Arenicola* which are commonly everted with more force during initial penetration, for instance 5000 Pa by *Sipunculus nudus*.

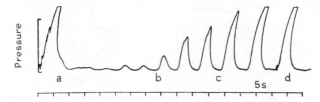

Fig. 8.9 Recording of pressure (scale: 0–4900 Pa) produced in trunk coelom of *Arenicola marina* during burrowing. (a) head on sand followed by proboscis eversion; (b, c and d) 2, 3 and 4 chaetigerous annuli respectively beneath the surface. Flat top of last two pulses due to limitation of pen's travel (from Trueman, 1975).

This worm can enter the sand rapidly when immersed in seawater and the manner in which it obtains sufficient anchorage to avoid being pushed backwards, whilst using pressures of this order is not fully understood.

Study of the burrowing mechanism of *Astropecten* has shown how turgid tube feet can penetrate the substratum and become anchored so as gradually to drag its arms into the sand (Heddle, 1967).

8.6.3 Movement through the substratum by worms

Worms move through the substratum by means of a series of movements. These may consist of excavation of a burrow by the prostomial horns and peristalsis of the body wall as in *Polyphysia*, which lives in flocculent mud (Elder, 1973); or penetration of sand by the proboscis followed by a step-like movement as in *Arenicola* (Seymour, 1971), *Sipunculus* (Trueman and Foster-Smith, 1976) or *Priapulus* (Hammond, 1970). In the three latter genera the coelom, or trunk coelom in *Arenicola*, functions as a single hydraulic organ with antagonistic circular and longitudinal muscle fibres in the body wall generating pressure in the coelomic fluid so as to effect locomotory movements.

The advantage of a single large coelomic cavity may be clearly observed in *Nephtys*, a polychaete worm with well-developed parapodia and powerful extrusion of the proboscis. The intersegmental septa are lost in the 34 anterior segments (Clark and Clark, 1960), so providing space for the retracted proboscis in a large coelomic cavity about which contraction of the longitudinal muscles generates the force (10 000 Pa, Trevor, personal communication) required for eversion.

Each digging cycle in *Arenicola* consists of a major movement of the trunk into the sand followed by an interval during which the worm elongates and everts its proboscis so as to extend the burrow. Each movement of the trunk

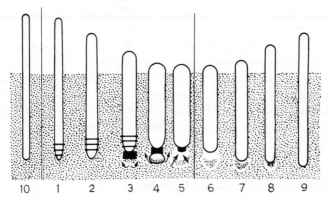

Fig. 8.10 Diagram showing principal events of a single flange–proboscis sequence in the burrowing of *Arenicola*. Stipple represents sand; coarse stipple, sand softened by water drawn through in direction of arrows. See text (from Seymour, 1971).

into the burrow is accompanied by a pressure pulse in the trunk coelom characteristically of about 10 000 Pa. This pressure pulse corresponds with dilation of the anterior segments of the lugworm forming a terminal anchor so as to allow contraction of the longitudinal trunk muscles to pull the worm along the burrow (Trueman, 1975). Between these pulses proboscis eversion occurs accompanied by the erection of flanges on the anterior segments and the protraction of chaetae to form a penetration anchor. As burrowing continues the digging cycles lengthen in duration and a new activity termed the 'flange-proboscis' (f-p) sequence, occurs. In this activity the proboscis and flanges combine as a digging tool as detailed in Fig. 8.10. The sequence commences with the worm shortening, the trunk moving into the sand (1) followed by the anterior end being pulled back with the flanges raised (2). Proboscis eversion (3–4) scrapes the sand away from the end of the burrow and its withdrawal (5) draws water through the sand into the cavity so formed. A recovery phase (6–10) follows with elongation of the trunk. This f-p sequence has four

217

functions in burrowing: penetration and removal of the sand, softening of the sand, and the drawing of the trunk into the burrow. These movements may be repeated up to five times in each digging cycle (Seymour, 1971).

Sipunculus nudus presents a similar pattern of activity to *Arenicola* when moving through sand. However, during each digging cycle there are two pressure pulses, the primary (18 000 Pa) causing a very powerful eversion of the proboscis to about one third of the overall trunk length and the secondary of much lower amplitude (2000 Pa) occurring as the proboscis is retracted. During retraction the tip of the proboscis is dilated so as to form a terminal anchor and the worm is drawn along the burrow by powerful proboscis retractor muscles (Trueman and Foster-Smith, 1976).

Generation of high pressure pulses in the coelom of a worm involves tension being developed in the trunk musculature. Determinations of muscle tension in the circular fibres in *Arenicola* gives values of 6×10^3 N m^{-2} at resting pressures of 200–400 Pa and 3×10^5 N m^{-2} at maximal pulse pressures, this powerstroke being undoubtedly derived from the contraction of the relatively massive longitudinal fibres (Trueman, 1975). In an earthworm, the longitudinals can develop a pressure in the coelomic fluid of ten times that of the circular fibres and the longitudinal muscles probably never exert their maximum tension in a worm unless the body wall is supported by the burrow (Chapman, 1950).

8.6.4 Movement through substrata by molluscs

The Mollusca are another major group of soft-bodied animals which inhabit marine sand, muds and gravel, members of the Bivalvia being particularly well-adapted for this life. Burrowing is essentially similar in all bivalves and is summarized in Fig. 8.11 (Trueman and Ansell, 1969; Trueman, 1975).

Bivalvia are not the only molluscs to adopt an infaunal life but other examples are all from the highly specific adaptations of small molluscan groups, e.g. the scaphopods and naticid gastropods (Trueman and Ansell, 1969). Prosobranch snails of the genus *Bullia* are another example which has successfully adapted to burrowing by specialization, showing a number of points of evolutionary convergence with Bivalvia (Trueman and Brown, 1976).

8.7 Burrowing by hard-bodied animals

Hard-bodied animals, such as crabs or heart urchins, generally have multiple appendages which are employed in scraping out a cavity in the sand whilst other spines or limbs are used to hold the animal in position (Fig. 8.7c). Movement into the cavity is achieved either passively by means of the weight of the animal, as in the starfish *Asterias*, or actively by the action of other appendages as in some crabs (Trueman and Ansell, 1969). These animals are not pulled into

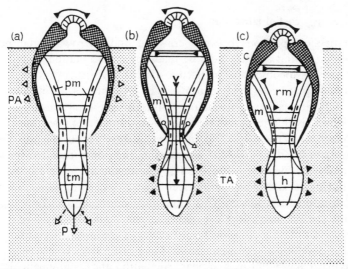

Fig. 8.11 Diagrams of principal stages in the burrowing of a bivalve mollusc. (A) valves press against sand by opening thrust of hinge ligament to provide a penetration anchor (PA) while foot extends by probing (P); (B) contraction of adductor muscles (am) ejects water from the mantle cavity (m) to soften the sand around valves (c) and causes a high pressure pulse in the haemocoel of the foot (h) so producing dilation to form a terminal anchor (TA). (C) contraction of retractor muscles (rm) then pulls shell down to anchored foot. tm, Transverse pedal muscle; pm, protractor muscle; ⟩——⟨, tension in ligament, adductor and retractor muscles (from Trueman and Ansell, 1969).

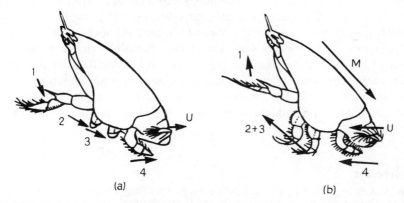

Fig. 8.12 Diagram of *Emerita* from the side showing position during burrowing with the limbs in (a) the recovery and (b) the power stroke. Arrows indicate movement. 1–4 thoracic legs; M, motion of crab, U, uropod (from Trueman, 1970).

the sand by means of longitudinal or retractor muscles so that terminal anchorage is not required. The details of the use of appendages in burrowing are recorded for few animals but the amphipod, *Talorchestia* (Reid, 1938) and the mole crab, *Emerita* (Trueman, 1970) (Fig. 8.12) are good examples.

219

8.8 The energy cost of burrowing

Quantitative comparisons between crawling and burrowing and between different burrowing animals may be made by determination of the energy requirement for locomotion expressed as J Kg^{-1} dry tissue wt.

Energy cost may be determined from oxygen consumption during locomotion but it is difficult to achieve satisfactory results using a respirometer. An alternative method, used in respect of the clam *Donax* spp. and mole crabs (*Emerita* spp.) (Ansell and Trueman, 1973), involves measurement of the rate of movement (*U*) and the drag (*D*), *DU* representing the power required for burial. Drag cannot be measured directly, but the maximum force exerted by an animal during burial can be readily ascertained by attaching a thread from the animal to a force transducer. This force must normally overcome drag and represents the maximal drag experienced. The snail, *Bullia digitalis*, has a digging period of 45 s duration for 10 cycles, a mean velocity of 6.7 × 10^{-4} m s^{-1} and develops a maximum tensile force of 0.17 N. The energy cost of burial is 6.6 J Kg^{-1} and by a similar calculation the requirement for locomotion over a sandy surface is about a tenth of this (Trueman and Brown, 1976). The energy cost of burial by the tropical clam, *Donax incarnatus*, of 22 J Kg^{-1} (Ansell and Trueman, 1973) is much greater than for *Bullia* and is probably related to its speed of burial for the digging period is only of 4–5 s duration.

Account may be taken of the efficiency of propulsion in *B. digitalis* which is likely to be of the order of 0.2 by analogy with previous calculations (Ansell and Trueman, 1973). Converted into joules the total energy for burial by this snail is thus 2.5 × 10^{-2} J over 45 s. When suspended in a respirometer with the foot expanded and making some movements *Bullia* has an energy requirement of 1.6 × 10^{-1} J over the same period. The cost of burial is only about one sixth of the latter and this suggests that, given the structural adaptations which enable burial to be effected, the cost is relatively small. This is especially so when the advantages of avoidance of prolonged exposure on a sandy beach are considered.

References

Ansell, A.D. and Trueman, E.R. (1973) The energy cost of migration of the bivalve *Donax* on tropical sandy beaches. *Mar. Behav. Physiol.* **2**, 21–32.
Chapman, G. (1950) Of the movement of worms. *J. exp. Biol.* **27**, 29–39.
Clark, R.B. (1964) *Dynamics in Metazoan Evolution.* Oxford: Clarendon Press.
Clark, R.B. and Clark, M.E. (1960) The ligamentary system and the segmental musculature of *Nephtys*. *Q. Jl. Microsc. Sci.* **101**, 149–176.
Elder, H.Y. (1972) Connective tissues and body wall structure, of the polychaete *Polyphysia crassa* (*Lipobranchius jeffreysii*) and their significance. *J. mar. biol. Ass. U.K.* **52**, 747–764.
Elder, H.Y. (1973) Direct peristaltic progression and the functional significance of the dermal connective tissues during burrowing in the polychaete *Polyphysia crassa* (Oersted). *J. exp. Biol.* **58**, 637–655.

Gray, J. (1968) *Animal Locomotion*. London: Weidenfeld & Nicolson.

Hammond, R.A. (1970) The burrowing of *Priapulus caudatus*. *J. Zool.*, *(Lond.)* **162**, 469–480.

Heddle, D. (1967) Versatility of movement and the origin of the Asteroids. *Symp. zool. Soc. Lond.* Millott, N., Ed., **20**, 125–142.

Hughes, G.M. and Mill, P.J. (1974) Locomotion: Terrestrial. In: *The Physiology of Insecta*, Rockstein, M., Ed., 2nd edn., **3**, Chapter 5. London: Academic Press.

Hyman, L.H. (1940) *The Invertebrates. I. Protozoa through Ctenophora*. New York: McGraw-Hill.

Hyman, L.H. (1959) *The Invertebrates. V. Smaller Coelomate Groups*. New York: McGraw-Hill.

Jones, H.D. (1973) The mechanism of locomotion of *Agriolimax reticulatus* (Mollusca: Gastropoda). *J. Zool.*, *(Lond.)* **171**, 489–498.

Jones, H.D. (1975) Locomotion. In: *Pulmonates*, Fretter, V. and Peake, J., Eds., **1**, Chapter 1. London: Academic Press.

Jones, H.D. and Trueman, E.R. (1970) Locomotion of the limpet, *Patella vulgata* L. *J. exp. Biol.* **52**, 201–216.

Lissmann, H.W. (1945) The mechanism of locomotion in gastropod molluscs. I. Kinematics. *J. exp. Biol.* **21**, 58–69.

Lissmann, H.W. (1946) The mechanism of locomotion in gastropod molluscs. II. Kinetics. *J. exp. Biol.* **22**, 37–50.

Miller, S.L. (1974a) Adaptive design of locomotion and foot form in prosobranch gastropods. *J. exp. Mar. Biol. Ecol.* **14**, 99–156.

Miller, S.L. (1974b) The classification, taxonomic distribution, and evolution of locomotor types among prosobranch gastropods. *Proc. Malacol. Soc. Lond.* **41**, 233–272.

Morton, J.E. (1964) Locomotion. In: *Physiology of Mollusca*, Wilbur, K.M. and Yonge, C.M., Eds., **1**, Chapter 12. London: Academic Press.

Nair, N.B. and Ansell, A.D. (1968) Characteristics of penetration of the substratum by some marine bivalve molluscs. *Proc. malac. Soc. Lond.* **38**, 179–197.

Pearl, R. (1903) The movements and reactions of fresh-water planarians: a study in animal behaviour. *Q. Jl. Microsc. Sci.* **46**, 509–714.

Reid, D.M. (1938) Burrowing methods of *Talorchestia deshayesii* (Audouin) (Crustacea, Amphipoda). *Ann. Mag. nat. Hist.* (II) **1**, 155–157.

Roberts, M.J. (1971) On the locomotion of cyclorrhaphan maggots (Diptera). *J. nat. Hist.* **5**, 583–590.

Seymour, M.K. (1969) Locomotion and coelomic pressure in *Lumbricus terrestris* L. *J. exp. Biol.* **51**, 47–58.

Seymour, M.K. (1970) Skeletons of *Lumbricus terrestris* L. and *Arenicola marina* (L.) *Nature, Lond.*, **228**, 383–385.

Seymour, M.K. (1971) Burrowing behaviour in the European lugworm *Arenicola marina* (Polychaete: Arenicolidae). *J. Zool.*, *(Lond.)* **164**, 93–132.

Trueman, E.R. (1970) The mechanism of burrowing of the mole crab, *Emerita. J. exp. Biol.* **53**, 701–710.

Trueman, E.R. (1975) *The locomotion of soft-bodied animals*. London: Edward Arnold.

Trueman, E.R. and Ansell, A.D. (1969) The mechanism of burrowing into soft substrata by marine animals. *Ocean. mar. Biol.* **7**, 315–366.

Trueman, E.R. and Brown, A.C. (1976) Locomotion, pedal retraction and extension, and the hydraulic systems of *Bullia* (Gastropoda: Nassaridae). *J. Zool.*, *(Lond.)* **178**, 365–384.

Trueman, E.R. and Foster-Smith, R.L. (1976) The mechanism of burrowing of *Sipunculus nudus* L. *J. Zool.*, *(Lond.)* **179**, 373–386.

Webb, J.E. (1969) Biologically significant properties of submerged marine sand. *Proc. R. Soc., Lond.* B, **174**, 355–402.

Webb, J.E. (1973) Swimming in *Amphioxus, J. Zool.*, *(Lond.)* **170**, 325–338.

Wilson, C.B. (1900) The habits and early development of *Cerebratulus lacteus* (Verrill). A contribution to physiological morphology. *Q. Jl. Microsc. Sci.* **43**, 97–198.

9 Swimming

R. McN. Alexander

9.1 Introduction

Just as Chapter 7 was mainly about mammals, this chapter is mainly about teleost fish. Their techniques of swimming present particularly challenging problems to zoologists and hydrodynamicists, and great advances have been made in understanding them in the past few years. Other swimmers such as worms, beetles and squids are considered more briefly. Protozoa are discussed in Chapter 11.

9.2 Basic hydrodynamics

Chapters 7 and 8 made extensive use of elementary statics, dynamics and hydrostatics. These subjects are generally taught in schools and it was assumed that they would be already familiar to most readers. This chapter and Chapter 10 (on flight) demand a knowledge of hydrodynamics which cannot be expected of the average reader. This section is designed as an introduction to the parts of hydrodynamics which will be needed. To achieve brevity, an approach has been adopted which is far from rigorous. Further information can be found in Alexander (1968) and in textbooks of hydrodynamics such as Prandtl and Tietjens (1957).

Viscosity is one of the properties of fluids which we have to consider (high viscosity is the property which makes treacle hard to stir). Consider a layer of fluid of thickness d, sandwiched between two parallel plates of area, S (Fig. 9.1a). The lower plate is fixed and the upper one moved parallel to it with velocity u. A force F (proportional to the velocity) is needed, to overcome the viscosity of the fluid. It is found that

$$F = \mu S u/d, \tag{9.1}$$

where μ is a constant (for a given fluid at a given temperature). It is known as the viscosity of the fluid and has values 1.8×10^{-5} N s m^{-2} and 1.0×10^{-3} N s m^{-2} for air and water, respectively, both at 20°C.

222

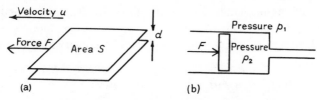

Fig. 9.1 These diagrams are explained in the text.

As well as forces due to viscosity we will have to consider forces required to accelerate or decelerate fluid. By Newton's Second Law of Motion the force F required to give a mass m an acceleration du/dt is given by

$$F = m.du/dt \qquad (9.2)$$

Since the momentum of a mass m travelling with velocity u is mu, equation 9.2 states that the force equals the rate of change of momentum.

Many animals considered in this chapter and Chapter 10 make use of this principle for propulsion. They obtain the thrust which propels them by giving the fluid backward momentum. Suppose an animal obtains the thrust F required to propel it with velocity u, by giving some of the fluid a backward velocity $-u'$. If the rate of change of momentum is to be F the mass of fluid accelerated in unit time must be F/u'. Hence the kinetic energy given to the fluid in unit time is $\frac{1}{2}(F/u')u'^2 = Fu'/2$. This is the power consumed in driving the fluid backwards and is, in a sense, wasted power. The power used usefully, driving the animal through the water, is Fu. Hence we can define an efficiency, known as the Froude efficiency

$$\text{Froude efficiency} = \frac{\text{useful power}}{\text{useful power} + \text{power lost to fluid}}$$
$$= u/(u + \tfrac{1}{2}u'). \qquad (9.3)$$

This efficiency is higher if the animal accelerates a large mass of fluid (per unit time) to a low velocity, than if it obtains the same thrust by accelerating a small mass to a high velocity.

Work is done on a fluid when it moves from a region of high pressure to one of low pressure, for instance when driven by a pump. Consider the pump shown in Fig. 9.1b. The piston of area S works against a pressure difference ($p_1 - p_2$) so the force F required to drive it is $S(p_1 - p_2)$. If a volume V of fluid is moved the piston must travel a distance V/S and the work which is done (force × distance) must be $V(p_1 - p_2)$.

This expression can be used to obtain a relationship between the pressure of a fluid and its velocity. Consider a chunk of fluid of mass m and density ρ in a situation where flow is steady (this means that though the velocity of a particular chunk of fluid may change the velocity at any particular location is constant). At time t_1 the chunk has velocity u_1 and pressure p_1 and is at a height h_1. At time t_2 these quantities are u_2, p_2 and h_2. Between t_1 and t_2 the kinetic energy of our chunk changes from $\frac{1}{2}mu_1{}^2$ to $\frac{1}{2}mu_2{}^2$. Its gravitational potential energy changes from mgh_1 to mgh_2. The volume of the chunk is m/ρ so the work done on it must be $(p_1 - p_2)m/\rho$. Hence by conservation of energy

$$(p_1 - p_2)m/\rho = \frac{1}{2}m(u_2{}^2 - u_1{}^2) + mg(h_2 - h_1),$$
$$u_1{}^2/2 + gh_1 + p_1/\rho = u_2{}^2/2 + gh_2 + p_2/\rho. \tag{9.4}$$

This is Bernoulli's equation. It has been assumed in the derivation that the fluid is incompressible so that there are no changes of elastic strain energy. Air and (to a much smaller extent) water are compressible, but errors involved in making the assumption are negligible in most problems in hydrodynamics, except those concerned with supersonic flight. Work done against viscosity is also ignored so the equation must not be applied in situations where viscosity is important: for instance it must not be applied to flow in a boundary layer (this term will be explained later).

The study of hydrodynamics is complicated by the way in which the pattern of flow of fluid around an object depends not only on the shape of the object but also on its size and velocity and the properties of the fluid. Hence equations which give a good account of the forces on a swimming eel, for instance, may be entirely inappropriate to a spermatozoon which swims by geometrically similar movements. We will often need to know in these chapters whether data obtained by engineers in entirely different contexts can be applied to animals, whether, for instance, data on aeroplane wings can be applied to the tail flukes of dolphins or the wings of insects.

Such questions can be answered by applying the theory of dynamical similarity (Duncan, 1953). The concept of geometrical similarity should be familiar to most readers: two triangles, for instance, are similar if their angles are the same so that the smaller could be made to match the larger simply by magnifying it. The concept of dynamical similarity is an extension of the same idea: two motions are dynamically similar if they can be matched by changes of size and time scale.

In Chapter 7 we considered mammals of height h running with mean velocity $\bar{u}$, taking strides of length λ. We found that mammals of different size tended to use similar gaits and to have equal values of λ/h, if they had equal values of $\bar{u}^2/gh$ (g is the acceleration of free fall). More generally, it can be shown that motions in which gravitational accelerations are important can only be dynamically similar if they have equal values of u^2/gl, where u is a velocity and l is a linear dimension. The velocities must be measured at corresponding stages in the two motions, and corresponding linear dimensions must be used. u^2/gl is a dimensionless number, known as the Froude number. It is important in discussions of the hydrodynamics of ships and of animals such as ducks which swim at the surface of the water because the resistance to motion is largely due to gravitational forces on the waves set up by the motion (see Prange and Schmidt-Nielsen, 1970).

When an animal of the same density as water swims under water, gravitational forces are of no importance but viscous forces are important. It can be shown that in such cases, two motions can only be dynamically similar if they have equal values of $\rho u l / \mu$, where ρ and μ are the density and viscosity of the fluid. $\rho u l / \mu$ is another dimensionless number and is known as the Reynolds number. l may be any characteristic length provided corresponding lengths are used for objects being compared but it is customary to use the overall length of a body, measured in the direction of motion.

The ratio ρ / μ is about 10^6 s m^{-2} for water and about 7×10^4 s m^{-2} for air at atmospheric pressure, both at temperatures around 20°C. For motion in water (which is the subject of this chapter) the Reynolds number is thus 10^6 (velocity in m s^{-1}) $\times$ (length in m).

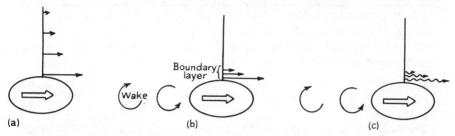

(a) (b) (c)

Fig. 9.2 Diagrams of a body moving to the right at three different Reynolds numbers Thin arrows represent velocities of the surrounding fluid.

When a body moves through a fluid the pattern of flow around it depends on the Reynolds number. Figure 9.2 shows the principal types of pattern. In Fig. 9.2a the fluid parts as the body approaches it and closes up smoothly behind it. Because of viscosity, the body carries fluid along with it. Fluid immediately in contact with the body has the same velocity as the body, and the velocity gradually diminishes with distance from the body. Fluid far in front of the body, far to either side of it and far behind it is at rest. In Fig. 9.2b the bulk of the fluid remains at rest: though the fluid in contact with the body moves with the velocity of the body, fluid a short distance away is almost stationary so there is only a thin boundary layer of moving fluid. Flow in this boundary layer is regular, and parallel to the surface of the body: this is described as laminar flow. There is also a wake of moving, swirling fluid behind the body. Figure 9.2c is like Fig. 9.2b, except that flow in the boundary layer is turbulent (irregular).

Flow patterns like the one shown in Fig. 9.2a occur at Reynolds numbers below 1. As the Reynolds number rises above 1 a gradual change occurs to the pattern shown in Fig. 9.2b and this pattern is maintained until the Reynolds number reaches a critical value between about 2×10^5 and 2×10^6 when it changes abruptly to the pattern shown in Fig. 9.2c. The critical value is higher for streamlined bodies such as torpedoes than for bluff (unstreamlined)

225

bodies such as spheres. This chapter is mainly concerned with flow patterns like Fig. 9.2b. The smallest and slowest swimmers to be discussed are copepods about 2 mm long which swim at around 5 cm s^{-1}, that is at Reynolds numbers around $10^6 \times 5 \times 10^{-2} \times 2 \times 10^{-3} = 100$. Smaller, slower swimmers with Reynolds numbers well below 1 are discussed in Chapter 11. A dolphin 2 m long swimming at 5 m s^{-1} (a moderate speed, Lang and Norris, 1966) would have a Reynolds number of $10^6 \times 5 \times 2 = 10^7$ so a turbulent boundary layer would be expected. However, for all but the largest and fastest fish Reynolds number remains below 10^6 and even in the case of the dolphin the boundary layer may remain laminar (see Section 9.4.1).

As fluid enters a boundary layer its velocity changes. The rate of change of momentum must equal the force acting on the fluid, which can also be calculated from considerations of viscosity. In this way the thickness of the boundary layer can be calculated. The precise value obtained depends on how the thickness is defined (for the outer limit of the boundary layer is not a sharp one). The usual definition gives a thickness d at a distance x from the front edge of a body moving with Reynolds number R

$$d = 3.4xR^{-1/2}. \tag{9.5}$$

The force which resists the motion of a body through a fluid is called drag. It acts backwards along the direction of motion. The amount of drag is given by the equation

$$\text{Drag} = \tfrac{1}{2}\rho S u^2 C_D, \tag{9.6}$$

where ρ and u have the same meanings as before, S is some area and C_D is a dimensionless number called the drag coefficient. The value of the drag coefficient depends on the shape of the body and on the Reynolds number. It also depends on what area S is used as the basis of its definition. Sometimes the wetted area S_w is used and sometimes the frontal area S_f which is the area of a full-scale front view of the body (this is generally the area of the largest cross-section). When the body is a hydrofoil or aerofoil, the plan area S_p (the area of a full-scale plan) is usually used.

Drag has two components, friction drag due to the viscosity of the boundary layer and pressure drag due to the changed momentum of the fluid in the wake. For a flat plate moving edge-on nearly all the drag is friction drag but for one moving at right angles to its surface the drag is much larger and nearly all of it is pressure drag. It can generally be assumed that the friction drag on a body will be $\tfrac{1}{2}\rho S_w u^2$ (1.3 $R^{-1/2}$), so the friction drag coefficient based on wetted area is $1.3R^{-1/2}$. This can be converted to a coefficient based on frontal area by multiplying by S_w/S_f. The pressure drag coefficient depends on the shape of the body but tends to be more or less constant for bodies of given shape over a wide range of Reynolds numbers (about 300 to 2×10^5). Based on frontal area, it is about 1.0 for long slender cylinders travelling at right angles to their length and 0.5 for spheres.

Streamlined bodies are bodies designed to leave as little disturbance as possible in their wake. They are torpedo-shaped, rounded in front and tapering gently behind (the taper is particularly important). A well streamlined body experiences friction drag but relatively little pressure drag. The drag coefficient based on frontal area is commonly less than 0.05 and the coefficient based on wetted area is little more than $1.3R^{-1/2}$.

Now consider an asymmetrical body moving through a fluid or a symmetrical one moving asymmetrically (for instance a plate tilted at an angle to its direction of motion). The force exerted by the fluid on such a body is not in general parallel to the direction of motion: it has both a drag component acting backwards along the direction of motion and a lift component at right angles

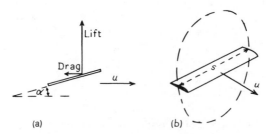

Fig. 9.3 Diagrams of hydrofoils, explained in the text.

to the direction of motion (Fig. 9.3a). Despite its name, lift does not necessarily act upwards. Hydrofoils and aerofoils are structures designed to give maximum lift for minimum drag and are used as aeroplane wings, propeller blades, etc. The lift and drag on a hydrofoil depend on its angle of attack (α, Fig. 9.3a) as well as on its speed. Lift is zero at some small value of α (typically when $\alpha = 0$) and increases as α increases until it reaches a maximum which often occurs when $\alpha \simeq 20°$. At greater angles of attack lift diminishes again due to the phenomenon known as stalling. The greater the angle of attack the greater the drag. For any given angle of attack, a lift coefficient C_L and a drag coefficient C_D (both based on plan area) are defined by the equations

$$\left.\begin{array}{l} \text{Lift} = \frac{1}{2}\rho S_p u^2 C_L. \\ \text{Drag} = \frac{1}{2}\rho S_p u^2 C_D. \end{array}\right\} \tag{9.6a}$$

The maximum attainable lift coefficient for ordinary hydrofoils is about 1.5 for Reynolds numbers around 10^6 and 1.0 for Reynolds numbers around 10^3. (The length used in calculating the Reynolds number of a hydrofoil is the chord, that is the distance from the front edge to the rear edge.)

Aerofoils and hydrofoils obtain lift by accelerating fluid at right angles to their path. Consider a hydrofoil of span s moving horizontally with velocity u (Fig. 9.3b). A vertical circle of diameter s, area $\pi s^2/4$ has been drawn through its tips. In unit time a mass $\rho \pi s^2 u/4$ of fluid passes through this circle. It can be shown that it is a reasonable approximation to assume that only this fluid is affected by the passage of the hydrofoil and that all of this fluid is given a downward velocity v. If so the fluid is given momentum at a rate $\rho \pi s^2 uv/4$. This must equal the lift, L, so v must equal $4L/\rho \pi s^2 u$. Also the rate at which the fluid is given kinetic energy is $\rho \pi s^2 uv^2/8 = 2L^2/\rho \pi s^2 u$. Some of the work done against drag in propelling the hydrofoil must be used to give this kinetic energy to the fluid. The part of the drag concerned is called the induced drag and since rate of working = (drag) × (velocity)

$$\text{Induced drag} = 2L^2/\rho \pi s^2 u^2. \tag{9.7}$$

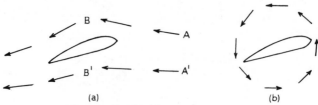

Fig. 9.4 Diagrams of circulation round a hydrofoil, in section.

The aspect ratio A of a hydrofoil is the ratio of its span to its mean chord (so it is high for a long slender wing). Thus $s^2 = S_p A$ and

$$\text{Induced drag} = 2L^2/\rho \pi S_p A u^2. \tag{9.7a}$$

Hence a hydrofoil of high aspect ratio can supply the same lift for less drag than one of the same plan area but lower aspect ratio.

Figure 9.4a shows a hydrofoil moving through a fluid from left to right. The arrows show the movements of the fluid relative to the hydrofoil. The pressure and velocity are the same at A and at A' but fluid from A passes over the hydrofoil while fluid from A' passes under it. If there is upward lift the pressure over the hydrofoil at B must be less than the pressure below it at B'. Hence by Bernoulli's equation (9.4) the velocity must be greater at B than at B'. The motion of the fluid relative to the hydrofoil can thus be repre-sented as a flow from right to left with an anticlockwise circulation (shown in Fig. 9.4b) superimposed on it. When a hydrofoil starts abruptly from rest the lift does not reach its full value until it has travelled about one chord length, because this distance is needed to build up the circulation.

It will be necessary to add a little to this rudimentary outline of hydrodyn-amics in Chapter 10.

9.3 Swimming by typical teleosts

9.3.1 Mechanics and energetics

This section of the chapter is about typical fish with fairly large caudal fins such as trout (*Salmo*). It is not concerned with tunnies (*Thunnus*) and similar fish which have stiff caudal fins in the form of hydrofoils of rather high aspect ratio: they are discussed in a later section.

Figure 9.5 shows a slender, rather eel-like teleost swimming. Waves of bending travel backwards along the body, increasing in amplitude towards the tail.

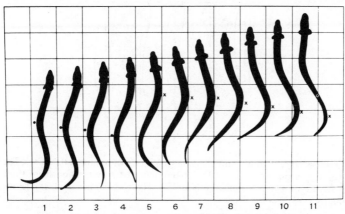

Fig. 9.5 Frames taken at intervals of 0.05 s, from a film of a butterfish (*Centronotus gunnelus*) swimming. The grid lines are 1 in (25 mm) apart and successive frames have been displaced laterally by one square (from *Animal Locomotion*, Sir James Gray, by permission of G. Weidenfeld & Nicolson Ltd., London, and W.W. Norton & Co. Inc., New York).

Figure 9.6 shows a trout swimming. There are similar waves of bending but they are less obvious than the side-to-side movements of the tail which they cause.

When a fish swims at constant velocity in a straight line its momentum is constant and the net momentum of the water must also be constant (by conservation of momentum). Any momentum given to the water in one direction must be balanced by equal momentum in the opposite direction. The tail drives water alternately to the left and to the right, giving the water equal momentum in each direction. The energy cost of swimming can be estimated by the following simple arguments which were devised by Sir James Lighthill (1969, 1970, 1971) and are presented much more rigorously in his papers of 1970 and 1971 than they are here.

Consider a fish swimming with velocity u, passing waves of bending posteriorly along its body with velocity c (Fig. 9.7a). At time t the posterior end of the body has a transverse component of velocity w and the water being left

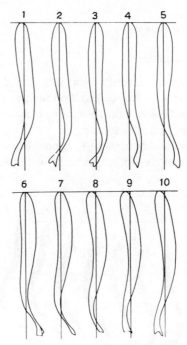

Fig. 9.6 Outlines traced from a film of *Salmo gairdneri* swimming at 0.5 m s⁻¹. The interval between successive positions was 0.03 s. (Reproduced by permission from the Fisheries Research Board of Canada Bulletin No. 190, Hydrodynamics and energetics of fish propulsion, by P.W. Webb.)

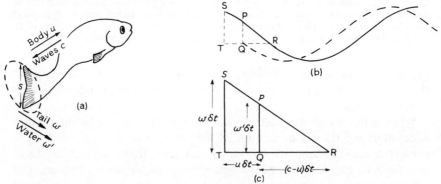

Fig. 9.7 Diagrams of a typical teleost swimming. Further explanation is given in the text.

behind by it has transverse velocity w'. It can be shown that it is a reasonable approximation to assume that this velocity is imparted to all the water which passes through the circle drawn round the caudal fin and that water which passes outside the circle is unaffected. The circle is in a transverse plane. Its diameter is equal to the span s of the caudal fin so its area is $\pi s^2/4$. The mass of water of density ρ which passes through it in unit time is $\rho\pi s^2 u/4$ so the rate at which momentum is being given to the water is $\rho\pi s^2 uw'/4$. By Newton's Second Law of Motion, this is the force the tail must exert. The power output of the tail is this transverse force multiplied by the transverse component of the velocity of the tail so

$$\text{Total power} = \rho\pi s^2 uw' w/4. \tag{9.8}$$

Mass $\rho\pi s^2 u/4$ moving with velocity w' has kinetic energy $\rho\pi s^2 uw'^2/8$ so power must be expended at this rate, giving kinetic energy to the water. This power is wasted, in the sense that it is not used to overcome the drag on the body. The remainder of the power is useful power, used against drag.

$$\text{Useful power} = (\rho\pi s^2 uw'/4)(w - \tfrac{1}{2}w'), \tag{9.9}$$

and

$$\text{Froude efficiency} = (w - \tfrac{1}{2}w')/w. \tag{9.10}$$

We need to evaluate w'. Figure 9.7b shows the median axis of our fish at time t and by a broken line, at the slightly later time $(t + \delta t)$. Water at Q which is just being left behind by the tail at time $(t + \delta t)$ was at P at time t. Hence the distance PQ is $w' . \delta t$. The tail has moved a transverse distance ST in time t so $\text{ST} = w . \delta t$. It has moved forward TQ so $\text{TQ} = u . \delta t$. At time $(t + \delta t)$ Q is at the stage in the wave motion which R was at a time t. Since the velocity of the waves is c relative to the fish or $(c - u)$ relative to the ground, $\text{QR} = (c - u)\delta t$. If δt is small enough RPS is a straight line and RPQ, RST are similar triangles (Fig. 9.7c). Hence

$$w' = w(c - u)/c,$$

and

$$\text{Total power} = \rho\pi s^2 uw^2(c - u)/4c, \tag{9.8a}$$

$$\text{Useful power} = \rho\pi s^2 uw^2(c^2 - u^2)/8c^2, \tag{9.9a}$$

$$\text{Froude efficiency} = (c + u)/2c. \tag{9.10a}$$

Equations 9.8a and 9.9a give power output at time t. The power output fluctuates over a cycle of swimming movements and we can get mean values by replacing w^2 by the mean value of w^2, $[\overline{w^2}]$. If the tail moves from side to side in simple harmonic motion of amplitude a and frequency n it moves $2a$ to the left and $2a$ to the right in time $1/n$ so the mean value of w is $4an$. However, $[\overline{w^2}]$ is not $16 a^2 n^2$ but rather greater, because of the large effect of the higher values of w. It can be shown by calculus that it is $2\pi^2 a^2 n^2$. Hence

$$\text{Mean total power} = \rho\pi^3 s^2 a^2 n^2 u(c - u)/2c \tag{9.8b}$$

$$\text{Mean useful power} = \rho\pi^3 s^2 a^2 n^2 u(c^2 - u^2)/4c^2 \tag{9.9b}$$

231

The drag on the body of the fish is $\frac{1}{2}\rho S_w u^2 C_D$ (Equation 9.6) where C_D is the drag coefficient based on the wetted area S_w. Fish have streamlined bodies so it would seem reasonable at first sight to expect C_D to be little more than $1.3R^{-1/2}$: $2R^{-1/2}$ seems likely. The power needed to overcome the drag is the drag multiplied by the velocity so

$$\text{Expected power requirement} \simeq \rho S_w u^3 R^{-1/2}. \tag{9.11}$$

These equations can be applied to real data. Webb (1971a) made observations specifically for this purpose on 28 cm Rainbow trout (*Salmo gairdneri*) swimming in a water tunnel. The fish were stationary, swimming against a

Table 9.1 Calculations of the power output of 28 cm Rainbow trout, *Salmo gairdneri*, swimming at various speeds. These fish had mass 0.22 kg and wetted area 3.1 × 10^{-2} m². Modified from Webb (1975)

Swimming velocity, u (m s^{-1})	0.10	0.21	0.34	0.52
Wave velocity, c (m s^{-1})	0.45	0.49	0.69	0.86
Span of caudal fin, s (cm)	4.2	4.8	5.6	6.1
Frequency of tail beat, n (s^{-1})	2.1	2.3	3.2	4.0
Amplitude of tail beat, a (cm)	1.1	2.0	1.9	2.1
Mean total power (W)	0.0011	0.0090	0.030	0.084
Mean useful power (W)	0.0007	0.0064	0.023	0.068
Froude efficiency	0.61	0.71	0.75	0.81
Reynolds number	28 000	59 000	95 000	146 000
Expected power requirement (W)	0.0001	0.0008	0.0025	0.0074

current, so u was the velocity of the water passing them in the tunnel. This is a little greater than the velocity the water would have if the fish were removed because the body of the fish occupies a proportion of the cross-section of the tunnel, making the water flow faster in the remainder of the section. The fish were filmed in side view and simultaneously in top view, using a mirror. Values of a and n were obtained from the top views and values of s from the side views. The wavelength of the waves of bending must equal c/n: it was measured from the top views and used to calculate c.

Data obtained in this way are shown in Table 9.1. The amplitude (a) of the tail beat was low at very low speeds but more or less constant at speeds of 0.2 m s^{-1} and over. The fish spread the caudal fin rather more at high speeds, increasing the span s. The wavelength of the waves of bending was about three-quarters of the length of the fish, at all speeds. The frequency n of the tail beat increased as speed increased so that san/u was very nearly constant.

Table 9.1 shows that the Froude efficiency ranged from about 0.6 to about 0.8. Equation 9.10a shows that the Froude efficiency must always lie between 0.5 and 1.0 (since u must be positive and less than c). Large values of s help to make the Froude efficiency high because they increase the mass of water accelerated in unit time (see p. 223).

The useful power calculated from Equation 9.9b might be expected to match the power requirement calculated from Equation 9.11, but Table 9.1 and Fig. 9.8 show that it is 7–9 times higher. Part of the discrepancy may be due to flow in the water tunnel being turbulent but this should not have much effect: flow in the boundary layer should nevertheless be laminar at the Reynolds

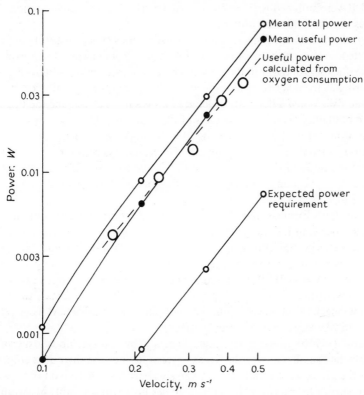

Fig. 9.8 Graphs of power against speed of swimming for 28 cm *Salmo gairdneri*. Data from Tables 9.1 and 9.2.

numbers in question. Part of the discrepancy may be because of energy losses due to the movements of the tail wagging the anterior part of the fish (Lighthill, 1970). These losses are minimized by the shape of the fish. The anterior parts of the body, which bend little, have a large area in side view and are not easily moved transversely through the water. The caudal peduncle is slender so that its transverse movements do not exert large forces.

These considerations can only explain a small part of the discrepancy between useful power and expected power requirement, shown by Table 9.1. Attempts have been made to check Equation 9.11 by measuring drag on the bodies of dead fish. For instance Webb (1975) attached dead Rainbow trout to a hydrodynamic balance in his water tunnel. He used fish of the same size as for the observations recorded in Table 9.1, and measured drag at the same range of speeds. The Reynolds numbers were of the order of 10^5 so drag coefficients (based on wetted area) could be expected to be around $2(10^5)^{-1/2} = 0.006$. However, the drag coefficients which were measured were much higher, probably because the fish fluttered in the stream of water like a flag. Even when the fish were stiffened with wires and fixed as rigidly as possible in the tunnel, values around 0.02 were obtained.

Lighthill (1971) has suggested that the discrepancy between useful power and expected power requirement may be due to an effect of swimming movements on the thickness of the boundary layer. The same effect could result from passive fluttering as from active swimming and may explain the high drag coefficients of dead fish. When a flat plate of length l moves in its own plane the thickness of the boundary layer (if laminar) is of the order of $3.4R^{-1/2}l$ (Equation 9.5). When it moves at the same velocity at right angles to its plane the boundary layer on its front face is much thinner. The transverse components of the motion of the body of the fish may thin the boundary layer. If it is reduced to a fraction q of the value for a rigid body the friction drag must (from Equation 9.1) be increased by a factor $1/q$.

Webb (1971b) measured the oxygen consumption of trout, swimming at various speeds in his water tunnel. Metabolism using 1 cm^3 oxygen releases chemical energy amounting to about 20 J, whatever food is being oxidized. If a fish uses M cm^3 oxygen s^{-1} at rest and N cm^3 s^{-1} when swimming it presumably needs $20(N - M)$ W for swimming at the speed in question. This is metabolic power (rate of consumption of chemical energy). To calculate the corresponding mechanical power it must be multiplied by the efficiency of conversion of chemical energy to mechanical work.

Webb (1971b) investigated this efficiency by an ingenious experiment. He attached small plates to the backs of trout, to increase their drag (Fig. 9.9). Since the plates were arranged transversely nearly all the drag on them would be pressure drag and would be unlikely to be affected appreciably by swimming movements. The extra drag caused by each plate at each speed was determined by experiments with the plates attached to model fish. Table 9.2 shows measured rates of oxygen consumption of trout swimming with and without the plates. The extra mechanical power needed to propel a plate is the drag on the plate multiplied by the velocity. The extra metabolic power can be calculated from the difference between the rates of oxygen consumption swimming with and without the plate at the same speed. Hence the efficiency can be calculated. Table 9.2 shows that it increases gradually from about 0.06 at 0.17 m s^{-1} to about 0.12 at 0.45 m s^{-1}. This is the efficiency of conversion of chemical energy to useful work. The Froude efficiency was probably in the region of 0.75 in each case (see Table 9.1) so the efficiency of conversion of chemical energy to work of any sort (useful or not) was probably about $0.06/0.75 = 0.08$ at 0.17 m s^{-1} and 0.16 at 0.45 m s^{-1}. These values may be compared to the

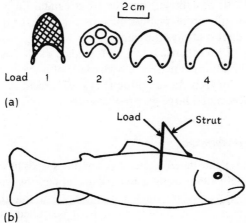

Fig. 9.9 (a) The plates which Webb attached to trout to increase their drag. (b) A trout with a plate attached (from Webb, 1971a).

efficiency of working of men on bicycle ergometers which is about 0.22 at the optimum pedalling frequency and lower at other frequencies (Dickinson, 1929).

The broken line in Fig. 9.8 shows the mechanical power requirement for swimming, calculated from Webb's (1971b) measurements of oxygen consumption assuming the efficiencies indicated by Table 9.2. The values are reasonably similar to the values of useful power calculated from Equation 9.9b.

Table 9.2 Calculations of the efficiency of conversion of chemical energy to work against drag by 28 cm Rainbow trout, *Salmo gairdneri*. Extra drag was imposed by attached plates (Fig. 9.9) and metabolic power was calculated from measurements of oxygen consumption. Data from Webb (1971a, b)

Swimming velocity (m s^{-1})	Extra drag (N)	Extra mechanical power (W)	Metabolic power (W)	Extra metabolic power (W)	Efficiency
0.17	—	—	0.17	—	—
	0.031	0.005	0.26	0.09	0.06
0.24	—	—	0.22	—	—
	0.060	0.014	0.40	0.18	0.08
0.31	—	—	0.27	—	—
	0.033	0.010	0.38	0.11	0.09
	0.067	0.021	0.56	0.29	0.07
0.38	—	—	0.33	—	—
	0.030	0.011	0.41	0.08	0.14
	0.052	0.020	0.53	0.20	0.10
0.45	—	—	0.40	—	—
	0.040	0.018	0.55	0.15	0.12

McCutchen (1970) has shown that the trout caudal fin is constructed in such a way that it will automatically adjust its curvature so that the distribution of forces along it is always the same.

The swimming of no other fish has been studied as thoroughly as that of trout but the observations which have been made of movements (Bainbridge, 1958, 1963) and of oxygen consumption (Fry, 1971) seem to show that trout are reasonably typical of a large group of teleosts.

9.3.2 Swimming performance

Some of the most reliable data on teleost swimming speeds have been obtained by Bainbridge (1960, 1962) using a remarkable apparatus which he called a fish wheel. This was an annular trough of diameter 2.3 m which could be rotated by a powerful motor about its (vertical) axis. It was filled with water and a fish swam in it. As it swam the wheel was rotated at whatever speed was necessary to keep the fish stationary. Gates across the trough forced the water to revolve at the same velocity as the trough but each gate opened automatically as it approached the fish to allow the fish through. A speed meter was attached to the trough and connected to an oscilloscope so that if the experimenter kept the fish stationary he obtained a continuous record of its speed.

Bainbridge (1960, 1962) studied Rainbow trout, dace (*Leuciscus*) and goldfish (*Carassius*). He found that specimens 4–30 cm long could reach speeds around 10 body lengths per second (Fig. 9.10). Such speeds were only maintained for a second or so, and bursts of swimming lasting 20 s or more were seldom faster than 4 body lengths per second. Similar speeds (measured in lengths per second) have been recorded for various other species (Blaxter, 1969; Webb, 1975). Most of the data concerns fish less than 30 cm long but *Onchorhynchus* 50–100 cm long can exceed 6 m s^{-1} briefly.

Bainbridge (1960, 1962) found that 28 cm trout had top speeds around 2.8 m s^{-1}. At this speed the amplitude of the tail beat was 2 cm and the frequency was 18 s^{-1}. It will be assumed that the caudal fin was fully spread so that its span was 6 cm and that the wavelength of the waves of bending was three-quarters of the length of the fish, as at lower speeds. If so, by Equation 9.8b, the total mechanical power output was about 6 W. The mass of a 28 cm trout is about 0.22 kg of which about 0.13 kg is swimming muscle so the power output is about 50 W (kg muscle)$^{-1}$. This is only about one-fifth of the power output obtainable from pigeon wing muscles (Alexander, 1973; Weis-Fogh and Alexander, 1977), but pigeon muscles operate at about 40°C while trout muscles operate at the temperature of the water which in Bainbridge's (1962) experiments was only about 14°C.

Bainbridge found that though 28 cm trout could swim at 2.8 m s^{-1} for one second, five second bursts of swimming were seldom faster than about 1.5 m s^{-1}. A 1 s burst of swimming at 2.8 m s^{-1}, involving a power output of 6 W, would use 6 J. Similarly it can be calculated from Bainbridge's data that the power output at 1.5 m s⁻¹ would be about 1.4 W so that a 5 s burst would use 7 J. This suggests that the duration of short bursts may be limited by the stock of ATP and creatine phosphate, which can only be replenished relatively slowly. Men cannot maintain their top sprinting speed for more than 100 m or so because ATP and creatine phosphate are depleted (Margaria, 1968).

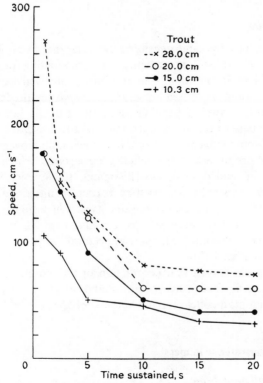

Fig. 9.10 The maximum speeds sustained for different periods of time by four Rainbow trout (*Salmo gairdneri*) (from Bainbridge, 1960).

Trout muscle contains 18 μmol creatine phosphate g⁻¹ (Goldspink, personal communication) which is several times too much for it to be the limiting factor. Data on the mechanical work obtainable from high-energy phosphate is given by Goldspink (1975).

Fish can be made to swim at constant speed in a water tunnel until they are exhausted. Thus it is possible to determine the maximum speed which can be maintained for a long period (for instance an hour). This is presumably the maximum speed which can be maintained by aerobic metabolism. For Webb's (1971b) trout it was 0.6 m s⁻¹ (2 body lengths s⁻¹) but small *Onchorhynchus* can maintain 4 body lengths s⁻¹ (Brett, 1964). Such speeds are probably maintained largely or entirely by activity of the red muscle fibres (see Chapter 1). The power output of Webb's trout at 0.6 m s^{-1} was about 0.15 W (Fig. 9.8), which is only about 7 W (kg red muscle)$^{-1}$.

Several species of teleost are known to be capable of accelerations over 40 m s^{-2} when starting from rest (Webb, 1975b). Weihs (1972, 1973) has discussed the mechanics of acceleration from rest and of turning.

9.3.3 Use of fins

Teleosts which have the same density as the water they live in make much use of their fins (without trunk movements) for compensating for errors of buoyancy and for slow locomotion. The movements involved have been studied by Harris (1953) and (in the sea horse, *Hippocampus*) by Breder and Edgerton (1942). Typically, waves of bending travel along the fin generating a thrust parallel to the base of the fin, in the same way as forces are generated by undulatory swimming movements. For a few teleosts, fin movement is the principal means of propulsion. They include *Hippocampus*, which uses its dorsal and pectoral fins and *Gymnarchus* (Lissmann, 1961) which uses the dorsal fin alone. *Cymatogaster* propels itself by its pectoral fins alone at speeds up to 3.4 body lengths s^{-1} but uses tail movements as well at higher speeds (Webb, 1973). Fins, particularly pectoral fins, are also used for braking. *Taurulus* has been filmed decelerating with spread pectoral fins, with a deceleration of 15 m s^{-1} (Alexander, 1970).

Arita (1971) has shown that contraction of the muscles of one face of a fin tends to make the fin concave towards that face so that a fin used in braking, for instance, adopts a concave, parachute-like shape.

9.4 Other underwater swimmers

9.4.1 Hydrofoil propulsion

There is a continuous gradation of body form between typical teleosts and the tunnies, which have stiff tails shaped as hydrofoils mounted on very slender caudal peduncles (Fig. 9.11). The theory of swimming presented for the trout is not appropriate to the tunny because of the abrupt transition between the slender caudal peduncle and the tall caudal fin (Lighthill, 1970).

Figure 9.12 shows how a tunny swims. The caudal fin, moving from side to side takes a sinuous path through the water. Its angle of attack is adjusted so that the resultant force on it has a forward component, whether it is moving to the left or to the right (Fierstine and Walters, 1968). The propulsive force is provided by the lift on the fin but work is done against the drag so it is advantageous to have as high a ratio of lift to drag as possible. The high aspect ratio of the caudal fin helps to achieve this by keeping induced drag reasonably low (Equation 9.7a). The caudal fin travels about 6 times its average chord

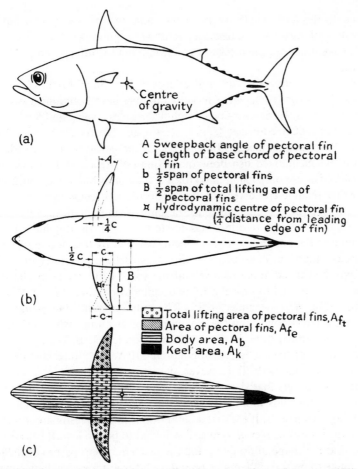

A Sweepback angle of pectoral fin
c Length of base chord of pectoral fin
b ½ span of pectoral fins
B ½ span of total lifting area of pectoral fins
¤ Hydrodynamic centre of pectoral fin (¼ distance from leading edge of fin)

▢ Total lifting area of pectoral fins, A_{f_t}
▨ Area of pectoral fins, A_{f_e}
▤ Body area, A_b
■ Keel area, A_k

Fig. 9.11 A wavy skipjack *Euthynnus affinis*, 45 cm long (from Magnuson, 1970).

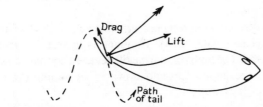

Fig. 9.12 A diagram of a tunny swimming, seen from above.

length in each half cycle of swimming movements (estimated from data in Fierstine and Walters, 1968). This is enough for the generation of circulation about the hydrofoil. Anticlockwise circulation (as seen from above) is required while the tail is moving to the right and clockwise circulation while it is

239

moving to the left. Whales swim in the same manner as tunnies but have horizontal tail flukes which beat up and down. Lighthill (1969, 1970) and Wu (1971) have discussed the mechanics of swimming by tunnies and similar fish, and whales.

These animals are generally large and swim fast. Small tunnies (*Euthynnus pelamis* and *Thunnus albacares*) about 0.5 m long have been filmed swimming at 6 m s^{-1} (Yuen, 1966). A trained dolphin (*Tursiops truncatus*) 1.9 m long was able to swim a 60 m course at 8.3 m s^{-1} (Lang and Norris, 1966). Speeds up to 21 m s^{-1} have been claimed for *Thunnus* and *Acanthocybium* from records of the speed at which hooked fish drew out a line (Walters and Fierstine, 1964). These speeds were only maintained for a few seconds but are so much higher than other reliable speed records that confirmation seems desirable. Speeds up to 13 m s^{-1} have been claimed for dolphins swimming near ships but may not be reliable because of difficulties due to parallax if the dolphin is far from the ship and the possibility that it is being assisted by the motion of the ship if it is near the ship. A dolphin which places itself appropriately in the bow wave will be carried along passively at the speed of the ship (Hertel, 1969).

The power needed to propel a 1.9 m dolphin swimming at 8.3 m s^{-1} would be about 1700 W (Alexander, 1975). The dolphin could only maintain this speed for a few seconds and the maximum power output which athletes can maintain for similar times is about 1400 W (Wilkie, 1960). The power output per unit mass of muscle is probably about 110 W kg^{-1} in each case (Weis-Fogh and Alexander, 1977). However, the estimate for the dolphin was based on the assumptions that the boundary layer was turbulent (as would be expected at a Reynolds number of 1.6×10^7) and that it was not thinned by swimming movements. If the first assumption is false the estimate will tend to be too high but if the second is false it will tend to be too low. It has been suggested that the shape of dolphins and the mechanical properties of their skin are such as to preserve a laminar boundary layer at unusually high Reynolds numbers (Kramer, 1965) but calculations of drag based on a film of a dolphin gliding to a halt gave values consistent with a fully turbulent boundary layer (Lang and Pryor, 1966).

The Manta ray (*Mobula*) has long pectoral fins and swims by flapping them like birds' wings (Klausewitz, 1964). They act as hydrofoils, generating lift which propels the ray. The limbs of marine turtles are also shaped as hydrofoils and the forelimbs are flapped up and down in swimming (Walker, 1971).

9.4.2 Worms

Two types of force act on an animal undulating in water. There are resistive forces due to the velocity of parts of the body: lift and drag as represented by Equations 9.6 are forces of this sort. There are also reactive forces due to changes of velocity, which impose accelerations on the nearby water. Reactive forces

240

predominate on the large caudal fins which propel typical fish, as was assumed in Section 9.3.1. They are relatively unimportant in the swimming of worms and snakes which have no caudal fin, especially if the Reynolds number is low. Only resistive forces are considered in this section.

Snakes swim by undulatory movements like those of slender fish (Fig. 9.5). Leeches also swim by undulation but bend their bodies dorsoventrally, not laterally. Taylor (1952) devised a theory of swimming applicable to both. He assumed that the force acting on a short length of the body at any instant would be the same as the force on an equal-sized piece of a long straight cylinder, moving at the same velocity relative to the water. Let waves of amplitude a and wavelength λ travel along the body with velocity c, propelling it at velocity u. Taylor obtained equations from which he was able to calculate the value of u/c which would be obtained, for given values of a/λ and Reynolds number. He analysed film of a snake and a leech swimming and found reasonable agreement with his theory. He also showed that for any Reynolds number there was a value of a/λ which minimized the power needed for swimming at given speed. The snake had about the ideal value of a/λ but the leech had a value which was only 0.7 of the ideal one.

Taylor (1952) went on to show that whereas a smooth worm must send waves backwards along its body to propel itself forward, a sufficiently rough one would have to send the waves forward. This seemed to explain how polychaetes with paddle-like parapodia projecting from their bodies swim forward by forward-moving waves. However, the parapodia are not rigidly fixed to the body so they do not function simply as 'roughness' as supposed in the theory. They swing backward while on the outside of a bend and forward again while on the inside, which helps to propel the body (Clark and Tritton, 1970).

9.4.3 Rowing

Hydrofoil propulsion (Section 9.4.1) uses lift on the hydrofoil to provide thrust. Rowing uses drag on the oars, which are moved backwards through the water. The most thorough study of rowing by animals is Nachtigall's (1960, 1965) study of water beetles. Most of the larger water beetles row with their middle and hind legs. Figure 9.13 shows the action of the hind legs of *Acilius*. These legs have long hairs which are hinged to them in such a way as to spread automatically in the power stroke (stages 1 to 4) and trail behind in the recovery stroke (5 to 8). The leg also twists in the recovery stroke so as to present a narrow edge to the water. In the power stroke the spread hairs form a large oar which can move a relatively large mass of water, making a high Froude efficiency possible. In the recovery stroke the configuration of the limb reduces to a minimum the drag which acts on it. Nachtigall took films of *Acilius* swimming and measured the drag on the trunk and on individual limbs in a water tunnel. He calculated that the useful power, used against drag on the trunk, was 0.45

241

of the total power output. The drag coefficient of the trunk based on frontal area was 0.23, but a well-streamlined body at the same Reynolds number would probably have a drag coefficient less than 0.1.

Morgan (1971) has studied the mechanics of swimming by pycnogonids, which propel themselves by movements of slender cylindrical limbs. Many Crustacea also swim by a rowing action (Lochhead, 1961). Vlymen (1970) has calculated the drag coefficient of the body of a copepod and the energy cost of locomotion, from observations of the deceleration between strokes.

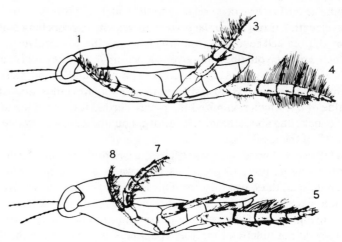

Fig. 9.13 Lateral views of *Acilius sulcatus* swimming, showing successive positions (numbered) of the left hind leg. Other legs have been omitted (from Nachtigall, 1960).

9.4.4 Jet propulsion

Cephalopods swim by jet propulsion, by squirting water from the mantle cavity (Trueman, 1975). Among them squid such as *Loligo* are particularly fast. *Loligo* of mass 100 g can accelerate from rest to 2 m s^{-1} by a single squirt, and maximum speeds of 8 m s^{-1} have been claimed for other squids. Squid travel backwards when travelling fast and look well-streamlined but Packard's (1969) measurements from cine films of *Loligo* gliding to a halt after a burst of speed indicate a rather high drag coefficient (based on frontal area) of about 0.3, for a 3 g animal.

The Froude efficiency of squid jet propulsion must be low. Compare a 220 g *Loligo* with the trout of the same mass considered in Section 9.3. This squid would have a mantle cavity capacity of about 130 cm^3 and could probably emit jets at a frequency of 1 s^{-1} (Trueman and Packard, 1968), so it could give backward momentum to about 130 cm^3 water s^{-1}. The trout gives momentum to a volume $\pi s^2 u/4$ of water in unit time: at a swimming speed of 0.5 m s^{-1}, this is 1400 cm^3 water s^{-1}. The squid must accelerate much less water than the trout to a higher velocity, to obtain the same thrust. Hence its Froude efficiency must be lower.

Trueman and Packard (1968) recorded pressures in the mantle cavities of tethered cephalopods. They recorded pressures up to 3×10^4 N m^{-2} above ambient in *Loligo* during jetting and calculated that this implied a stress of 1.5×10^5 N m^{-2} in the circular muscle fibres of the mantle. These muscle fibres shorten by about 30% in jetting. Ward (1972) has shown that they are probably re-extended, and the mantle cavity re-filled, by contraction of radial fibres which are interspersed among them.

Siekmann (1963) has discussed the hydrodynamics of squid jet propulsion. Trueman (1975) has reviewed the use of jet propulsion by animals other than cephalopods.

9.5 Buoyancy

The densities of fish which have no adaptations to give them buoyancy generally lie between 1060 and 1090 kg m^{-3}. They are thus considerably denser than either freshwater or seawater (density 1026 kg m^{-3}). Many of them spend much of their time on the bottom, where excess density helps to keep them in position: if the fish were not denser than the water, there would be no friction between it and the bottom. When a fish swims it must generate an upward hydrodynamic force unless its density is reduced to match that of the water. Most fish which swim perpetually have densities very close to that of the water due to a gas-filled swimbladder (in many teleosts; Alexander, 1966), or to low-density organic compounds such as squalene (in some selachians, Bone and Roberts, 1969; Corner, Denton and Forster, 1969) or wax esters (in Myctophidae and the coelacanth *Latimeria*, references in Alexander, 1972). However, some fish which swim perpetually are much denser than the water. They include many tunnies such as *Euthynnus affinis* (Fig. 9.11) which has a density of 1086 kg m^{-3} (Magnuson, 1970).

Sharks and tunnies which are denser than the water have pectoral fins shaped as hydrofoils and swim with them spread at a positive angle of attack, to generate upward lift. They depend largely on this lift to prevent them from sinking but since the pectoral fins are anterior to the centre of mass (Fig. 9.11) they cannot achieve equilibrium unless some of the lift is provided by hydrofoils posterior to the centre of mass. (This would not necessarily be true if the anterior parts of the fish were denser than the posterior parts so that the centre of mass was anterior to the centre of buoyancy.) Upward lift is provided by a heterocercal tail in selachians (Alexander, 1965; Simons, 1970) and possibly by the caudal peduncle in tunnies (Magnuson, 1970) as well as by the pectoral fins.

The pectoral fins cannot supply the required lift below a certain minimum speed. This speed can be calculated from Equation 9.6a, assuming a plausible maximum lift coefficient. Note that the plan area S_p which must be used is not the area of the fins alone, but the 'total lifting area' shown in Fig. 9.11 (see for instance Houghton and Brock, 1960). *Euthynnus affinis* has a high density and relatively small pectoral fins and was not observed by Magnuson (1973) to swim slower than 2 body lengths s^{-1}. *Thunnus obesus* is less dense and has larger pectoral fins and swims at speeds down to 1 length s^{-1}. Magnuson calculated that if the maximum lift coefficient of the pectoral fins was 0.8 *Euthynnus* should be able to swim at speeds down to 1.6 lengths s^{-1} and *Thunnus* down to 0.7 lengths s^{-1}. Fish which have the same density as the water have no minimum speed but can remain stationary in mid-water.

The relative merits of different means of avoiding sinking have been discussed (Alexander, 1972). It will be convenient to take as an example a 28 cm, 220 g fish as in Section 9.3. Suppose that, without buoyancy aids, it is 5 % denser than the water so that it needs about 0.1 N (10 g wt) hydrodynamic lift. A fish of that size would probably have pectoral fins of chord about 2 cm and would be likely to swim at speeds of the order of 0.5 m s^{-1}. Hence the Reynolds number of the pectoral fins would be of the order of $10^6 \times 0.5 \times 0.02 = 10^4$. At such Reynolds numbers the drag on a hydrofoil is likely to be at least 0.1 of the lift (see for instance the graph for a locust wing in Fig. 10.1) so the extra drag incurred by the fish, getting the lift it needed, would be at least 0.01 N. (If the lift were not required the fins could be held flat against the trunk and would add virtually nothing to the drag on the trunk.) If instead of using hydrofoils the fish had squalene or wax esters, the quantity required to match its density to that of the water would increase its volume by about 32 % and its wetted area by a factor of about $1.32^{0.67} = 1.20$. The drag on the fish at any swimming speed would be increased by 20 %. A swimbladder would be much smaller (because gas is much less dense than squalene or wax esters) and would only increase the drag by about 4 %.

The actual drag on a 28 cm trout can be estimated from Table 9.1, by dividing useful power by velocity. It has been used as the basis for the estimates of additional drag shown in Table 9.3. The means of avoiding sinking which imposes least additional drag will be the most economical of energy. It appears from the table that squalene or wax esters would be superior to hydrofoils at speeds below about 0.3 m s^{-1}, but not at higher speeds. A swimbladder is better than either throughout the range of speeds which is shown but would probably become inferior to hydrofoils at rather higher speeds.

Many tunnies are much larger than the fish we have been considering but the 31 cm Bullet mackerel (*Auxis rochei*) studied by Magnuson (1973) were about the same size. They had no swimbladder, and their density was 1086 kg m^{-3}. They swam continuously at an average speed of 0.7 m s^{-1} which may have been fast enough for hydrofoil lift to be more economical than a swimbladder. Many fish of similar size which have swimbladders spend a lot of time stationary but some including trout and dace swim more or less continuously against the current in rivers. The currents are probably generally a good deal slower than 0.5 m s^{-1}. Evolution has not offered selachians the option of a swimbladder but the ones which have squalene include *Cetorhinus*, the Basking shark, which swims extremely slowly.

The argument so far indicates that a swimbladder is likely to be the most economical means of preventing sinking, for all but the fastest teleosts. However, the energy cost of keeping the swimbladder inflated has been ignored. The pressure of the gas in the swimbladder of a submerged fish must exceed one atmosphere (and will be several hundred atmospheres at great depths). The total of the partial pressures of dissolved gases in the water will not exceed one atmosphere. Gases will therefore diffuse out of the swimbladder and must be replaced by a process of secretion which will require energy. Rough calculations (Alexander, 1972) lead to the conclusion that the swimbladder would

Table 9.3 Estimates of the cost in terms of drag, for a 28 cm fish, of three different means of avoiding sinking. The fish will incur the basic drag plus one of the three alternatives

Swimming speed (m s^{-1})	0.10	0.34	0.52
Basic drag (N)	0.007	0.07	0.13
Additional drag (N) due to:			
(a) Hydrofoils	0.01	0.01	0.01
(b) Squalene or wax esters	0.0014	0.014	0.03
(c) Swimbladder	0.0003	0.003	0.005

probably lose its advantage for small fish living at great depths (e.g. for 1 g fish living at 1000 m). There are fish with swimbladders at depths below 3500 m but at least most of them are moderately large.

The Myctophidae are small oceanic teleosts. Many of them spend the night near the surface and the day at depths around 300 m. Some of them have swimbladders which give them about the same density as the water by night, but must be collapsed by the high pressure at their daytime depths (it is extremely unlikely that they can secrete gas fast enough to keep their swimbladders distended). During the day, they must rely on hydrodynamic lift to avoid sinking further. In such circumstances the swimbladder may lose its advantage over wax esters. Some myctophids have no swimbladder gas but contain enough wax esters to give them about the same density as seawater (Alexander, 1972).

Many of the cephalopods are pelagic and many of them have densities about equal to that of the seawater they live in (Denton, 1974). Some owe their buoyancy to rigid chambered shells from which water is withdrawn osmotically, leaving spaces filled with gases at pressures below atmospheric. Such shells cannot withstand the high pressures which occur at great depths. The cuttlebone of *Sepia* would implode at about 200 m, and the shell of *Spirula* at about 1700 m. Many deep sea squid have their density matched to seawater by very large volumes of an ammoniacal solution of density 1010 kg m^{-3}. However,

Loligo and fast-swimming squid which live near the surface have no buoyancy aids, and are considerably denser than seawater. They must rely on hydrodynamic lift to avoid sinking. This may be advantageous for the same reason as in fast pelagic teleosts. *Octopus* spend most of their time on the bottom and are also denser than seawater.

References

Alexander, R.McN. (1965) The lift produced by the heterocercal tails of Selachii. *J. exp. Biol.* **43,** 131–138.

Alexander, R.McN. (1966) Physical aspects of swimbladder function. *Biol. Rev.* **41,** 141–176.

Alexander, R.McN. (1968) *Animal Mechanics,* Sidgwick and Jackson, London.

Alexander, R.McN. (1970) Mechanics of the feeding action of various teleost fishes. *J. Zool., (Lond.)* **162,** 145–156.

Alexander, R.McN. (1972) The energetics of vertical migration by fishes. *Symp. Soc. exp. Biol.* **26,** 273–294.

Alexander, R.McN. (1973) Muscle performance in locomotion and other strenuous activities. In: *Comparative Animal Physiology,* Bolis, L., Maddrell, S.H.P. and Schmidt-Nielsen, K., Eds., pp. 1–22. North Holland, Amsterdam.

Alexander, R.McN. (1975) *The Chordates,* Cambridge University Press, London.

Arita, G.S. (1971) A re-examination of the functional morphology of the soft rays in teleosts. *Copeia,* **1971,** 691–697.

Bainbridge, R. (1958) The speed of swimming of fish as related to size and to the frequency and amplitude of the tail beat. *J. exp. Biol.* **35,** 109–133.

Bainbridge, R. (1960) Speed and stamina in three fish. *J. exp. Biol.* **37,** 129–153.

Bainbridge, R. (1962) Training, speed and stamina in trout. *J. exp. Biol.* **39,** 537–555.

Bainbridge, R. (1963) Caudal fin and body movements in the propulsion of some fish. *J. exp. Biol.* **40,** 23–56.

Blaxter, J.H.S. (1969) Swimming speeds of fish. *FAO Fisheries Report* **62,** Vol. 2, 69–100.

Bone, Q. and Roberts, B.L. (1969) The density of elasmobranchs. *J. mar. biol. Ass. U.K.* **49,** 913–938.

Breder, C.M. and Edgerton, H.E. (1942) An analysis of the locomotion of the seahorse, *Hippocampus,* by means of high speed cinematography, *Ann. N.Y. Acad. Sci.,* **43,** 145–172.

Brett, J.R. (1964) The respiratory metabolism and swimming performance of young sockeye salmon. *J. Fish. Res. Board Can.* **21,** 1183–1226.

Clark, R.B. and Tritton, D.J. (1970) Swimming mechanisms in nereidiform polychaetes. *J. Zool., (Lond.)* **161,** 257–271.

Corner, E.D.S., Denton, E.J. and Forster, G.R. (1969) On the buoyancy of some deep-sea sharks. *Proc. R. Soc., Ser. B,* **171,** 415–429.

Denton, E.J. (1974) On buoyancy and the lives of modern and fossil cephalopods. *Proc. R. Soc., Ser. B,* **185,** 273–299.

Dickinson, S. (1929) The efficiency of bicycle-pedalling, as affected by speed and load. *J. Physiol.* **67,** 242–255.

Duncan, W.J. (1953) *Physical Similarity and Dimensional Analysis,* Edward Arnold, London.

Fierstine, H.L. and Walters, V. (1968) Studies of locomotion and anatomy of scombroid fishes. *Mem. S. Calif. Acad. Sci.* **6,** 1–31.

Fry, F.E.J. (1971) The effect of environmental factors on the physiology of fish. In: *Fish Physiology,* Hoar, W.S. and Randall, D.J., Eds., **6,** 1–98. Academic Press, New York.

Goldspink, G. (1975) Biochemical energetics of fast and slow muscles. In: *Comparative Physiology: Functional Aspects of Structural Materials,* Bolis, L., Maddrell, S.H.P. and Schmidt-Nielsen, K., Eds., North Holland, Amsterdam.

Gray, J. (1968) *Animal Locomotion,* Weidenfeld and Nicolson, London.

Harris, J.E. (1953) Fin patterns and mode of life in fishes. In: *Essays in Marine Biology,* Marshall, S.M. and Orr, P., Eds., Oliver and Boyd, Edinburgh.

246

Hertel, H. (1969) Hydrodynamics of swimming and wave-riding dolphins. In: *The Biology of Marine Mammals*, Andersen, H.T., Ed., Academic Press, New York.

Houghton, E.L. and Brock, A.E. (1960) *Aerodynamics for Engineering Students*. Edward Arnold, London.

Klausewitz, W. (1964) Der Lokomotionsmodus der Flugelrochen (Myliobatoidei). *Zool. Anz.* **173**(2), 111–120.

Kramer, M.O. (1965). Hydrodynamics of the dolphin. *Adv. Hydrosci.* **2**, 111–130.

Lang, T.G. and Norris, K.S. (1966) Swimming speed of a Pacific bottlenose porpoise. *Science*, **151**, 588–590.

Lang, T.G. and Pryor, K. (1966) Hydrodynamic performance of porpoises (*Stenella attenuata*). *Science* **152**, 531–533.

Lighthill, M.J. (1969) Hydromechanics of aquatic animal propulsion. *Ann. Rev. Fluid Mech.* **1**, 413–446.

Lighthill, M.J. (1970) Aquatic animal propulsion of high hydromechanical efficiency. *J. Fluid Mech.* **44**, 265–301.

Lighthill, J.J. (1971) Large-amplitude elongated-body theory of fish locomotion. *Proc. R. Soc. Lond. Ser. B*, **179**, 125–138.

Lissmann, H.W. (1961) Zoology, locomotory adaptations and the problem of electric fish. In: *The Cell and the Organism*. Ramsay, J.A. and Wigglesworth, V.B., Eds., pp. 301–307, Cambridge University Press, Cambridge.

Lochhead, J.H. (1961) Locomotion. In: *The Physiology of Crustacea*, Waterman, T.H., Ed., **2**, 313–364, Academic Press, New York.

Magnuson, J.J. (1970) Hydrostatic equilibrium of *Euthynnus affinis*, a pelagic teleost without a gas bladder. *Copeia*, **1970**, 56–85.

Magnuson, J.J. (1973). Comparative study of adaptations for continuous swimming and hydrostatic equilibrium of scombroid and xiphoid fishes. *Fish. Bull.* **71**, 337–356.

Margaria, R. (1968) Capacity and power of the energy processes in muscle activity: their practical relevance in athletics. *Int. Z. angew. Physiol.* **25**, 352–360.

McCutchen, C.W. (1970) The trout tail fin: a self-cambering hydrofoil. *J. Biomech.* **3**, 271–281.

Morgan, E. (1971) The swimming of *Nymphon gracile* (Pyconogonida): the mechanics of the leg-beat cycle. *J. exp. Biol.* **55**, 273–287.

Nachtigall, W. (1960) Uber Kinematik, Dynamik & Energetik des Schwimmens einheimischer Dytisciden. *Z. vergl. Physiol.*, **43**, 48–118.

Nachtigall, W. (1965) Locomotion: swimming (hydrodynamics) of aquatic insects. In: *The Physiology of Insecta*, Rockstein, M., Ed., **2**, 255–281, Academic Press, New York.

Packard, A. (1969) Jet propulsion and the giant fibre response in *Loligo*. *Nature*, **221**, 875–877.

Prandtl, L. and Tietjens, O.G. (1957) *Applied Hydro- and Aeromechanics*, 2nd edn. Dover Books, New York.

Siekmann, J. (1963) On a pulsating jet from the end of a tube, with application to the propulsion of certain aquatic animals. *J. Fluid Mech.* **15**, 399–418.

Simons, J.R. (1970) The direction of the thrust produced by the heterocercal tails of two dissimilar elasmobranchs: the Port Jackson shark, *Heterodontus portus jacksoni* (Meyer) and the Piked dogfish, *Squalus megalops* (Mackleay). *J. exp. Biol.* **52**, 95–107.

Taylor, G. (1952) Analysis of the swimming of long narrow animals. *Proc. R. Soc. Lond. Ser. A*, **214**, 158–183.

Trueman, E.R. and Packard, A. (1968) Motor performances of some cephalopods. *J. exp. Biol.* **49**, 495–507.

Trueman, E.R. (1975) *The Locomotion of Soft-bodied Animals*. Arnold, London.

Vlymen, W.J. (1970) Energy expenditure of swimming copepods. *Limnol. Ocean.* **15**, 348–356.

Walker, W.F. (1971) Swimming in sea turtles of the family Cheloniidae. *Copeia*, **1971**, 229–233.

Walters, V. and Fierstine, H.L. (1964) Measurements of swimming speeds of yellowfin tuna and wahoo. *Nature* **202**, 208–209.

Ward, D.V. (1972) Locomotory function of the squid mantle. *J. Zool., Lond.*, **167**, 487–499.

Webb, P.W. (1971a) The swimming energetics of trout. I. Thrust and power output at cruising speeds. *J. exp. Biol.* **55**, 489–520.

Webb, P.W. (1971b) The swimming energetics of trout. II. Oxygen consumption and swimming efficiency. *J. exp. Biol.* **55**, 521–540.

Webb, P.W. (1973) Kinematics of pectoral fin propulsion in *Cymatogaster aggregata*. *J. exp. Biol.* **59**, 697–710.

Webb, P.W. (1975a) Hydrodynamics and energetics of fish propulsion. *Bull. Fish. Res. Board Can.* **190**, 1–158.

Webb, P.W. (1975b) Acceleration performance of Rainbow trout *Salmo gairdneri* and Green sunfish *Lepomis cyanellus. J. exp. Biol.* **63**, 451–465.

Weihs, D. (1972) A hydrodynamical analysis of fish turning manoeuvres. *Proc. R. Soc. Lond. Ser. B*, **182**, 59–72.

Weihs, D. (1973) The mechanism of rapid starting of slender fish. *Biorheol.* **10**, 343–350.

Weis-Fogh, T. and Alexander, R.McN. (1977) The sustained power output obtainable from striated muscle. In: *Scale Effects in Animal Locomotion.* Pedley, T.S., Ed., Academic Press, London.

Wilkie, D.R. (1960) Man as an aero engine. *J. R. Aero. Soc.*, **64**, 477–481.

Wu, T.Y.-T. (1971) Hydromechanics of swimming of fishes and cetaceans. *Adv. appl. Mech.* **11**, 1–63.

Yuen, H.S.H. (1966) Swimming speeds of yellowfin and skipjack tuna. *Trans. Am. Fish. Soc.* **95**, 203–209.

10 Flight

R. McN. Alexander

10.1 Introduction

This chapter would have been written by Professor Torkel Weis-Fogh, had it not been for his untimely death. It will be clear from the text that a very large part of our understanding of animal flight is due to him.

It will be assumed in this chapter that any reader who was not previously familiar with the rudiments of aerodynamics will by now have read Section 9.2 of the chapter on swimming.

This chapter is mainly about insects and birds but includes passages on bats and pterosaurs. It starts with a general discussion of wings and their aerodynamic properties. This is followed by sections on hovering and gliding. Hovering is practised extensively by insects, and to a lesser extent by small birds and bats. The animals' trunk being stationary, or nearly so, simplifies aerodynamic analysis. Gliding and soaring are common modes of flight for many of the larger birds. The wings being stationary relative to the trunk simplifies analysis. Only at the end of the chapter will fast flapping flight be discussed.

10.2 Wings

Let us start by examining the properties of wings. The drag on a well-designed wing consists mainly of friction drag and induced drag. In all but the largest flying animals the Reynolds number R is less than the value (about 10^6) at which a turbulent boundary layer would be expected. Hence the friction drag coefficient based on the wetted area S_w should be $1.3R^{-1/2}$ (see p. 226). It is customary to define the drag coefficients of aerofoils in terms of plan area S_p ($\simeq \frac{1}{2}S_w$) and the friction drag coefficient based on plan area is $2.6R^{-1/2}$. From Equations 9.6a and 9.7a the part of the drag coefficient attributable to induced drag is $C_L^2/\pi A$, where C_L is the lift coefficient and A the aspect ratio. Hence the total drag coefficient C_D should be given by

$$C_D \simeq 2.6R^{-1/2} + (C_L^2/\pi A). \tag{10.1}$$

Any given aerofoil can have a wide range of lift coefficients corresponding to different angles of attack. The relationship between lift and drag coefficients predicted by Equation 10.1 is shown by broken lines in Fig. 10,1, for particular values of R and A. The continuous lines show the observed relationship for real aerofoils, with the same values of R and A. The aerofoils perform less well than Equation 10.1 predicts (i.e. they suffer more drag for given lift) for several reasons. Even at low angles of attack there is a little pressure drag which was ignored in Equation 10.1. The value given for induced drag in Equations 9.7a and 10.1 is the minimum attainable value for an ideally designed aerofoil, so real values are likely to be a little higher. Finally and most important, wings stall: if the angle of attack is increased beyond about 20° eddies form behind the wing, the lift falls somewhat and pressure drag increases greatly.

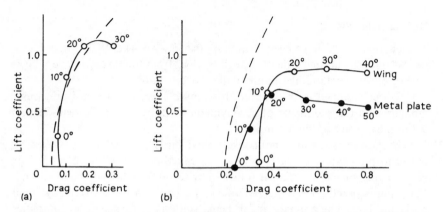

Fig. 10.1 Graphs of lift coefficient against drag coefficient for various aerofoils. Continuous lines show experimental data and broken ones have been calculated (for the same aspect ratio and Reynolds number) from Equation 10.1. Angles of attack are indicated alongside points on the experimental curves. (a) refers to a locust (*Schistocerca*) hind wing of aspect ratio 2.8 at a Reynolds number of 4000 (Jensen, 1956). (b) refers to *Drosophilia* wings and thin, flat metal plates, both of aspect ratio about 2.7 and both at a Reynolds number of 200 (Vogel, 1967b).

Figure 10.1 confirms what Equation 10.1 predicts, that drag coefficients are inevitably high (for given lift coefficient) when the Reynolds number is low. A flat plate (Fig. 10.2a) suffers less drag than a cambered one (Fig. 10.2b) at very low angles of attack but a cambered plate achieves a higher lift coefficient before stalling. Aeroplane wings are not flat plates but have relatively thick streamlined sections (Fig. 10.2d: note that this is a cambered version of the symmetrical streamlined section shown in Fig. 10.2c). This makes higher lift coefficients possible at Reynolds numbers above about 10^5, but cambered plates are superior to streamlined sections at lower Reynolds numbers (Schmitz, 1960; Alexander, 1968). The wings of moderate to large birds weighing 0.5 kg or more operate at Reynolds numbers over 10^5 so that a streamlined section

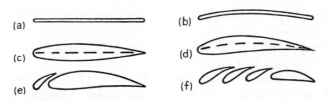

Fig. 10.2 Sections of aerofoils, cut at right angles to the span. (a) is a flat plate, (b) a cambered plate, (c) a symmetrical streamlined section, (d) a cambered streamlined section, (e) a slotted aerofoil and (f) a multi-slotted aerofoil (from Alexander, 1968).

might be advantageous, but for small birds and bats and for insects a cambered plate is best.

Vogel (1967a) showed by photography that the wings of the fruit fly *Drosophila* are cambered in the downstroke and flat in the upstroke and Jensen (1956) showed more complicated changes during the wing beat of the locust. Insect wings are thin membranes, stiffened by pleating and by a network of veins (Fig. 10.3a. Hertel, 1966; Rees, 1975a). Smooth wings perform better than rough or, presumably, pleated ones, at high Reynolds numbers. In general,

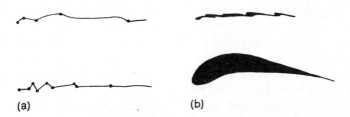

(a)   **(b)**

Fig. 10.3 Sections through the wings of (a) a hoverfly, *Syrphus balteatus* (re-drawn from Rees, 1975a) and (b) a pigeon, *Columba livia* (re-drawn from Nachtigall and Weiser, 1966). The upper section is distal to the lower one in each case.

any irregularity which projects through the boundary layer is harmful. Rees (1975b) tested models of a hoverfly (*Syrphus*) wing, in the range of Reynolds number (450–900) which is used in flight. He found that their aerodynamic properties were more or less identical with those of a smooth aerofoil, shaped to match the envelope through the high points of the pleats. The depth of the deepest pleats in *Syrphus* wings is about 0.08 times the chord length, about equal to the thickness of the boundary layer halfway across the chord.

The lift coefficients of metal plates fall when the angle of attack is increased above about 20°, but the lift coefficient of *Drosophila* wings is almost constant (at about 0.8) for angles of attack between 20 and 50° (Vogel, 1967b), and those of butterfly wings fall only a little at high angles of attack (Nachtigall, 1967).

The wings of bats are thin cambered membranes. The camber can be adjusted by muscles and, in the fruit bat *Rousettus*, is particularly marked when the bat is gliding slowly and needs a high lift coefficient (Pennycuick, 1971).

251

The bones of the wing and the muscles which support them form ridges which project above the general wing surface, which may not be harmful and may even be advantageous. Schmitz (1960) has shown that at Reynolds numbers around 5×10^4 a spoiler wire near the leading edge of a wing can improve its performance (*Rousettus* glides in this range of Reynolds numbers). Hertel (1966) suggested that the anterior veins in the wings of large dragonflies may also act as spoiler wires.

Bird wings have bones and muscles along the proximal part of the leading edge, but the feathers are arranged so as to give the wing a streamlined section (Fig. 10.3b). In the distal part of the wing the primary feathers form a thin cambered plate.

Ordinary aerofoils do not generally develop lift coefficients above 1.5, even at high Reynolds numbers. Much higher lift coefficients, up to about 2.0, can be obtained from slotted wings (Fig. 10.2e). At low angles of attack the lift coefficient is about the same as for similar aerofoils without slots, but it continues to increase as the angle of attack increases beyond the value at which the ordinary aerofoil would stall (Prandtl and Tietjens, 1957). Even higher lift coefficients are obtainable from multiply-slotted wings (Fig. 10f). Note that a multiply-slotted aerofoil is in effect a stack of aerofoils of greater total area than an ordinary aerofoil of similar size. Also, these component aerofoils are set at a lower angle of attack than the multiply-slotted aerofoil as a whole.

The alula or bastard wing is a group of feathers borne by the thumb, which is abducted from the anterior edge of the wing at low flying speeds (Fig. 10.6) and presumably acts as a slot. However, the maximum lift coefficients of bird wings tested in a wind tunnel by Nachtigall and Kempf (1971) were increased only a little by abduction of the alula. The primary feathers of many birds separate in the upstroke in slow flight, forming a multiply-slotted aerofoil (Fig. 10.7, which will be explained further). The primaries of vultures and some other large birds separate at the wing tip whenever the wing is fully extended (see for instance McGahan, 1973a, b; Oehme and Kitzler, 1975).

10.3 Hovering

10.3.1 Hovering animals and their movements

This section of the chapter deals with slow forward flight, in which the velocity of the wing tips is much higher than that of the trunk, as well as with stationary hovering.

Hummingbirds and most insects hover with the trunk vertical or steeply tilted and the wings beating more or less horizontally. The movements of the moth shown in Fig. 10.4 are typical. In the forward stroke (morphologically the downstroke) the dorsal surface of the wing is uppermost but for the backward stroke (upstroke) the wings turn upside-down. It has a positive angle of attack in both strokes, so upward lift is produced in both. This type of hovering

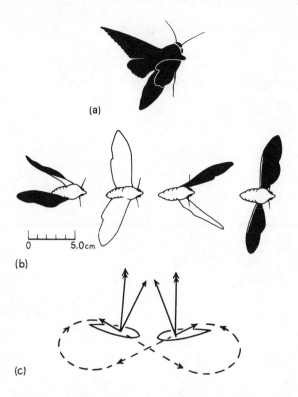

(a)

(b)

(c)

Fig. 10.4 (a) A large moth, *Manduca sexta*, hovering. Drawn from a flash photograph. (b) Tracings from a film taken vertically from above, of the same species hovering. Successive tracings are separated by intervals of about 12 ms. The undersides of the wings are shown black. (c) A diagram showing a section of the wing at two stages of the wing beat. (a) and (b) are from Weis-Fogh (1973).

merges imperceptibly into slow forward flight as practised, for instance, by Diptera (Nachtigall, 1966).

Figure 10.5 shows a minute wasp, also flying by horizontal wing beats with the wings turning ventral side up for the upstroke. The wings clap together at the top of the upstroke (frames 0 and 17) and they are severely bent at frames 9 and 10. The significance of these peculiarities will be discussed in Section 10.3.7. Some other insects including *Drosophila* clap their wings together at the top of the upstroke and so do pigeons when they are taking off (Weis-Fogh, 1973).

Hoverflies (Syrphinae) and dragonflies (Odonata) hover with their bodies horizontal, and their wings beat in a steeply inclined plane.

A different technique of hovering and slow flight is used by birds other than hummingbirds (Figs. 10.6 and 10.7). In the downstroke (Fig. 10.6) the alula is

253

Stroke 2 ends   Stroke 3 starts

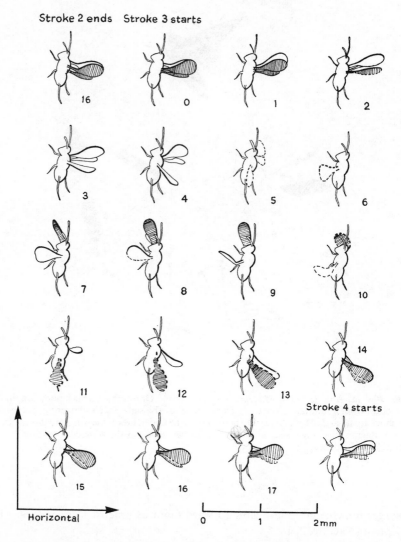

Fig. 10.5 Tracings of successive frames of a film taken at 7150 frames s^{-1} of a minute wasp, *Encarsia formosa*, flying slowly backwards and upwards (from Weis-Fogh, 1973).

abducted. The primaries are bent upwards at their tips (Fig. 10.6c) which shows that upward lift acts on them. The wings swing forward to a position parallel to each other in front of the body (Fig. 10.6d). The upstroke involves sharp bending of the wrist (Fig. 10.7b) followed by a flick which straightens the wrist again and brings the wings to their initial position over the back. The primaries separate at this stage (Fig. 10. 7c) and are bent upwards, which shows that upward lift acts on them again.

254

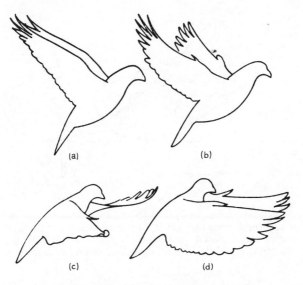

Fig. 10.6 Outlines traced from photographs taken at intervals of 0.01 s, of a pigeon flying slowly. The sequence is continued in Fig. 10.7 (from Brown, 1963).

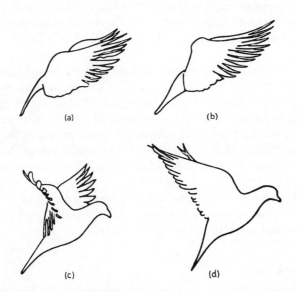

Fig. 10.7 These outlines complete the sequence started in Fig. 10.6 (from Brown, 1963).

Figure 10.8 shows how the primary feathers of the pigeon wing produce upward lift in the downstroke and in the upstroke. The angle of attack becomes very high during the upstroke but this need not cause stalling because the primaries separate to form a multi-slotted wing. The primaries are asym-

255

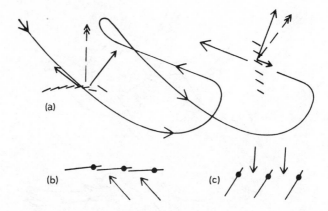

Fig. 10.8 (a) A diagram showing the path of the primary feathers of a pigeon in slow flight, and the forces which act on them in the downstroke and in the upstroke. (b), (c) Diagrams showing why the primaries overlap in the downstroke and separate in the upstroke (from Alexander, 1968).

metrical, with a broader vane posterior to the shaft than anterior to it, so any difference in pressure between the two faces of a primary tends to twist the shaft. The primaries overlap in such a way that their tendency to twist makes them form a continuous lamina in the downstroke but separates them in the upstroke.

Most insects and small birds and bats can and do hover. Hummingbirds (mass 2–20 g) can hover for long periods; a 3 g *Calypte* once hovered continuously for 50 min in an experiment (Lasiewski, 1963). Larger birds can only hover briefly and apparently depend on anaerobic metabolism to make even this possible. 0.4 kg pigeons can hover or climb vertically for about 1 s (Pennycuick, 1968b) and substantially larger birds cannot hover at all. Inability to hover implies inability to take off in still air from a standing start on level ground. Many large birds such as the Kori bustard (*Ardeotis kori*, 12 kg, see Pennycuick, 1969) and swans (*Cygnus* spp., up to 12 kg) need a taxiing run over land or water and some of the large birds of prey gain the speed needed to become airborne by diving from a cliff.

10.3.2 A simple model

The next few pages of this chapter present a quantitative analysis of the mechanics of hovering, based on a brilliant paper by Weis-Fogh (1973). We will use the model shown in Fig. 10.9, which is supposed to be hovering by horizontal wing beats like the moth shown in Fig. 10.4. The model has mass m, and has rectangular wings which it beats with frequency n (Weis-Fogh's model allowed more realistic wing shapes, but a rectangular wing is specified here for the sake of simplicity). These wings have length r and chord c and

256

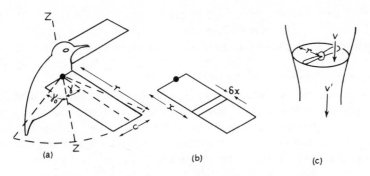

Fig. 10.9 The schematic hovering animal described in the text.

moment of inertia I and at time t they make an angle γ with the transverse axis ZZ. They swing to maximum angles γ_0 in front of and behind ZZ, according to the equation

$$\gamma = \gamma_0 \sin 2\pi nt \tag{10.2}$$

We will need realistic values for the various parameters, for hovering animals of different size. Figure 10.10 shows that for a very wide range of sizes of hovering animal, the length of the wings tends to be proportional to (body

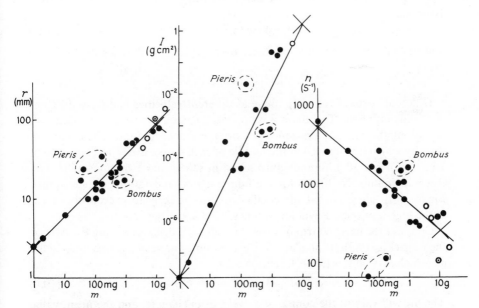

Fig. 10.10 Graphs of the length (r) and moment of inertia (I) of the wings and of their frequency of beating in hovering flight (n) against body mass (m) for various insects (●), hummingbirds (○) and a bat (⊙). The crosses refer to the hypothetical 10 g hummingbird and 1 mg mosquito referred to in the text (data from Weis-Fogh, 1973).

mass)$^{0.39}$ their moment of inertia to (body mass)^2 and their frequency to (body mass)$^{-0.33}$. The relationships are not strict ones. Some animals such as the bumblebee *Bombus* have short wings for their mass and beat them at unusually high frequency. Others such as the butterfly *Pieris* have long wings and beat them at unusually low frequency.

To show the effects of differences of size we will refer repeatedly in the next few pages to two hypothetical hovering animals, each with the wing length, moment of inertia and frequency typical for its size (Fig. 10.10). The larger of these will be supposed to have a mass of 10 g and wings of length 85 mm, moment of inertia 1.8 g cm ^{2} and frequency 25 s^{-1}. It could well be a rather large hummingbird. The smaller will be supposed to have a mass of 1 mg and wings of length 2.3 mm, moment of inertia 1.5×10^{-8} g cm^2 and frequency 500 s^{-1}, about the same as a mosquito. For a great many hovering animals the average chord length is about one quarter of the wing length (see Weis-Fogh, 1973) so we will assume $c = r/4$ in each case. We will also assume that $\gamma_0 = 1$ radian, the value given by Weis-Fogh for most of the hovering animals he described.

10.3.3 The lift coefficient

Differentiation of Equation 10.2 gives the angular velocity of the wing, ω.

$$\omega = 2\pi n \gamma_0 \cos 2\pi nt, \tag{10.3}$$

Note that

$$\omega^2 = 4\pi^2 n^2 \gamma_0^2 \cos^2 2\pi nt \tag{10.3a}$$

and since the mean value of $\cos^2 2\pi nt$ over a cycle is $\frac{1}{2}$, the mean value of ω^2 is

$$[\overline{\omega^2}] = 2\pi^2 n^2 \gamma_0^2. \tag{10.3b}$$

Different parts of the wing move at different velocities and have different Reynolds numbers. To obtain a representative value consider a section half way along the wing, at a distance $\frac{1}{2}r$ from the axis of rotation. The maximum velocity which it will reach in the middle of each stroke is $\pi n \gamma_0 r$. This is about 7 m s^{-1} for our 10 g hummingbird and 4 m s^{-1} for our 1 mg mosquito. The Reynolds number is 10 000 for the hummingbird and 150 for the mosquito. Figure 10.1 suggests that lift coefficients are unlikely to exceed about 1.3 at the higher of these Reynolds numbers, and 0.9 at the lower.

Consider the narrow strip of wing, δx wide, at a distance x from the base of the wing (Fig. 10.9). Its plan area is $c . \delta x$. At time t its velocity is ωx and the lift δL which acts on it is (from Equation 9.6a)

$$\delta L = \tfrac{1}{2}\rho . c\delta x . \omega^2 x^2 . C_L. \tag{10.4}$$

The lift will vary in the course of a cycle, as ω changes, and the mean value over a cycle can be obtained by replacing ω^2 with $[\overline{\omega^2}]$, from Equation 10.3b. This gives

$$[\overline{\delta L}] = \rho c \pi^2 n^2 \gamma_0^2 . x^2 \delta x . C_L. \tag{10.4a}$$

The mean lift on the whole wing can be obtained by integration and it must equal half the body weight (since the body is supported by two wings). If the lift coefficient is constant along the wing

$$\bar{L} = \tfrac{1}{2}mg = \rho c \pi^2 n^2 \gamma_0^2 C_L \int_0^r x^2\, dx$$

$$= \rho c \pi^2 n^2 \gamma_0^2 C_L r^3/3.$$
$$C_L = 1.5\, mg/\rho c \pi^2 n^2 \gamma_0^2 r^3. \tag{10.5}$$

This gives lift coefficients of 1.4 for our 10 g hummingbird and 0.5 for our 1 mg mosquito. Each is feasible though fairly high for the Reynolds number in question. Since $\gamma_0 = 1$ rad and $c = r/4$, the wing tip moves 8 chord lengths and the mid-point of the wing 4 chord lengths, in the course of a stroke. This is ample for the generation of circulation (p. 228).

Weis-Fogh (1973) calculated lift coefficients between 0.5 and 1.3 for most of the animals he considered. However, he obtained values around 2 for hover-flies and a butterfly and well over 2 for the tiny wasp *Encarsia*. The hoverflies have very low values of γ_0 (0.7 rad or less) and the butterfly and wasp have much lower values of n than Fig. 10.10 indicates as normal for their size. The techniques which these animals use to obtain high lift coefficients are discussed in Section 10.3.7.

10.3.4 Aerodynamic power requirements

An animal hovering by beating its wings horizontally is essentially a helicopter and the power it needs for aerodynamic purposes can be estimated by simple helicopter theory. An animal with wings of length r is more or less equivalent to a helicopter with a rotor of radius r (Fig. 10.9c).

A helicopter rotor obtains the lift needed to support the craft by giving downward momentum to the air. Air passes through the plane of the rotor with velocity v. It continues to accelerate below the rotor (this will be proved) and eventually reaches a velocity v', well below the rotor. The air well above the rotor is stationary with pressure p, but when this air arrives immediately above the rotor its velocity is v so by Bernoulli's equation (Equation 9.4) its pressure must be $p - \tfrac{1}{2}\rho v^2$. Bernoulli's equation cannot be applied to flow through the rotor because the rotor does work on the air but it can be applied again to flow below it. Since the air well below the rotor has pressure p and velocity v' the pressure immediately below the rotor, where the velocity is v, must be $p + \tfrac{1}{2}\rho(v'^2 - v^2)$. Hence the pressure difference across the plane of the rotor is $\tfrac{1}{2}\rho v'^2$ and the upward force on the rotor is $\tfrac{1}{2}\rho v'^2 \pi r^2$.

The mass of air which flows through the rotor in unit time is $\rho \pi r^2 v$. This air eventually attains velocity v' so the rate at which the air is given momentum is $\rho \pi r^2 v v'$. By Newton's Second Law of Motion, this must equal the upward force already calculated. The upward force must also equal the weight mg of the helicopter so

$$mg = \tfrac{1}{2}\rho v'^2 \pi r^2 = \rho \pi r^2 vv',$$
$$v = \tfrac{1}{2}v' = (mg/2\rho\pi r^2)^{1/2}. \tag{10.6}$$

v is thus 1.3 m s^{-1} for our 10 g hummingbird and 0.5 m s^{-1} for our 1 mg mosquito, so it is considerably less in each case than the peak velocity of the midpoint of the wing (7 and 4 m s^{-1}, respectively).

The mass $\rho\pi r^2 v$ of air which passes through the rotor in unit time is accelerated to a final velocity v' (=$2v$) so the rate at which kinetic energy is given to the air is $2\rho\pi r^2 v^3$. This is the minimum power requirement for hovering, $P_{\min}$. Using the value of v from Equation 10.6 we find

$$P_{\min}/m = (mg^3/2\rho\pi r^2)^{1/2}. \tag{10.7}$$

This is 13 W kg^{-1} for the 10 g hummingbird and only 5 W kg^{-1} for the 1 mg mosquito. However, this is a theoretical minimum value which can never be achieved in practice. Weis-Fogh (1972) obtained more realistic values by applying propeller theory to the wings, and later by a simpler approach (1973). The next few paragraphs present a cruder version of his approach which still gives similar conclusions.

We start with a calculation which will be shown to need refinement. Consider again the narrow strip of wing δx wide at a distance x from the base of the wing. By an argument similar to the one used to derive Equation 10.4a the mean drag on this strip is

$$[\overline{\delta D}] = \rho c\pi^2 n^2 \gamma_0^2 x^2 \, \delta x . C_D.$$

The moment exerted by this drag about the base of the wing is

$$[\overline{\delta M}] = x[\overline{\delta D}] = \rho c\pi^2 n^2 \gamma_0^2 x^3 \, \delta x . C_D.$$

The mean total moment exerted by the drag on the whole wing can be obtained by integration:

$$\bar{M} = \rho c\pi^2 n^2 \gamma_0^2 C_D . \int_0^r x^3 \, dx,$$
$$= \rho c\pi^2 n^2 \gamma_0^2 C_D r^4/4.$$

In each cycle of wing movements the wing rotates through an angle $4\gamma_0$ against this mean moment ($2\gamma_0$ in the downstroke and $2\gamma_0$ in the upstroke), and there are n cycles in unit time so an initial estimate of the power P_a required to overcome drag on *two* wings is

$$P_a = 8\gamma_0 n\bar{M} = 2\rho c\pi^2 n^3 \gamma_0^3 C_D r^4. \tag{10.8}$$

The wing of our 10 g hummingbird operates at a Reynolds number around 10^4, with a mean lift coefficient of 1.3. Since the Reynolds number is a little above that of the locust wing in Fig. 10.1 a slightly better performance could be hoped

260

for, but the drag coefficient would probably be about 0.25. If so, the value of P_a/m given by Equation 10.8 is 11 W kg⁻¹, or about the same as the value given by Equation 10.7. The wing of our 1 mg mosquito operates at a Reynolds number of only 150, with a lift coefficient of 0.6. Its properties would probably be similar to those of the *Drosophila* wing (Fig. 10.1) so the drag coefficient would probably be about 0.4. If so, P_a/m is 21 W kg⁻¹, considerably more than Equation 10.7 suggested.

These values are too low, because the downward flow of air past the wings was ignored. The velocity of the air relative to a strip of wing has a downward component v and a backward component ωx so the resultant velocity is tilted at an angle ψ to the horizontal (Fig. 10.11). The lift δL and drag δD are tilted

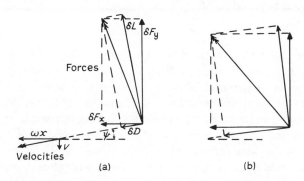

Fig. 10.11 Diagrams showing the effect of the downward air velocity v on the forces on the wing. (a) is drawn to scale to represent the situation at the mid-point of the wing of the hypothetical 10 g hummingbird discussed in the text. (b) similarly refers to the mid-point of the wing of the 1 mg mosquito.

at the same angle to the vertical and horizontal, respectively. Their resultant has a vertical component δF_y and a horizontal component δF_x. It is against this horizontal component rather than the drag that the wing muscles must work, if the wings are beating horizontally.

Figure 10.11a is drawn to scale to represent the situation at the mid-point of the wing of our 10 g hummingbird. The maximum value of ωx there is 7 m s⁻¹ and as most of the lift is produced while the wing is near its maximum velocity, this value of ωx is represented. v is 1.3 m s⁻¹, so v is shown with 1.3/7 of the value of ωx. The lift coefficient has been estimated as 1.3 and the drag coefficient as 0.25 so δL has been shown 1.3/0.25 times δD. The geometrical construction shows that $\delta F_x = 2\delta D$. Rather larger factors would apply nearer the base of the wing and smaller ones nearer the tip, but it seems clear that the power required to overcome aerodynamic forces must be nearer 20 W kg⁻¹ than the 11 W kg⁻¹ calculated from Equation 10.8.

Figure 10.11b shows the situation at the mid-point of the wing of the 1 mg mosquito. δF_x is only about 1.2 δD, so a much smaller correction is needed to the aerodynamic power calculated from Equation 10.8. This power is probably about 25 W kg^{-1}, little different from the value for the 10 g hummingbird.

For both animals $\delta F_y \simeq \delta L$ so the errors involved in using Equation 10.5 (which ignored v) to calculate lift coefficient are negligible.

10.3.5 Inertia, elasticity and power

If the moment of inertia of each wing is I the total kinetic energy of a pair of wings, each moving with angular velocity ω, is $I\omega^2$. The maximum value of ω, achieved in the middle of a wing stroke, is $2\pi n\gamma_0$. This value is attained twice in each cycle of wing movements, so the total work done accelerating the wings in each cycle is $2I(2\pi n\gamma_0)^2 = 8\pi^2 In^2\gamma_0^2$. This work is done with frequency n so the power required is

$$P_i = 8\pi^2 In^3\gamma_0^2. \tag{10.9}$$

This will be called the inertial power. On this reckoning P_i/m is 22 W kg^{-1} for our 10 g hummingbird and 15 W kg^{-1} for our 1 g mosquito. However, we have ignored so far the mass of air in the boundary layer which has to be accelerated with the wings and adds to their moment of inertia (Vogel, 1962). This is negligible in the case of the hummingbird but important in the case of the mosquito and when it is considered P_i/m becomes about 22 W kg^{-1} in each case. If the wings were accelerated and decelerated solely by muscle action the animals would do positive work at this rate, and negative work at this rate. Metabolic energy is needed both for positive and for negative work (Chapter 3).

The power needed for hovering cannot be calculated simply by adding together the estimates of aerodynamic and inertial power. Figure 10.12a shows the moment which the muscles must exert about the base of the wing to overcome aerodynamic forces, at each stage in a stroke. This moment is

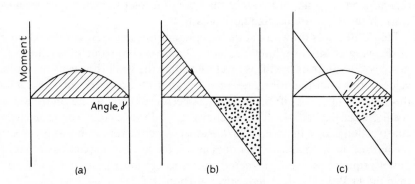

Fig. 10.12 Graphs of moments acting about the base of a wing, against the angle of the wing, in a single stroke. Further explanation is given in the text.

greatest in the middle of the stroke, when the wing is travelling fastest. The hatched area under the graph is aerodynamic work. Figure 10.12b shows the moment required to accelerate or decelerate the wing. If the motion is simple harmonic as shown in Equation 10.2, the graph of moment against angle must be a straight line. The hatched area represents positive work done accelerating the wing, and the stippled area negative work done decelerating it. This negative work need not all be undertaken by the muscles: aerodynamic forces can be used to slow the wing down so that the kinetic energy which has to be dissipated does useful aerodynamic work. Figure 10.12c shows the positive aerodynamic work and the negative inertial work which can be set against each other. The positive work which has to be done by muscles in the course of the stroke is not the sum of the areas hatched in Figs. 10.12a and 10.12b, but this sum less the area hatched in Fig. 10.12c. Figure 10.12 has been drawn with the aerodynamic work and the inertial work about equal, as for our 10 g hummingbird. It can be deduced from it that if $P_a = P_i$ the total power requirement is about $1.6\,P_a$.

So far it has been assumed that the joint between the wing and the body is a free-moving hinge, with only the muscles and the aerodynamic forces exerting moments about it. Suppose instead that the wing is attached by some elastic material so that when it is at the top of the upstroke there is an elastic moment tending to swing it down, and when it is at the bottom of the downstroke an elastic moment tends to swing it up. If the elastic connection has the right dimensions, the elastic moment at each wing position will match the inertial moment shown in Fig. 10.12b and the wing will have a natural frequency of vibration which matches the frequency of the wing beat. In this case the muscles will not have to do any work against inertial forces. They will simply have to do the aerodynamic work shown in Fig. 10.12a, plus a small amount to make good the losses which must occur due to imperfect elastic properties.

Hence our 10 g hummingbird and our 1 mg mosquito must each exert total mechanical power amounting to about 35 W kg^{-1} if their wings are attached by free-moving joints, but this can be reduced to about 25 W kg^{-1} by ideally designed elastic wing attachments.

The metabolic rates of hummingbirds and mosquitos have been measured. Lasiewski (1963) found that a hummingbird used 230 W kg^{-1} while hovering and 30 W kg^{-1} while resting. The difference of 200 W kg^{-1} was presumably used for flight. Similarly a mosquito used 110 W kg^{-1} for flight (Hocking, 1953). These are rates of consumption of chemical energy. We have calculated that the mechanical power required for flight is about 35 W kg^{-1} both for our typical hummingbird and for our typical mosquito, if neither make use of elastic wing attachments. If so the efficiency of conversion of chemical energy to mechanical work must be $35/200 = 0.18$ in the hummingbird. This is entirely feasible, and similar to the efficiencies obtainable from men on bicycle ergometers (Dickinson, 1929, and Chapter 3). Similarly the efficiency would be $35/110 = 0.32$ in the mosquito, which seems most improbably high. However, if the mosquito had elastic wing attachments, ideally arranged, its mechanical power requirement would be only 25 W kg^{-1} and the efficiency $25/110 = 0.23$, which is just feasible.

Weis-Fogh (1973) performed similar calculation for six hovering species, using actual dimensions and wing frequencies instead of the typical values used in this chapter. He calculated that without elastic wing attachments the efficiency would be 0.17 to 0.22 (and so feasible) in a hummingbird (*Amazilia*), a bee (*Apis*) and a fruit fly (*Drosophila*). It would be 0.40 to 0.58 (far beyond the bounds of feasibility) in a wasp (*Vespa*), a mosquito (*Aedes*) and a drone fly (*Eristalis*). It seems that at least some insects must depend on elastic structures to reduce the energy needed for flight.

These structures are known. They consist in part of ordinary cuticle and in part of a remarkable elastic protein called resilin which was discovered by Weis-Fogh (1960) (see also Chapter 1). Resilin can be stretched to three times its initial length, and snaps back immediately to its initial length when it is released. Its Young's modulus is similar to that of soft rubber (about 2×10^6 N m^{-2}) (Weis-Fogh, 1961) but it is greatly superior to ordinary rubber in one important property. When a rubber is deformed and allowed to recoil, the energy recovered in the recoil is generally no more than about 0.91 of the energy used to deform it. However, locust resilin returns up to 0.97 of the energy used to deform it (Jensen and Weis-Fogh, 1962).

Resilin is used in different ways in the thoraxes of different insects (see Jensen and Weis-Fogh, 1962) but there seem to be no detailed accounts of the functional morphology of the structures involved.

Weis-Fogh (1972) found no evidence of energy saving by elastic storage in hummingbirds but Pennycuick (1975) has suggested for other birds that elastic storage in the primary feathers (which are bent during the downstroke, Fig. 10.6) may be important in slow flight and hovering.

10.3.6 Scale effects in hovering

Large hovering animals are not geometrically similar to small ones. If they were, wing lengths would be proportional to (body mass)$^{0.33}$ and the moment of inertia of the wings to (body mass)$^{1.67}$. In fact we find that there is a marked tendency for wing length to be proportional to (body mass)$^{0.39}$ and the moment of inertia of the wings to (body mass)^2 (Fig. 10.10). Wing frequency tends to be proportional to (body mass)$^{-0.33}$. These tendencies have remarkable effects, as has been shown by the calculations for hypothetical 10 g and 1 mg hovering animals. In each case the lift coefficient required is fairly high for the Reynolds number in question, but entirely feasible. The minimum power requirement for hovering, calculated from considerations of momentum (Equation 10.7) is less, per unit body mass, for small animals than for large ones. However, the small animals operate at low Reynolds numbers and suffer the penalty of high drag coefficients so that the aerodynamic power requirement is probably about

25 W kg^{-1}, both for the 10 g animal and for the 1 mg one. The inertial power requirements are also fairly similar.

The flight muscles make up 17–35% of the body mass, in various insects (Weis-Fogh, 1952) and most birds (Greenewalt, 1962). The maximum power output available from fast aerobic muscles seems to be about 250 W kg^{-1} at frequencies above 10 s^{-1}, and less at lower frequencies (Weis-Fogh and Alexander, 1977). A typical flying animal with 25% flight muscle should be capable of a mechanical power output of 60 W (kg body mass)$^{-1}$, provided its wing beat frequency is 10 s^{-1} or more. This is ample for hovering, for animals with body mass between 1 mg and 10 g. Hovering becomes impossible for much larger animals for two reasons. First, the scaling laws which keep power requirements (per unit body mass) fairly constant over so great a range of sizes, cannot prevent the requirement from rising with further increase in size: the advantages of higher Reynolds number will not compensate for ever for the increase in $P_{min}/\dot{m}$ implied by Equation 10.7. Secondly, power output per unit muscle mass must fall, as wing beat frequency falls below about 10 s^{-1}.

The next few paragraphs are largely about a solution to a different scaling problem, which arises for exceptionally small insects.

10.3.7 Unconventional mechanisms of lift generation

It was shown in Section 10.3.3 that the hovering ability of a few insects depends on astonishingly high lift coefficients. The highest values of all are 2.4–3.2, calculated by Weis-Fogh (1973) for *Encarsia formosa* (Fig. 10.5). This wasp has a mass of only 25 μg and extrapolation from Fig. 10.10c suggests that it would have a wing beat frequency around 1700 s^{-1}. So high a frequency would probably not be physiologically feasible and the actual frequency is only about 400 s^{-1}. A low frequency implies a high lift coefficient (Equation 10.5).

Encarsia probably generates more lift than predicted by conventional aerodynamics, by a mechanism proposed by Weis-Fogh (1973; Lighthill, 1973). The wings are clapped together at the top of the upstroke (Fig. 10.13a; see also Fig. 10.5). The anterior edges of the forewings separate, so that the wings are flung open like a book (Fig. 10.13b, c) before they separate for the downstroke (Fig. 10.13d). During the fling, air must flow round the wings into the cleft between them so that when the wings separate there is a circulation of air round them. This implies lift (p. 228). Lighthill (1973) has calculated the strength of this circulation, and it seems clear that it is adequate to explain the ability of *Encarsia* to fly.

A refinement of the Weis-Fogh mechanism is shown in Fig. 10.13e. When one end of a structure is twisted suddenly the whole structure does not start to turn simultaneously: an elastic deformation is propagated along it at a velocity which depends on the density and shear modulus of the material and is probably about 51 m s^{-1} for insect wings. Fast though this is, it is enough to ensure that the bases of the wings start to open before their tips so that most of the air flows

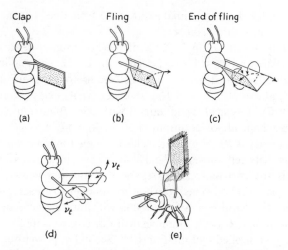

Fig. 10.13 Diagrams showing how *Encarsia* is believed to generate lift at the start of the downstroke. Each wing in this diagram represents the fore and hind wing of its side (from Weis-Fogh, 1973).

in between the anterior edges of the wings, as required to generate circulation, and not round the wing tips.

The Weis-Fogh mechanism is probably also important in butterfly flight, and may be used by pigeons at take-off (when they clap their wings together at the top of the upstroke).

Weis-Fogh (1973) has suggested, more speculatively, that the rapid rotation of the wing about its long axis which occurs at the bottom of the downstroke may also serve to generate circulation. The effect of this rotation is that the anterior edge which led the wing in the downstroke, also leads it in the upstroke (Fig. 10.5, frames 7 to 9). Weis-Fogh argued that this 'flip' mechanism was probably important in hoverflies and dragonflies, as well as in *Encarsia*.

10.4 Gliding

10.4.1 Gliding performance

The most impressive gliding animals are large birds. Vultures (*Gyps*, etc.) remain airborne for hours at a time by soaring in thermals, making very little use of wing movements. Some species regularly soar over distances of 100 km or more, between nest sites and foraging areas (Pennycuick, 1972). Storks (*Ciconia*) migrate by soaring between northern Europe and tropical Africa, travelling round the eastern end of the Mediterranean (Pennycuick, 1972). Albatrosses (*Diomedea*) soar for hours over the sea using a different soaring technique but hardly ever flapping their wings (Jameson, 1958). However, smaller birds such as swifts (*Apus*) also soar, and so do a few insects. Nachtigall

266

(1967) watched the butterfly *Iphiclides podalirius* soaring near a cliff, where the wind was deflected upwards. He observed glides without wing beats lasting up to 30 s. A specimen which he watched for almost an hour spend 80 % of the time gliding.

Figure 10.14a shows an animal gliding. If its velocity is constant it must be in equilibrium under the action of its weight *mg*, the lift *L* (which acts mainly on the wings) and the drag *D* (which acts partly on the wings and partly on the rest of the body). This is only possible if the velocity of the animal (relative to the air) is inclined downwards at an angle θ from the horizontal. In still air, the bird must lose height. In rising air, it may keep its level, or even rise. Most techniques of soaring depend on rising air (Section 10.4.2).

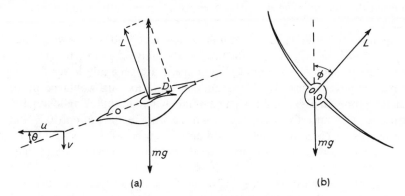

(a)   (b)

Fig. 10.14 Diagrams of animals gliding.

In our discussion of gliding, as in our discussion of hovering, we will use as examples two animals of very different size. Instead of using hypothetical animals of typical dimensions, however, we will use two species for which there is exceptionally good aerodynamic data. The larger example is one of the largest vultures, *Gyps ruppellii* (Pennycuick, 1971, 1972). Its average mass is 7.6 kg, its wing span 2.4 m and its wing area (S_p) 0.83 m². The span is about what would be predicted for this mass, by extrapolation from Fig. 10.10a. For birds of all sizes, as for hovering animals, wing length and wing span tend to be proportional to (body mass)$^{0.39}$ (Greenewalt, 1962; Alexander, 1971). Our smaller example is the locust *Schistocerca gregaria* (Jensen, 1956), which generally travels by flapping flight but also glides and soars. Its average mass is 2 g, its wing span 12 cm and its wing area 30 cm² (Fig. 10.10a shows a wing length of 4.5 cm and so a span of about 10 cm as typical for hovering animals of this mass).

In Fig. 10.14 equilibrium requires:

$$L = mg \cos \theta \tag{10.10}$$

and if θ is fairly small so that $\cos \theta \simeq 1$

$$L \simeq mg. \tag{10.10a}$$

267

From this and Equation 9.6a, at velocity u

$$mg \simeq \tfrac{1}{2}\rho S_p u^2 C_L. \tag{10.11}$$

The lower the speed, the higher the lift coefficient C_L must be. There is therefore a minimum speed u_{min}, below which gliding is impossible, which requires the maximum attainable lift coefficient, $C_{L\,max}$,

$$u_{min} \simeq (2\,mg/\rho S_p C_{L\,max})^{1/2}. \tag{10.12}$$

A vulture's wing could probably achieve lift coefficients up to about 1.5, in which case the minimum speed would be about 9.6 m s^{-1}. Locust wings, working at lower Reynolds numbers, have maximum lift coefficients of 1.2 (Jensen, 1956) which implies a minimum speed of 2.9 m s^{-1}. The Reynolds number of the wings at the minimum speed is about 230 000 for the vulture and 5000 for the locust.

Equation 10.12 shows that the minimum speed depends on the wing loading (mg/S_p) which is 90 N m^{-2} for the vulture and only 6.5 N m^{-2} for the locust. Wing loading and so minimum gliding speed are generally lower for small animals (for geometrically similar animals wing loading would be proportional to (body mass)$^{0.33}$) but some animals have unusual wing loadings for their size. The butterfly *Pieris brassicae* has a wing loading of only 0.7 N m^{-2} (Hertel, 1966) corresponding to a minimum speed of 1 m s^{-1}. The huge pterosaur *Pteranodon* had a wing span of 8 m and a wing area of 6 m^2 but its mass has been estimated at only 18 kg, giving a wing loading of only 30 N m^{-2} and a minimum speed of 7 m s^{-1} (Bramwell and Whitfield, 1970; Bramwell, 1971). This animal was far too large for slow, near-hovering flight to be possible, so it could not have taken off from rest on level ground in still air. It seems clear from its shape that it could not have made a fast taxiing run. However, because its minimum speed was so low it could probably gain enough lift to become airborne, simply by spreading its wings and facing into a mild breeze.

Consider Fig. 10.14a again. Let the downward component of the animal's velocity, relative to the air, be v ($=u\tan\theta$ where u is the horizontal component). For equilibrium

$$D = mg\sin\theta \simeq mgv/u. \tag{10.13}$$

(The approximation holds only if θ is small). We can estimate the sinking speed v, if we can calculate the drag. The drag on the trunk and wings will be estimated separately.

The drag on wingless trunks of locusts and vultures has been measured (Weis-Fogh, 1956; Pennycuick, 1971). The drag coefficient C_{Df} based on frontal area S_f was found to be about 0.4 for the vulture and nearly 1 for the locust, far more than for streamlined bodies. Hence it must have been mainly pressure drag and it can be assumed that the drag coefficient was independent of speed. This component of the drag can be estimated as $\tfrac{1}{2}\rho u^2 S_f C_{Df}$, where $S_f C_{Df}$ is 1.3×10^{-2} m^2 for the vulture and 8×10^{-5} m^2 for the locust (Weis-Fogh, 1956; Pennycuick, 1971). For geometrically similar animals, $S_f C_{Df}$

should be proportional to (body mass)$^{0.67}$, and a locust-sized vulture should have a value of only 5×10^{-5} m^2. The higher value for the real locust is partly due to its long legs.

The drag on the wings will be estimated using Equation 10.1. This Equation is appropriate for the locust but possibly not for the vulture as flow over the vulture wing might be turbulent. Fortunately, the possible error makes little difference to the estimates of total drag. Hence

$$D = \tfrac{1}{2}\rho u^2 \{S_f\, C_{Df} + S_p[2.6R^{-1/2} + (C_L^2/\pi A)]\}. \tag{10.14}$$

Equation 10.11 gives the value of C_L. By putting the expression for D into Equation 10.13 we get an equation giving the sinking speed v.

$$v = (\rho u^3/2mg)(S_f\, C_{Df} + 2.6\, S_p\, R^{-1/2}) + (2mg/\rho \pi A S_p\, u). \tag{10.15}$$

Figure 10.15a shows graphs of sinking speed against forward speed, calculated for the locust and the vulture from Equation 10.15. Note that sinking speed increases downwards on the graph. Also shown is a similar graph for the

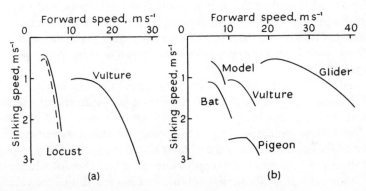

Fig. 10.15 Graphs of sinking speed v against forward speed u for gliding animals. (a) shows theoretical graphs for a 2 g locust and a 7.6 kg vulture. (b) shows the actual performance of a full-sized glider, a model glider, a 120 g bat (*Rousettus*), a 400 g pigeon (*Columba*) and a 1.8 kg vulture (*Coragyps*) (data from Pennycuick, 1968a; 1971 and Parrott, 1970).

locust calculated from Jensen's (1956) measurements of the drag on an intact locust. The main difference between the graph calculated from the equation and this more realistic one is that the latter shows high sinking speeds at forward speeds near the minimum. This difference arises because the equation ignores pressure drag on the wings, which becomes large at low speeds when the wings are nearly stalled.

Gyps africanus is similar to *G. ruppellii* in size and shape. Pennycuick (1971) followed specimens in a motor glider and estimated that their mean sinking speeds were about 1.5 m s^{-1} when gliding with forward air speeds between 13 and 20 m s^{-1}. The vulture curve in Fig. 10.15a is probably reasonably realistic.

269

Figure 10.15b shows more precise information on the gliding performance of birds and a bat. The animals were trained to glide in tilted wind tunnels. The jet was directed at an angle upwards and the animal glided in such a way as to remain stationary. The angle was altered and the minimum angle at which gliding occurred without flapping was determined for each airspeed. The vulture was a much smaller species than *Gyps ruppellii*. It and the bat and also a falcon (Tucker and Parrott, 1970) had minimum sinking speeds around 1 m s^{-1} and so did the fulmars (*Fulmarus*) studied in the field by Pennycuick (1960). The pigeon sank much faster.

Large animals almost inevitably have larger wing loadings than small ones. This implies that their minimum gliding speeds are higher than for small animals and also that they can (and must) glide faster, for similar sinking speeds. The minimum gliding gradient is lower for them than for smaller animals. A small glider like a locust cannot make progress against any but the lightest winds, and it cannot travel as far as larger gliders for given loss of height.

Birds can reduce their wing area and so increase their wing loading. This also increases the forward speed for any given sinking speed. They lose height less rapidly when gliding fast if they partly fold their wings, and there is an optimum wing area for any given forward speed. Pigeons (Pennycuick, 1968) and falcons (Tucker and Parrott, 1970) greatly reduce their wing area at high speeds, but the bat wing can only be kept taut over a very limited range of areas (Pennycuick, 1971).

Animals often need to glide at a steeper gradient than the minimum for the speed in question. Pennycuick (1971) has shown how vultures achieve this by reducing the wing area below the optimum for their speed and by lowering their feet and so increasing drag. They seem also sometimes to reduce the camber of the proximal part of the wing by allowing the secondary feathers to rise, and so make the distal part supply more of the lift. This disturbance of the distribution of lift along the wing should increase induced drag. Pennycuick and Webbe (1959) described how fulmars use their wings, tail and feet to control gliding manoeuvres.

Baudinette and Schmidt-Nielsen (1974) measured the metabolic rate of gulls (*Larus*) gliding in a wind tunnel. They found it was only about double the resting rate, and much lower than for flapping flight.

10.4.2 Soaring

Soaring techniques enable birds to remain airborne for long periods without flapping their wings, and so at relatively little energy cost.

One important technique uses thermals, which are rising columns or bubbles of warm air, due to solar heating of the earth's surface. Strong thermals are generally formed only over land, and only from mid-morning until evening. Pennycuick (1972) made fascinating observations from a motor glider of the soaring behaviour of vultures, storks and other birds over East Africa. These birds depend mainly on thermals to keep them airborne for many hours at a time.

If it is to gain much height in a thermal, a bird must remain in the rising air for some time. Typically, it will circle for a while in the thermal (gaining height) glide on towards its destination (losing height) and then find another thermal and rise again (Fig. 10.16).

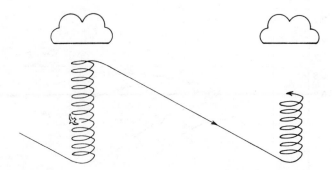

Fig. 10.16 Thermal soaring. Cumulus clouds tend to form at the tops of thermals (from Pennycuick, 1972).

Thermal soaring depends on ability to glide in fairly small circles. Figure 10.14b shows forces acting on a circling bird. It must bank at an angle ϕ so that the lift has a horizontal component which provides the necessary centripetal force, mu^2/r

$$L \sin \phi = mu^2/r. \tag{10.16}$$

The lift cannot exceed $\frac{1}{2}\rho S_p u^2 C_{L\,max}$, where $C_{L\,max}$ is the maximum attainable lift coefficient. Also, $\sin \phi$ can never be as high as 1. Hence

$$\frac{1}{2}\rho S_p u^2 C_{L\,max} > mu^2/r,$$
$$r > 2m/\rho S_p C_{L\,max}. \tag{10.17}$$

This sets an unattainable lower limit to the circling radius. The limit is proportional to the wing loading, and is 9 m for the 7.6 kg vulture considered in Section 10.4.1 (assuming $C_{L\,max} = 1.5$).

A circling bird has a higher sinking speed v than if it were gliding along a straight path at the same forward speed u. This is because the additional lift required for circling implies additional induced drag. Pennycuick (1971) calculated the probable circling performance of *Gyps africanus*, assuming a straight-glide performance very like that shown for *Gyps ruppellii* in Fig. 10.15a. He obtained the family of curves, each for a different angle of bank ϕ, shown in Fig. 10.17. The envelope of these curves shows the minimum attainable sinking speed, for each radius. Its asymptotes are the minimum radius given by Equation 10.17 (in this case 8 m) and the minimum sinking speed attainable

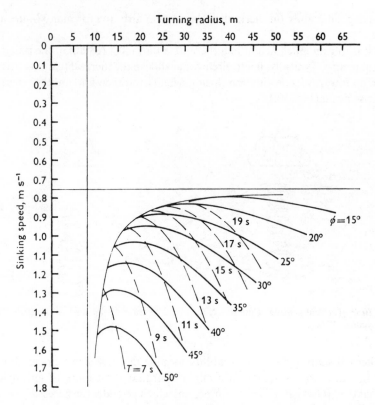

Fig. 10.17 Estimated graphs of sinking speed against radius of circling for the vulture *Gyps africanus*, for particular angles of bank ϕ (continuous lines) and for particular times per revolution (broken lines) (from Pennycuick, 1971).

in a straight glide (0.76 m s^{-1}). The rate of gain of height in a thermal is the difference between the upward velocity of the air and the downward velocity v of the bird relative to the air. The optimum circling radius is a compromise between the advantages of the faster up currents near the centre of the thermal and the lower values of v attainable when the radius is large. *Gyps africanus* seems usually to circle with radii between 15 and 25 m.

Many birds, particularly sea birds which nest on cliffs, soar in the upward currents of air which result from slopes or cliffs deflecting the wind upwards. The bird may be able to remain motionless if the wind is faster than its minimum gliding speed: otherwise it must soar along the slope, or tack backwards and forwards along it. Pennycuick and Webbe (1959) described the slope-soaring behaviour of the fulmar, *Fulmarus*. The kestrel *Falco tinnunculus* often slope-soars while searching for prey. McGahan (1973a) has described slope-soaring by the condor, *Vultur*.

The albatross (*Diomedea*) probably depends partly on slope soaring, where the wind is deflected upwards by waves (Wilson, 1975). It also uses a different technique of soaring which enables it to keep airborne over the sea for hours, hardly ever flapping its wings (Jameson, 1958). It glides downwind gaining speed and so kinetic energy, and turns into the wind when close to the surface of the sea. It rises again, converting some kinetic energy into potential energy. Though this reduces its velocity relative to the ground its speed relative to the air may increase, because the wind speed increases with height above the sea. Having reached a height of 12 m or so, it turns downwind again. It inevitably drifts gradually downwind, and is carried around Antarctica by the prevailing westerlies. Walkden (1925) showed that the faster an albatross could glide, the weaker would be the wind gradient needed for soaring.

Albatrosses have much smaller wing areas than vultures of similar mass, and so glide faster but cannot circle with as small radius. Their aspect ratio is much higher so they must suffer less induced drag and should lose height less fast in a straight glide. Pennycuick (1971) compared his estimates of circling performance of vultures with similar estimates for an albatross of the same mass. He concluded that the albatross should be able to climb faster than a vulture circling with the same radius in the same thermal, provided that radius exceeded 17 m. The large wing area of vultures may be an adaptation to soaring in small, early morning thermals, or it may be necessary to enable them to take off from the ground.

10.5 Fast flapping flight

Fast flapping flight has been studied more thoroughly for the locust than for any other animal (Weis-Fogh, 1956; Jensen, 1956). Weis-Fogh suspended a locust on a light pendulum in a wind tunnel. The wind was started, and the locust flew. The wind speed was adjusted so that the pendulum remained vertical, showing that the forward thrust produced by the wings balanced the drag on the body (the drag on the pendulum was balanced by an electromagnet). The pendulum was suspended from a balance, and the locust was filmed when the balance reading showed that the upward force generated by the wings was approximately equal to the weight of the locust. The forces produced by the locust were then the same as if it had been flying free at a forward speed equal to the wind speed. Figure 10.18 has been drawn from one of these films.

Weis-Fogh (1956) measured the drag on the wingless body of a locust, in the flying position, at various wind speeds. Jensen (1956) measured the lift and drag on individual wings, at various angles of attack (Fig. 10.1 includes data obtained by him). He mounted the wings near the wall of the wind tunnel so that the wind travelled faster over their tips than their bases, as nearly as possible as in flight. He used this data to calculate the forces acting on the wings in the position shown in each successive frame of Weis-Fogh's film. The mean upward force calculated in this way for a complete cycle of wing movements was approximately equal to the weight of the locust. The mean forward force was approximately equal to the drag on the wingless trunk. This indicates that

273

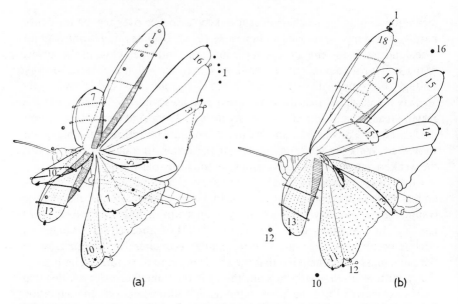

Fig. 10.18 Tracings from a film of a tethered locust *Schistocerca gregaria* flying in a wind tunnel. (a) shows the downstroke and (b) the upstroke. The upper surfaces of the wings are stippled and the flap along the hind edge of the fore wing is hatched. The transverse lines on the fore wing are fine hairs glued on to show the angle of the wing (from Weis-Fogh, 1973).

the aerodynamic force acting on a wing at any instant in flight is probably approximately equal to the force which acts on the same wing when it is held stationary at the same angle of attack and the same air speed.

In the downstroke the fore wings have an angle of attack such that the lift on them is much greater than the drag and the resultant aerodynamic force probably acts upwards and slightly forwards, helping both to support the animal's weight and to provide the thrust needed to propel it forwards (Fig. 10.19). In the upstroke the angle of attack is negative and might be expected to give rise to downward lift, but at this stage in the stroke the wing is bent so as to have a rather Z-shaped section and wind tunnel tests show that this gives it rather peculiar aerodynamic properties. The lift is probably small and acts in the opposite to the expected direction, and the resultant force on the wing is a small, more or less horizontal retarding force. The hind wings produce a large upward force in the downstroke, and some thrust during both wing strokes.

In fast flapping flight the induced drag and the drag on the trunk can be expected to be about the same as in gliding at the same forward speed, and the friction drag on the wings to be slightly more (because their velocity relative to the air is a little higher). This is, essentially, the basis of Pennycuick's (1975) theory of bird flight. Hence the aerodynamic power requirement for flight should be about the same as the rate of loss of potential energy in gliding at the same speed. Figure 10.15a shows that a gliding locust would probably

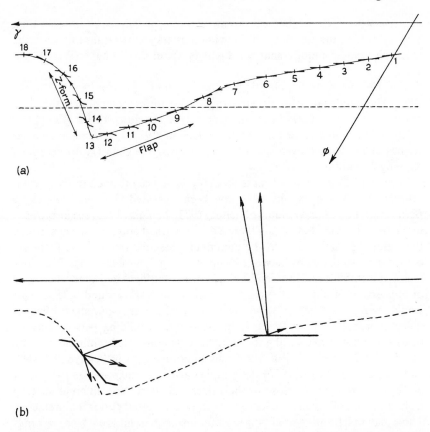

(a)

(b)

Fig. 10.19 (a) The movement relative to the air of a chordwise section through the mid-point of a locust wing, drawn from the same film as Fig. 10.18 (from Jensen, 1956). (b) A diagram showing forces believed to act on the fore wing at two stages in its cycle of movements.

sink at 0.55 m s⁻¹ when its forward speed was 3.5 m s⁻¹, and faster at other forward speeds. The rate of loss of potential energy would be 5.4 W kg⁻¹. Jensen (1956) calculated from his analysis of forces on the wings that the net aerodynamic power requirement for flight at this speed was 8 W kg⁻¹ but that this was the total of a larger positive power and some negative power (The wing muscles do negative work during the upstroke when they are being extended by the lift on the wings). The inertial power requirement can be calculated in the same way as for hovering, taking account of the wing movements not being sinusoidal. The total power requirement calculated as explained in the account of Fig. 10.12 is 14 W kg⁻¹ positive power and 6 W kg⁻¹ negative power. Weis-Fogh (1961) has shown that the cuticle of the locust thorax and resilin structures in it can save by elastic storage a large proportion of the energy which would otherwise have to be dissipated by muscles doing negative work. The mechanical power requirement may be little more than 8 W kg^{-1}. If so the efficiency of the wing muscles seems rather low since the metabolic power requirement for flight is about 76 W kg^{-1} (Weis-Fogh, 1964).

The principle of fast flapping flight is the same for birds as for locusts (Brown, 1963). The aerodynamic force on the wings has an upward component throughout the cycle, but a forward component only in the downstroke. Bilo (1971, 1972) Oehme and Kitzler (1974) and McGahan (1973b) have made detailed studies of the wing movements of various birds. Vaughan (1970) described flapping flight of bats.

Various birds and bats of mass 35–780 g have been trained to fly in wind tunnels, and their metabolic rates have been measured (Tucker, 1968, 1972; Bernstein, Thomas and Schmidt-Nielsen, 1973; Thomas, 1975; see also Chapter 6). In some cases, but not all, there is a well-marked optimum speed at which the metabolic rate is lowest. The minimum metabolic rate in level flight exceeded the resting rate by between 50 and 90 W kg^{-1} in each species. Measurements of metabolic rate in ascending and descending flight indicated efficiencies of conversion of chemical energy to mechanical work around 0.25, as one would expect (see Chapter 3), so the mechanical power requirement for level flight is probably 12–23 W kg^{-1}. Investigations of gliding performance of birds and a bat of mass 120 g to 1.8 kg have demonstrated minimum sinking speeds of 1.0–2.5 m s^{-1} (Fig. 10.15 and additional data in Pennycuick, 1960 and Tucker and Parrott, 1970). This implies aerodynamic power requirements of 10 to 25 W kg^{-1}, so close to the values calculated (for different species) from the metabolic data as to leave little scope for inertial power requirements. These may well be quite small. Pigeons, for instance, of mass 0.4 kg have wings of moment of inertia 1.8×10^{-4} kg m^2 (Pennycuick, 1968b). When they glide their sinking speed is least at a forward speed of about 14 m s^{-1} (Fig. 10.15b). In level flapping flight at this speed they flap their wings with frequency 5.5 s^{-1} and amplitude (γ_0) 0.8 rad. Equation 10.9 shows that the inertial power requirement is only 4 W kg^{-1}. Most of the kinetic energy given to the wing for the downstroke could be used to do aerodynamic work at the end of the stroke so only 2–3 W kg^{-1} need be added to the aerodynamic power requirement to obtain an estimate of the total power requirement.

References

Alexander, R.McN. (1968) *Animal Mechanics*. Sidgwick and Jackson, London.
Alexander, R.McN. (1971) *Size and shape*. Arnold, London.
Baudinette, R.V. and Schmidt-Nielsen, K. (1974) Energy cost of gliding flight in herring gulls. *Nature*, **248**, 83–84.
Bernstein, M.H., Thomas, S.P. and Schmidt-Nielsen, K. (1973) Power input during flight of the fish crow, *Corvus ossifragus*. *J. exp. Biol.*, **58**, 401–10.
Bilo, D. (1971) Flugbiophysik von Kleinvögeln. I. Kinematik und Aerodynamik des Flugelabschlages beim Haussperling (*Passer domesticus* L.) *Z. vergl. Physiol.*, **71**, 382–454.

Bilo, D. (1972) Flugbiophysik von Kleinvögeln. II. Kinematik und Aerodynamik des Flugelaufschlages beim Haussperling (*Passer domesticus* L.) *Z. vergl. Physiol.* **72**, 426–437.

Bramwell, C.D. (1971) Aerodynamics of *Pteranodon. Biol. J. Linn. Soc.* **3**, 313–328.

Bramwell, C.D. and Whitfield, G.R. (1970) Flying speed of the largest aerial vertebrate. *Nature* **225**, 660–661.

Brown, R.H.J. (1963) The flight of birds. *Biol. Rev.* **38**, 460–489.

Dickinson, S. (1929) The efficiency of bicycle-pedalling, as affected by speed and load. *J. Physiol.* **67**, 242–255.

Greenewalt, C.H. (1962) Dimensional relationships for flying animals. *Smithson. misc. Collns.* **144**(2), 1–46.

Hertel, H. (1966) *Structure, Form, Movement.* Reinhold, New York.

Hocking, B. (1953) The intrinsic range and speed of flight of insects. *Trans. R. ent. Soc. Lond.* **104**, 223–345.

Jameson, W. (1958) *The wandering Albatross.* Hart-Davis, London.

Jensen, M. (1956) Biology and physics of locust flight. III. The aerodynamics of locust flight. *Phil. Trans. R. Soc. Ser. B* **239**, 511–552.

Jensen, M. and Weis-Fogh, T. (1962) Biology and physics of locust flight. V. Strength and elasticity of locust cuticle. *Phil. Trans. R. Soc. Ser. B* **245**, 137–169.

Lasiewski, R.C. (1963) Oxygen consumption of torpid, resting, active and flying humming-birds. *Physiol. Zool.* **36**, 122–140.

Lighthill, M.J. (1973) On the Weis-Fogh mechanism of lift generation. *J. Fluid Mech.* **60**, 1–17.

McGahan, J. (1973a) Gliding flight of the Andean condor in nature. *J. exp. Biol.* **58**, 225–237.

McGahan, J. (1973b) Flapping flight of the Andean condor in nature. *J. exp. Biol.* **58**, 239–253.

Nachtigall, W. (1966) Die Kinematik der Schlagflugelbewegungen von Dipteren. *Z. vergl. Physiol.* **52**, 155–211.

Nachtigall, W. (1967) Aerodynamische Messungen am Tragflugelsystem segelnder Schmetterlinge. *Z. vergl. Physiol.* **54**, 210–231.

Nachtigall, W. and Kempf, B. (1971) Vergleichende Untersuchungen zur flugbiologischen Funktion des Daumenfittichs (Alula spuria) bei Vogeln. *Z. vergl. Physiol.* **71**, 326–341.

Nachtigall, W. and Wieser, J. (1966) Profilmessungen am Taubenflugel. *Z. vergl. Physiol.* **52**, 333–346.

Oehme, H. and Kitzler, U. (1974) Uber die Kinematik des Flugelschlages beim Unbeschleunigten Horizontalflug. *Zool. Jb. Physiol.* **78**, 461–512.

Oehme, H. and Kitzler, U. (1975) Zur Geometrie des Vogelflugels. *Zool. Jb. Physiol.* **79**, 402–424.

Pennycuick, C.J. (1960) Gliding flight of the fulmar petrel. *J. exp. Biol.* **37**, 330–338.

Pennycuick, C.J. (1968a) A wind-tunnel study of gliding flight in the pigeon *Columba livia. J. exp. Biol.* **49**, 509–526.

Pennycuick, C.J. (1968b) Power requirements for horizontal flight in the pigeon *Columba livia. J. exp. Biol.* **49**, 527–555.

Pennycuick, C.J. (1969) The mechanics of bird migration. *Ibis,* **111**, 525–556.

Pennycuick, C.J. (1971a) Gliding flight of the white-backed vulture *Gyps africanus. J. exp. Biol* **55**, 13–38

Pennycuick, C.J. (1971b) Control of gliding angle in Rupell's griffon vulture *Gyps ruppellii. J. exp. Biol.* **55**, 39–46.

Pennycuick, C.J. (1972) Soaring behaviour and performance of some East African birds, observed from a motor-glider. *Ibis* **114**, 178–218.

Pennycuick, C.J. (1975) Mechanics of flight. In: *Avian Biology* **5**, 1–75. Academic Press, New York.

Pennycuick, C.J. and Webbe, D. (1959) Observations on the fulmar in Spitzbergen. *British Birds* **52**, 321–332.

Parrott, G.C. (1970) Aerodynamics of gliding flight of a black vulture *Coragyps atratus. J. exp. Biol.* **53**, 363–374.

Prandtl, L. and Tietjens, O.G. (1957) *Applied Hydro- and Aeromechanics.* Dover, New York.

Rees, C.J.C. (1975a) Form and function in corrugated insect wings. *Nature* **256**, 200–203.

Rees, C.J.C. (1975b) Aerodynamic properties of an insect wing section and a smooth aerofoil compared. *Nature* **258**, 141–142.

Schmitz, F.W. (1960) *Aerodynamik des Flugmodells* ed. 4. Lange, Duisburg.

Thomas, S.P. (1975) Metabolism during flight in two species of bats, *Phyllostomus hastatus* and *Pteropus gouldii. J. exp. Biol.* **63**, 273–294.

Tucker, V.A. (1968) Respiratory exchange and evaporative water loss in the flying budgerigar. *J. exp. Biol.* **48**, 67–87.

Tucker, V.A. (1972) Metabolism during flight in the laughing gull, *Larus atricilla. Am. J. Physiol.* **222**, 237–245.

Tucker, V.A. and Parrott, G.C. (1970) Aerodynamics of gliding flight in a falcon and other birds. *J. exp. Biol.* **52**, 345–367.

Vaughan, T.A. (1970) Flight patterns and aerodynamics. In: *Biology of Bats*, Wimsatt, W.E., Ed., **1**, 195–216. Academic Press, New York.

Vogel, S. (1962) A possible role of the boundary layer in insect flight. *Nature* **193**, 1201–2.

Vogel, S. (1967a) Flight in *Drosophila*. II. Variations in stroke parameters and wing contour. *J. exp. Biol.* **46**, 383–392.

Vogel, S. (1967b) Flight in *Drosophila*. III. Aerodynamic characteristics of fly wings and wing models. *J. exp. Biol.* **46**, 431–443.

Walkden, S.L. (1925) Experimental study of the soaring of albatrosses. *Nature* **116**, 132–134.

Weis-Fogh, T. (1952) Fat combustion and metabolic rate of flying locusts (*Schistocerca gregaria* Forskal) *Phil. Trans. R. Soc. Ser. B* **237**, 1–36.

Weis-Fogh, T. (1956) Biology and physics of locust flight. II. Flight performance of the desert locust (*Schistocerca gregaria*). *Phil. Trans. R. Soc. Ser. B* **239**, 459–510.

Weis-Fogh, T. (1960) A rubber-like protein in insect cuticle. *J. exp. Biol.* **37**, 889–907.

Weis-Fogh, T. (1961a) Thermodynamic properties of resilin, a rubber-like protein. *J. mol. Biol.* **3**, 520–531.

Weis-Fogh, T. (1961b) Power in flapping flight. In: *The Cell and the Organism*, Ramsay, J.A. and Wigglesworth, V.B., Eds., pp. 283–300. Cambridge University Press.

Weis-Fogh, T. (1964) Biology and physics of locust flight. VIII. Lift and metabolic rate of flying locusts. *J. exp. Biol.* **41**, 257–271.

Weis-Fogh, T. (1972) Energetics of hovering flight in hummingbirds and *Drosophila. J. exp. Biol.* **56**, 79–104.

Weis-Fogh, T. (1973) Quick estimates of flight fitness in hovering animals, including novel mechanisms for lift production. *J. exp. Biol.* **59**, 169–230.

Weis-Fogh, T. and Alexander, R.McN. (1977) The sustained power output obtainable from striated muscle. In: *Scale effects in Animal Locomotion*, Pedley, T.J., Ed., Academic Press, London.

Wilson, J.A. (1975) Sweeping flight and soaring by albatrosses. *Nature* **257**, 307–308.

11 Locomotion of Protozoa and single cells

D. V. Holberton

11.1 Introduction

Protozoa solve the problems faced by whole organisms from the resources of a single cell. Like multicellular forms, their locomotion ultimately derives from forces developed between elements the size of macromolecules or assemblies of macromolecules. However, unlike many multicellular forms the process is coupled to these intracellular events in a very direct way.

We understand in some detail how forces arise in striped muscle from the interaction of fibrous proteins. We are far less well acquainted with the physiology of Protozoan motility. Most Protozoa are motile, but many are smaller than muscle fibres, and their motile organelles even smaller. Only a few are readily cultured in a laboratory. The principle of 'organic design' (Grimstone, 1959) suggests to us that constraints on the radiation of biochemical systems are relatively severe, and that the emergence of diverse contractile or motile mechanisms will be paced by the rate of evolution of functional molecules.

For these reasons, when structural studies seem to justify the comparison, it has been easier in the first instance to explain the primitive movements of cells by transferring arguments from how muscle shortens, than to experiment at first hand. At such times the premises made most often, which may have some claim to generality, are: that cytoplasmic filaments actively slide over one another, that cross-bridges carry out work cycles involving the splitting of ATP, and that calcium plays a controlling role.

All the more intriguing, then, when a quite separate mechanochemistry is found to govern the fast contractions of certain ciliates. 'Myonemal' responses—stalk shortening of peritrichs and body shortening of the heterotrichs *Spirostomum* and *Stentor*—are novel (to us) in drawing on the chemical potential of calcium rather than on ATP as the immediate energy source for developing tension (Amos, 1971; Weis-Fogh and Amos, 1972; Hawkes and Holberton, 1975; Holberton and Ogle, 1975). The engine in these cases seems to be the collapse on calcification of an extended and relatively simple protein aggregate, and not an active shear between actins and myosins or their analogues (Amos, Routledge and Yew, 1975).

279

Not all motile responses of single cells (myonemal contraction for instance) result in locomotion – when the organism as a whole makes forward progress with reference to a frame fixed by the observer. The repertoire includes purely intracellular motions of cytoplasm: cyclosis, particle transport along axopodia, streaming and membrane transport accompanying feeding through the tentacles of suctorians (Bardele, 1972), or the cytopharynges of gymnostomous ciliates (Wessenberg and Antiba, 1970; Tucker, 1972); the sinuous and peristaltic movements of gregarines (Stebbings, Boe and Garlick, 1974); the bending of flagellate axostyles (McIntosh, 1973). It is probable that some of these movements engage molecular machinery similar to (or the same as) that which drives the locomotory responses described in this chapter. When the mechanochemistry of all of these responses is better understood it may prove expedient to catalogue them according to mechanism (Sleigh, 1972): whether, for instance, dynamic polymerization of aggregation or proteins into tubules or fibres is involved; whether sliding of filaments or tubules is motivating; whether ATP splitting is intrinsic to the mechanism, etc.

In the meantime, considerable grey areas remain when we attempt to write a formal mechanics of cell locomotion in terms of the behaviour of cell components. Despite our deepened insight into the molecular biology of motility, the progralemma suggested by Wolpert in 1965 (*a propos* amoeboid movement) is still apt.

Wolpert states 'We need to know the nature and site of action of the forces (that produce movement), the structures that generate the forces and the structures on which they act, how the forces are generated, the energy supply, and the mechanism by which the forces are controlled.'

11.2 Mechanics of locomotion by cilia and flagella

11.2.1 The movements of cilia and flagella

Cilia and flagella are thin motile extensions of the cytoplasm of the protozoon bearing them. The vigorous bending movements that these organelles are seen to undergo are caused by configurational changes within the fibrous protein core, or axoneme.

In general, flagellar movements are flat or three-dimensional waves of bending that travel along the shaft with the effect of producing a net water displacement parallel to the wave axis. Flagellates with one or a few flagella are either pulled or pushed through the fluid, depending on whether waves travel from the tip of the flagellum towards its base, or from the base towards the tip. Ciliates are usually larger than flagellates, and are flat or ovoid organisms propelled by rows of beating cilia which vect water across the surface of the protozoon. Each cilium completes a characteristic asymmetrical cycle of bending. During the effective, or power, stroke the shaft is straight and stiff, and swings through an arc of movement by bending solely at the base. This movement is followed by the recovery stroke in which the bend, in moving toward the tip of the shaft, draws the now limp shaft back close to the surface of the cell. The recovery stroke may be completed in the same plane as the effective stroke (e.g. *Spirostomum* body cilia), but it is recognized that in many cases the envelope of bending is three-dimensional – the travelling bend of the recovery movement swings the tip of the cilium out in a sideways arc around the base of the cilium (Fig. 11.16).

Some authors prefer to acknowledge the fundamental identity of ciliary and flagellar structure and mechanism by adopting the single term *undulo-podium* (Frey-Wyssling, 1965; Jahn and Votta, 1972). This avoids problem-atical intermediate cases, such as when cilia beat helically (Preston, Jahn and Fonseca, 1970), or those flagellates propelled by oar-like (tonsate) strokes of their flagella (e.g. *Entosiphon*, *Chlamydomonas*, *Polytoma*) or by numbers of flagella beating in unison (e.g. *Koruga*, *Deltotrichonympha*). However, Sleigh (1974) has pointed out that for ease of analysis the need remains to distinguish different patterns of movement, even if these do not divide conveniently along taxonomic lines.

Cilia are closely and regularly packed on the surface of a ciliate protozoon. Their individual oscillations are entrained in a way that minimizes hydro-dynamic interference and imparts a steady relative velocity to the fluid in a direction opposite to that in which the organism as a whole is seen to move. Visible waves of co-ordinated beating spread across a field of cilia. The wave-front marks a row of cilia beating in register (synchrony). Cilia aligned by sight lines drawn in any other direction across the field are out of phase (metachrony), and it is clear that the greatest phase difference between adjacent cilia will coincide with the direction of wave travel – the mainline of meta-chrony. Metachronism is not a property confined to the reciprocating beat pattern of true cilia. Fields of helical oscillators may also give rise to systems of travelling waves, for example the mantle of flagella covering the anterior parts of the polymastigote flagellates *Koruga* and *Deltotrichonympha*, and the large number of symbiotic spirochaetes clinging to the surface of *Mixotricha* (Cleveland and Cleveland, 1966; Machemer, 1974).

The effect of the medium on helical or plane waves travelling along a flagel-lum can be resolved into transverse and longitudinal vectors of thrust, and the forward propulsion of the flagellate is the reaction to the sum of the longitud-inal components. If the flagellum bears an even number of half waves, trans-verse thrusts cancel out (Fig. 11.1). If the flagellum is fractionally longer or shorter than the arc wave-length, an additional vector force arises from that part of the shaft whose motion is not balanced by bending activity, of opposite sense, elsewhere. The extra force component rotates with the bending phase and becomes more influential for shorter flagella where other thrust com-ponents are smaller. Machemer (1974) has argued that the unbalanced vector is the principal force component from a very short oscillator the length of a cilium, and if the axis of oscillation is inclined to the protozoon body, a sig-

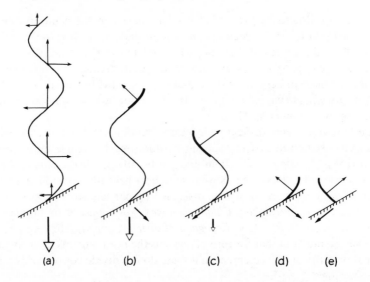

Fig. 11.1 Fluid reaction to thrusts of a helical oscillator with an integral number of bending waves (a) is, at the base of the organelle, a motive force (open-headed arrow) parallel to the axis of motion. (b) and (c): an additional force at the base (solid arrow) is the reaction to the unbalanced thrust vector from a shaft propagating incomplete waves. The unbalanced force rotates with the bending phase and is more significant for a shorter flagellum where axial thrusts are reduced. (d) and (e): for an oscillator of cilium length, fluid reaction is entirely to the unbalanced vector. If the insertion of the organelle is angled to the cell surface, as drawn here, the organism may be propelled in a direction parallel to the surface by a force that is 'effective' for only part of the cycle of motion (e) (after Machemer, 1974).

nificant motive force is generated parallel to that surface. In this simple ciliary motion, the direction in which water is moved by a beating cilium can be reconciled with a fundamental (primitive?) wave pattern that is helical. In most ciliates an effect of this sort arising purely from the length of the cilium is amplified by polarization of the beat cycle in time and space. Real cilia tend to increase their angular velocity, straighten, and flatten the circular contour of the beat pattern during the effective stroke (cf. *Paramecium* cilium, Fig. 11.16).

11.2.2 Hydrodynamics of swimming micro-organisms

In this section, we seek an explanation in basic fluid-mechanical terms of the self-propulsion of flagellated and ciliated cells. Mathematical models hold more than theoretical interest, they predict properties of movement, such as the swimming speeds for various beat patterns or wave geometries, and these may be verified by direct observation. All models include a level of approximation, but accurate models give confidence in estimates of important quantities used in their formulation, such as the force exerted on the fluid during movement.

Local fluid resistive forces give rise to bending moments at points along the axoneme and must enter into any biophysical explanation of bend propagation (see Section 11.2.5). Energy is dissipated in overcoming viscous forces, and is the least energy debt that must be met by the chemical input to the mechanically coupled reactions of the axonemal machine (Holwill, 1974).

The propulsion of small flagellates by one or a few widely spaced flagella can be reasonably described hydrodynamically by simple models in which flagella independently propagate regular waves. On the other hand, the close packing of cilia on the surface of a ciliate, often within a few ciliary diameters of each other, presents problems of hydrodynamic interference which (though we can be sure that the organism copes) are serious for the mathematician. Indeed, this principle has been invoked as a causative factor of synchrony and metachrony (Sleigh, 1962). To some extent it will not be valid to treat cilia within a field of cilia as independent oscillators, relating propulsion to the sum of their individual directional thrusts. The purpose of modelling might be better served by taking the extreme view ('the envelope model') that the metachronal wavefront is a moving boundary exerting its own thrust on the fluid environment, and that the influence of ciliary activity beneath the boundary is negligible by comparison. Since the locomotion of many ciliates is in a direction, and with a rotation, that follows the ciliary power stroke rather than the mainline of metachrony (Fig. 11.6; Machemer, 1972), the envelope approach would appear to be limited. The swimming of these organisms may be fitted better by the 'sublayer model' which allows for the effect of the cell surface and the interactions between cilia, on the velocity field in the fluid beyond the cell, even though the mathematics of this model are developed in the first instance for cilia in a flat field. Some aspects of these models are outlined below, they are extensively formulated by Blake (1971a, 1971b, 1971c, 1972, 1973), and the basic ideas and the problems attaching to them are reviewed by Blake and Sleigh (1974).

Fluid dynamics

To the fluid-dynamicist, micro-organisms moving through water inhabit a realm governed by low Reynolds number hydrodynamics. A fish swimming in the same medium occupies a different, high Reynolds number niche. Reynolds number (Re) is a dimensionless ratio that compares the inertial forces to the viscous forces acting on a body moving through a fluid. Formally, it is found by relating characteristics of the fluid and the body in the following way:

$$\text{Re} = \frac{\rho UL}{\mu} = \frac{\text{fluid density} \times \text{velocity} \times \text{size}}{\text{dynamic viscosity}} \qquad (11.1)$$

In relative terms, a fish is large, its rapid tail movements impart momentum to the fluid and generate inertial forces proportional to the rate of change of

283

momentum (Lighthill, 1969). Protozoan flagella, on the other hand, are small (0.2 μm in diameter) and move slowly in water (10–10^4 μm s^{-1}). The appropriate Reynolds number will be of the order 10^{-6}–10^{-3}, which means that their interaction with the fluid is dominated by viscous forces. For an explanation of their motion we must turn to the body of theory governing viscous flow problems and summarized in the Navier–Stokes equations. Perhaps the most familiar analytical solution of these equations is Stokes' own solution for the force, F, acting on a falling sphere of radius a:

$$F = -6\pi a U \mu. \tag{11.2}$$

Fluid in contact with the sphere moves at sphere velocity U (the condition of 'no slip' at the surface of a solid body), but since the fluid is at rest an infinite distance from the sphere, there must exist a region around the sphere where fluid moves at intermediate velocities. The origin of the force acting on the sphere is the frictional stress set up between adjacent fluid elements moving at different velocities in this field.

Equation (11.2) could also be expressed:

$$F = -C_H U a \tag{11.3}$$

where C_H is known as a drag coefficient, a parameter which allows for the shape of the sphere, scaled in the equation by the radius a. Consider now a long, thin body moving through a fluid. Intuitively it seems that the drag will depend on the orientation of the body to the axis of movement. By experiment it has been found that needle-shaped bodies fall in viscous media roughly half as fast sideways as they do end-on (White, 1946). This empirical result has only recently been confirmed analytically (Weinberger, 1972). The sideways, or normal, drag on a cylinder is, in fact, less than twice the longitudinal drag on the same cylinder moving parallel to its long axis at the same speed, but approaches this value if the cylinder is allowed to stretch to an infinite length. In terms of drag coefficients, the longitudinal and normal forces on a cylinder of length L and radius d are given by:

$$F_L = -C_L V_L L, \tag{11.4}$$

and

$$F_N = -C_N V_N L. \tag{11.5}$$

When the longitudinal velocity (V_L) is the same as the normal velocity (V_N),

$$C_N \to 2C_L, \quad \text{as} \quad L \to \infty \tag{11.6}$$

and it can be shown (Hancock, 1953; Gray and Hancock, 1955; Cox, 1970) that:

$$C_L = \frac{2\pi\mu}{\ln(2L/d) - 0.5} \tag{11.7}$$

Two important steps in explaining the propulsion of cells by flagella have been: firstly, to compare sections of an undulating flagellum to cylinders moving at wave speeds (Taylor, 1952; Hancock, 1953); and secondly, the use of surface coefficients of force (Equations 11.4 and 11.5) in a balance of forces argument that gives a useful approximate solution for swimming speeds.

Swimming of smooth flagella bearing flat waves

The 'resistive theory' of locomotion was developed originally by Gray and Hancock (1955) to account for propulsion by smooth flagella beating in uniform plane sine waves. The usefulness of the theory is such that it has since been extended by a number of authors to include propulsion by other flagellar waveforms and other motile structures, and the results of these calculations are listed in Table 11.1.

Active mechanochemical transformation of the axoneme deforms the flagellar filament into waves of bending of a characteristic amplitude and wavelength that travel at a wavespeed V_w, relative to the cell body. The organism as a whole moves with a speed, U, in the positive x direction of the moving

Table 11.1 Expressions for the self-propulsion of flagellated cells

---

1. Plane sine waves on a smooth flagellum propelling a spherical body (Gray and Hancock, 1955).

$$\bar{U}/V_w = \frac{\frac{1}{2}b^2 k^2}{1 + b^2 k^2 - (1 + \frac{1}{2}b^2 k^2)^{1/2} [\ln(d/2\lambda) + \frac{1}{2}](3a/n\lambda)}$$

---

2. Helical sine wave on a smooth flagellum propelling a spherical body (Holwill and Burge, 1963).

$$\bar{U}/V_w = \frac{b^2 k^2}{1 + 2b^2 k^2 - (1 + b^2 k^2)^{1/2} [\ln(d/2\lambda) + \frac{1}{2}](3a/n\lambda)}$$

---

3. Arc-line wave on a smooth flagellum propelling a body of unspecified shape (Brokaw, 1965).

$$\frac{\bar{U}}{V_w} = \frac{\lambda}{L}\left(1 - \frac{1 + C_H/nLC_L}{2 + C_H/nLC_L + (1/L)(2\rho \sin \theta \cos \theta - 2\rho\theta - \lambda \cos \theta)}\right)$$

---

4. Arc-line wave on an isolated smooth flagellum (Brokaw, 1965).

$$\frac{\bar{U}}{V_w} = 1 - \frac{LC_L}{LC_N + (C_N - C_L)(2\rho \sin \theta \cos \theta - 2\rho\theta - \lambda \cos \theta)}$$

---

5. Helical waves on an isolated smooth flagellum, or a spirochaete rotating with 'self-spin' (Holwill and Sleigh, 1967; Chwang, Winet and Wu, 1974).

$$\frac{\bar{U}}{V_w} = \frac{b^2 k^2(1 - [C_L/C_N])}{b^2 k^2 + [C_L/C_N]}$$

Parameters: $\bar{U}$ = forward velocity; V_w = apparent wave velocity (frequency × wavelength); b = wave amplitude; λ = wavelength; k = wave number $(2\pi/\lambda)$; n = number of waves along the axis of motion; L = flagellar length; d = radius of flagellum; a = radius of a spherical body; C_L = surface coefficient of resistance along the flagellum; C_N = surface coefficient of resistance normal to the flagellum; C_H = surface coefficient of resistance of a body; ρ = radius of curvature of a flagellar bend of arc shape; 2θ = angle subtended by the arc length.

Cartesian co-ordinate system superimposed on the flagellate of Fig. 11.2a. According to Newton's Laws, the sum of the forces acting will be zero. In this case, the components of force (dF) in the x direction arising from the motion of every section, ds, of the flagellum (Fig. 11.2b), must be summed over the length (L) of the flagellum and equated with the drag that resists movement

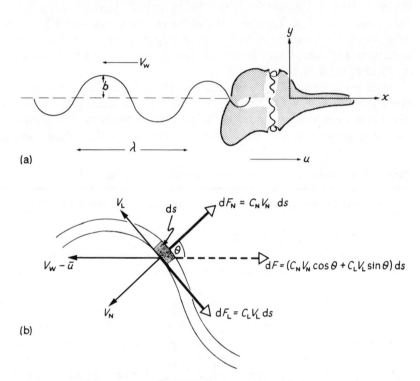

(a)

(b)

Fig. 11.2 (a) Parameters of the self-propulsion of *Ceratium* by flat bending waves of the posterior flagellum. (b) Longitudinal and normal resistive forces (open arrows) on a short element, ds, of a flagellum bending with wave speed V_w. Depending on wave shape, the longitudinal and normal velocities (V_L and V_N) may be derived from the wave speed to solve for the propulsive force, dF, along the axis of motion.

of the body. If the body is assumed spherical (Equation 11.3) this leads to:

$$\int_{x=0}^{L} dF - C_H U a = 0. \tag{11.8}$$

The value of the integral will depend on the shape of the wave, but a general equation can be written for dF in terms of normal and longitudinal drag coefficients when the displacement of the centre-line of the flagellum in the y direction is a single-valued function of distance x and time t, for instance a travelling sine wave. This condition does not hold for all real waves – consider the meander-like waveform depicted in Plate 11.1.d. Once dF is defined, expressions can be derived for the time-averaged velocity ($\bar{U}$) of the system (Table 11.1), a useful parameter that can be checked by observation. Flagellar propulsion by sine waves is mathematically attractive but not common in nature. There is evidence that flagella beating in two dimensions are fitted more exactly by a waveform of circular arcs separated by straight sections (Brokaw and Wright, 1963; Brokaw, 1965; and see Plate 11.1). Brokaw has calculated the velocity of an organism propelled by a flagellum undulating in this way, and found a value close to the result given by the sine wave approximation. Holwill and Sleigh (1967) have written the expression for propulsion by flat sine waves in a useful form that does not assume a particular numerical relationship between the drag coefficients of the flagellum:

$$\frac{\bar{U}}{V_{\mathrm{W}}} = \frac{b^2 k^2 (1 - C_{\mathrm{L}}/C_{\mathrm{N}})}{(2aC_{\mathrm{H}}/n\lambda C_{\mathrm{N}})(1 + \tfrac{1}{2}b^2 k^2)^{1/2} + (2C_{\mathrm{L}}/C_{\mathrm{N}}) + b^2 k^2} \tag{11.9}$$

where the wave parameters of this equation are defined in Table 11.1. Gray and Hancock's original expression for a smooth cylindrical flagellum is obtained by substituting values of C_{L} and C_{N} from Equations (11.6) and (11.7). In evaluating drag coefficients, a small element of a much longer flagellum is correctly modelled by a cylinder of infinite length since neither body has ends. When this is done two important properties of propulsion by smooth flagella are illustrated.

(1) Since the denominator in Equation 11.9 is positive and never smaller than $b^2 k^2$, the term $(1 - C_{\mathrm{L}}/C_{\mathrm{N}})$ indicates that the highest swimming speed attainable by a flagellate is half the wave speed of its flagellum.

(2) The arguments leading to positive values of $\bar{U}/V_{\mathrm{W}}$ require that the propulsive force from the fluid is opposed to the direction of wave travel. The result: $C_{\mathrm{L}}/C_{\mathrm{N}} < 1$ confirms the correctness of this assumption, and what is apparent from casual observation, that the organism travels in the opposite direction to the direction of wave propagation.

Hispid flagella

The flattened flagella of a number of Protozoa appear in electron micrographs not at all smooth, but to carry hairs or mastigonemes in one (stichoneme) or more than one (pantoneme) lateral rows (Plate 11.2b). The effect of mastigonemes on locomotion of some flagellates is not always obvious, for instance the simple rowing strokes made by the stichoneme flagellum of *Bodo* (Brooker, 1965). The flagellar appendages (flimmer filaments) of *Euglena* are flexible and may wrap around the flagellum during activity (Pitelka and Schooley, 1955), and may make no greater contribution to propulsion than to increase the diameter of the flagellum. But in the chrysomonad *Ochromonas*, and probably the related sessile genera *Monas*, *Actimonas*, and *Poteriodendron* (Sleigh, 1964), the mastigonemes are stiff rods standing out 1 μm or so from the flagellar shaft. Jahn, Landman and Fonseca (1964) have used this fact to account for the behaviour of the flagellate. *Ochromonas* is observed to move in the same direction as waves propagate down its long flagellum. In qualitative terms it appears that the backward thrust from mastigonemes passing over forwardly-directed bends in the flagellum exceeds the combined forward thrusts of the flagellar shaft and those overlapping mastigonemes clustered to the inside of the bends (Fig. 11.3). Alternatively, if mastigonemes project

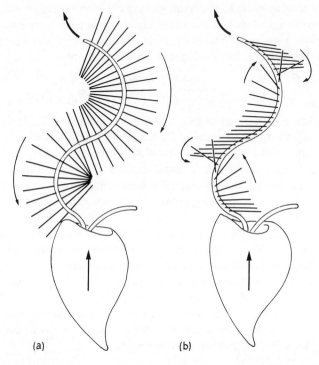

(a)   (b)

Fig. 11.3 Effect of mastigonemes in reversing the thrust from distally directed bending waves of the long flagellum of *Ochromonas*. (a) Mastigonemes in the bending plane carry out oar-like backward strokes at the crest of each wave, according to Jahn *et al.* (1964). (b) Whiplash effect described by Bouck (1972) from mastigonemes standing out of the bending plane.

out of the beat plane, as suggested by their morphological relationship to the central tubules of the axoneme, the reverse thrust may come from a whip-lash effect at the crown of each bend (Bouck, 1972). Holwill and Sleigh (1967) investigated rough flagella in detail and found a quantitative explanation of the reversed orientation of movement. Their analysis assumes that masti-gonemes are not independently motile and that they lie in the plane of the flagellar wave when arranged bilaterally, assumptions the validity of which is not certain. If there are two rows of q mastigonemes in a unit length of the flagellum, and each mastigoneme is regarded as a rigid cylindrical projection of length L, then surface coefficients of the conjugate mastigoneme-flagellum structure are:

$$C_N = C_N{}^C + 2qLC_L{}^P \qquad (11.10)$$

$$C_L = C_L{}^C + 2qLC_N{}^P \qquad (11.11)$$

where the superscripts C and P refer to the flagellar cylinder and the projections respectively. Evaluating coefficients gives $C_L/C_N = 1.8$ for the specific case covered by Equations 11.10 and 11.11, and the general condition: $C_L/C_N > 1$ for flagella with several rows of mastigonemes. Accordingly, from Equation 11.9 the ratio $\bar{U}/V_w$ takes a negative value, the circumstance under which the propulsive force acts in the same direction as wave travel.

Three-dimensional waves

Many flagellates rotate as they progress indicating that their flagella undulate in three dimensions, even where these cannot be seen directly. For example, Holwill (1965) describes the envelope of the body movements of *Crithidia oncopelti* as a 'twist surface' arising from combined rotation and side-to-side oscillation. Swimming *Giardia muris* are slowly rotated by the helical movements of two ventral flagella, though the organism may be driven forward principally by oar-like strokes of anterior and lateral flagella (Holberton, 1973). From high-speed films Lowndes (1941, 1944) early recognized that three-dimensional movements of the reflexed flagellum promote body rotations of a number of phytoflagellates and euglenoid flagellates. The trailing flagellum of *Tritrichomonas* attaches to a cytoplasmic flap of the body surface, the undulating membrane; nonetheless it has been observed to pass helical waves from base to tip (Jahn, reported in Holwill, 1966a). It is likely that many three-dimensional waves of protozoan flagella are not true helices, in most cases the cross-section of the wave is believed to be flattened to a greater or lesser extent (an exception is the anterior flagellum of *Ceratium*). This fact not withstanding, propulsion by helical waves can be analysed in a straightforward way (Holwill and Burge, 1963; Chwang and Wu, 1971; Schreiner, 1971) and remains a useful result, since numerical methods demonstrate that the swimming speed of an organism propelled by uniform flagellar waves of elliptical cross-section is bracketed by the results of calculations that assume the wave is planar or a circular helix, as might be expected (Coakley and Holwill, 1972).

It can be shown mathematically, and by physical models in water (Chwang and Wu, 1971; Taylor, 1952), that a flagellum with no body cannot progress by propagating helical waves. The reaction of the water to waves in three dimensions has, in addition to the thrust along the swimming axis, a torque component which causes an isolated flagellum to rotate around the axis at a rate that just offsets the rate of active wave travel. An attached body rotates

289

with the flagellum in the direction determined by the torque, but in so doing generates from the reaction of the water an opposite torque, the magnitude of which will depend on the size of the body and its speed of rotation. The balance of these two couples acting on the body decides the rate at equilibrium of rotation of the organism as a whole. This will always be less than the wave speed, allowing a forward thrust to be developed along the axis of motion. If wave parameters do not change, the effect of an increase in body size is twofold: allowing a greater developed thrust by slowing the rate of rotation, and additional drag along the axis of movement. Between extremes of no body, and a very large body, there must therefore be a range of body size which is optimal for effective propulsion by helical waves. According to Chwang and Wu, optimum body size is the hydrodynamic equivalent of a sphere whose radius is 15–40 times the flagellar radius, or between 1.5 μm and 4 μm for flagellates with typical smooth flagella. Since a number of Protozoa propelled by three dimensional flagellar waves are considerably larger, it is necessary to agree with Holwill (1974) that high hydrodynamic performance has not been an overriding factor in the evolution of body size by these organisms. The large size of *Euglena viridis* may be compensated by the coating of flimmer filaments that effectively doubles the diameter of its leading flagellum.

Screw-rotation of the flattened body of the flagellates *Rhabdomonas* and *Euglena* has been suggested as sufficient explanation for the forward movement of these organisms (Lowndes, 1944). The 'inclined plane theory' likens body twisting to the impeller action of a fan blade. Unfortunately, models constructed to demonstrate this mechanism have violated the principles of dynamical similarity in that their effects were governed by high Reynolds numbers, not at all appropriate to a micro-organism in water (see Holwill, 1974). Holwill (1966b) has demonstrated on theoretical grounds that, at least for *E. viridis*, this mechanism of locomotion is unlikely. On this point he parts company with some experimentalists who feel that films of *Rhabdomonas spiralis* and *Menoidium cultellus*, showing that the flagella are held out of the axis of movement, prove that the important thrusts are created by body rotation rather than directly by flagellar waves (Jahn and Votta, 1972a).

Observed swimming speeds of flagellates

Hopefully, the propulsive velocity of a flagellate protozoon may be predicted by applying the appropriate expression from Table 11.1. An assumption behind these expressions is that wave shapes conform to a regular and trivial mathematical function in two or three dimensions, whereas real waves captured on cine film, or by flash photography, are seen in many cases to be asymmetrical and non-sinusoidal, and to vary in amplitude and wavelength along the flagellum (Plate 11.1, and see Holwill, 1975, Plate 2). It is quite likely in some instances that the irregularities arise from unhappy conditions of observation, and the records are of moribund or traumatized organisms.

The properties of wave shapes other than a sine wave have been examined by Brokaw (1965) and by Silvester and Holwill (1972), with the result that there is confidence that approximating waves by sine functions does not misrepresent, by more than a few per cent, the swimming speeds of cells bearing waves that are really an arc-line configuration or meander-like. According to the numerical analysis undertaken by Holwill and Miles (1971), waves that increase or decrease exponentially in amplitude, or wavelength, as they progress would propel a flagellate at velocities that can be reasonably estimated

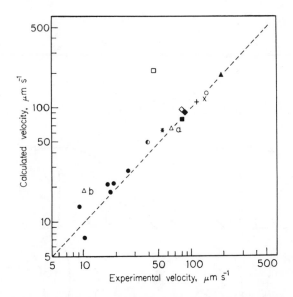

Fig. 11.4 Fit of swimming speeds of flagellated cells measured by various authors, to the theoretical velocities calculated from resistive theory using mean wave parameters. ● trypanosomatid flagellates; ×, □ bull sperm; ▲ sea urchin sperm; ■ *Euglena viridis*; ◐ *Chlamydomonas* sp.; o *Bodo saltans*; △a *Polytomella uvella*; △b isolated flagellum from *P. uvella*; + rabbit sperm; ◇ spore of *Blastocladiella emersonii*; ◆ *Allomyces*; * *Ochromonas malhamensis* (from Holwill, 1974).

by putting mean wave parameters in equations describing uniform waves, if the cell body is relatively large (a protozoon say, rather than a spermatozoon).

These are theoretical reasons for believing that resistive theory equations operate usefully over a range of waveforms that extends beyond the limits of their strict mathematical definition. Figure 11.4, taken from Holwill's (1974) review, demonstrates that in the few studies of flagellated cells where propulsive velocities and wave velocities have been recorded along with accurate measurements of wave parameters from film and photographs, there is an encouraging agreement between observation and calculations from resistive theory.

291

Ciliary metachronism

It is customary to distinguish four general systems of metachronism (Knight-Jones, 1954) although intermediate patterns have now been resolved by more detailed study of ciliates in which surface wave patterns have been preserved by rapid fixation. Metachronal waves travelling in the same direction as the power stroke are symplectic; they are antiplectic if they travel against the power stroke (Fig. 11.5). When the power stroke is to either side of the line of

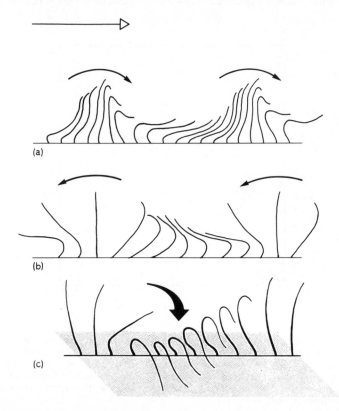

Fig. 11.5 Forms of metachrony. Metachronal wave (open arrow) travelling from left to right; effective stroke of cilia indicated by solid arrows. (a) Symplectic metachrony. (b) Antiplectic metachrony. (c) Dexioplectic metachrony.

wave propagation, metachrony is diaplectic, or, more specifically, dexioplectic or laeoplectic according to whether cilia beat to the right or left as the observer watches retreating metachronal waves. Of these, laeoplectic metachrony is yet to be observed on a ciliate protozoon. Symplectic metachrony is characteristic of *Opalina* and the polymastigote flagellates and is a form of wave movement that restricts the sweep of the cilia. By comparison antiplectic and diaplectic systems allow more freedom of movement for the individual cilia. During normal forward swimming, *Paramecium* shows dexioplectic metachrony, but Machemer (1972) describes a change towards symplexy when organisms are placed in viscous media (Fig. 11.6).

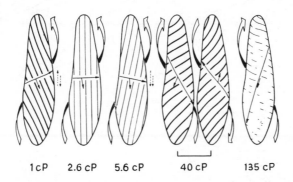

1 cP	2.6 cP	5.6 cP	40 cP	135 cP	

Fig. 11.6 Changes in the locomotion of *Paramecium* as the viscosity is raised. Effective beating of cilia (short arrow) and the orientation of metachronal wavefronts (direction of travel pointed by the long arrow) turn clockwise at higher viscosities. Normal swimming (flat arrow) in left-handed helices changes to right-handed swimming and, at 40 cP, may spontaneously reverse as the wave pattern inverts. True dexioplectic metachrony is progressively lost, and the pattern becomes symplectic and partly breaks down at 135 cP (from Machemer, 1972).

The ciliary envelope model

The envelope model was introduced in an earlier section. According to this model, thrust is developed by the wave motion of the bounding surface of ciliary activity. No allowance is made for water pumped back and forth across the boundary by changes in the spacing of adjacent cilia. For this reason it is more suited to the limited movements of cilia at high packing densities, such as in a symplectic system, and does not describe antiplectic motion successfully.

Some envelopes drawn in Fig. 11.7 from Table 11.2 account in a rough fashion for the locomotion of symplectic and antiplectic ciliary fields. For example, an envelope moving predominantly in a transverse direction (Fig. 11.7b) propels the organism in a direction opposite to wave travel. On the other hand, when ciliary tips move longitudinally and close to the cell surface (Fig. 11.7d), organism and metachronal wave move in the same direction. The first case models symplectic metachronism, and taken to a high enough order the wave equation can support the idea that ciliates like *Opalina* derive sufficient propulsive thrust from wavefronts alone (Sleigh, 1962; Parducz, 1967). But the second model always fails to predict the velocities achieved by antiplectic or diaplectic swimmers such as *Paramecium*, which generally move faster than metachronal wave speeds (cf. Table 11.2d).

293

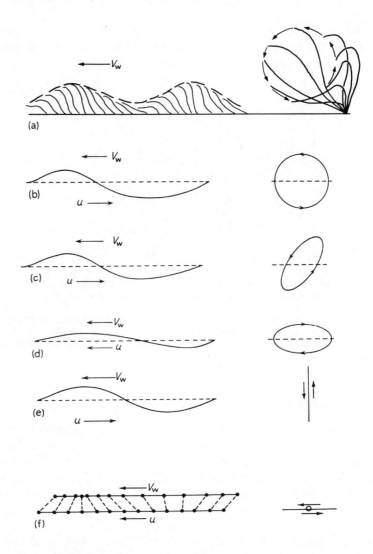

Fig. 11.7 Different flat ciliary envelopes, each moving with overall wave velocity V_w, are generated by combining a purely transverse or normal motion (e) and a longitudinal wave of compression in the plane of the boundary (f), these two motions separated in time by the phase difference ϕ (Table 11.2). The generating function (Blake, 1971) gives a non-dimensional approximate swimming speed:

$$\frac{U}{V_w} = \tfrac{1}{2}k^2(b^2 + 2\beta b \cos \phi - \beta^2)$$ (11.13)

where b and β are the transverse and longitudinal wave amplitudes.

Shown on the right is the path traced out by the tip of a cilium within the envelope. The direction in which the organism is propelled with velocity U is always opposite to the movement of the cilium over the upper part of its path (from Blake, 1971b, with modifications).

Table 11.2 Parameters for ciliary envelopes illustrated in Fig. 11.7

	βk	bk	ϕ	U/V_w	Efficiency $= 2kU^2/\bar{P}$
(b)	0.5	0.5	0°	0.25	25%
(c)	0.35	0.5	−45°	0.19	19%
(d)	−0.5	0.25	0°	−0.22	30%
(e)	0	0.5	0°	0.125	12.5%
(f)	0.5	0	0°	−0.125	12.5%

k is the wave number ($=2\pi/\lambda$), $\bar{P}$ is the time averaged rate of working over a unit area of ciliated surface.

The ciliary sublayer model

Blake (1972) has tried to estimate the mean velocity over a period of time of horizontal fluid layers within the ciliary sublayer, in the vicinity of a single cilium. The mean velocity profile in the region beyond the ciliary tips is then a measure of the speed of the organism relative to the fluid at rest at infinity (Blake, 1973, 1974). The strategy depends on writing, from resistive theory, equations for the influence on the fluid of each element of the ciliary shaft in imparting motion in directions set by the phases of the beat cycle. The equations must also include the effects of neighbouring cilia in stages of the cycle defined by the metachronal pattern, and the retarding effect on motion of the inert cell surface. The relations between these quantities were specified analytically, and then the equations were solved by numerical methods.

As a result of viscous interaction, a moving cilium carries with it layers of fluid at speeds that fall off with distance from the ciliary shaft. It is possible to define a 'near field' of fluid moving with the cilium the boundary of which is somewhat arbitrary, but will depend on the speed of movement of the cilium and its angle of attack (Fig. 11.8). The near field of the power stroke might roughly be described by an up-ended cone, since the tip of the straight cilium moves faster than the base and carries more fluid with it, and closer to the base fluid is held back by the cell surface. Blake allowed for the presence of the cell surface by setting the radius of the near field around each element of the cilium to half the height of that element above the surface. During the recovery stroke a much smaller near field moves with the cilium because the shaft lies closer to the cell surface, moves more slowly, and moves longitudinally rather than normally. For this reason some of the fluid moved with the power stroke is dragged off the cilium on its return. Over the period of one beat cycle (or along a metachronal row of cilia), fluid in the upper part of the sublayer will tend to move in one direction, while close to the cell surface the combined near field movements of effective and recovery strokes may cause oscillatory fluid motion. This prediction, that fluid behaviour changes at different depths in the sublayer, remains to be confirmed experimentally from Protozoan sources.

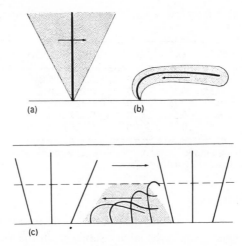

Fig. 11.8 The near field of fluid (stippled) influenced by a single cilium during: (a) the effective stroke, and (b) the recovery stroke. (c) Upper fluid layers over a metachronal wave are set in motion mainly by effective pointing cilia, while the near field effects of recovery stroke cilia are confined to a region (stippled) closer to the cell surface where the fluid may oscillate (after Blake and Sleigh, 1974).

From records of beat patterns of cilia of *Opalina* and *Paramecium*, Blake plotted the time-averaged velocity profile across the sublayer of these organisms. The profiles suggest that in antiplectic and diaplectic wavefronts there will be a weak backflow close to the surface of the cell. Table 11.3 compares

Table 11.3 Sublayer predictions compared to swimming speeds of ciliated organisms

Organism	Beat frequency (s^{-1}) $\sigma/2\pi$	Length of cilium (μm) L	Predicted velocity Non-dimensional velocity $U/\sigma L$	Velocity $(\mu m\ s^{-1})$ U	Observed velocity of propulsion $(\mu m\ s^{-1})$
Opalina	1–4	15	0.15–0.25	14–94	30–100
Paramecium	10–25	10	0.5–1.0	314–1570	500–2500
Pleurobrachia	5–15	600	0.75–1.25	1.4–7.1 ($\times 10^4$)	1–5 ($\times 10^4$)

(After Blake and Sleigh, 1974).

velocity at the outer limit of the sublayer with the normal range of swimming speeds of the organisms. Figure 11.9, taken from Blake's paper, traces the velocity profile through the sublayer and into the exterior flow field around a spherical model of *Paramecium*.

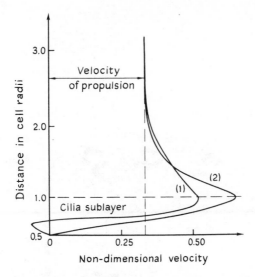

Fig. 11.9 The velocity of fluid in the ciliary sublayer and external flow field around *Paramecium*. (1) Profile calculated from sublayer theory applied to a spherical organism predicts a weak backflow (reflux) close to the cell surface. (2) Profile observed by Jahn and Votta (1972a) (from Blake, 1973).

11.2.3 The internal structure of flagella

The axoneme

Our knowledge of the architecture of the '9 + 2' core inside the membranous envelope of most flagella has been pieced together (Fig. 11.10) from thin sections of the cilia and flagella of many organisms (Plate 11.2a), and more recently from images of fibrillar components splayed out in films of electron-opaque stains (Warner, 1974; Hopkins, 1970).

The outer limit of the axoneme is a palisade of nine 'fibres' or doublet tubules. Each doublet is skewed somewhat across the imaginary circle joining all doublets in a cross-section. A doublet is built from one complete microtubule (A-subfibre), broadly similar to the 21–25 nm microtubules commonly found in many cytoplasms, that shares part of its wall with a second incomplete tubule (B-subfibre). Thirteen protofilaments have been counted in the wall of an A-subfibre and ten in the B-subfibre. Protofilaments are columns of the globular protein, tubulin, which is isolated from sheared flagella of *Chlamydomonas* as two species, α and β, in roughly equal amounts (Witman, Carlson and Rosenbaum, 1972).

At intervals of 16–22 nm along an A-subfibre, pairs of hooked arms reach out towards the B-subfibre of the next doublet. The outermost arm is longer than the inner arm, but both contain the ATPase dynein. Bridging along an

297

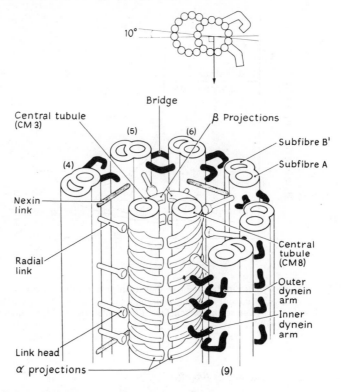

Fig. 11.10 Interpretation of axoneme structure. The base of the axoneme is toward the top of the page; outer doublets 1, 2 and 3 are removed. The upper diagram shows protofilament construction of doublet tubules; the arrow lies on an axoneme radius.

axoneme radius from the A-subfibre of each doublet to the central tubules is a series of hammer-headed links. In all sources so far examined, radial links space along the A-subfibre in one of two distinctive patterns, but in either case the repeat of the link pattern is in register with a multiple (six-fold) of the spacing of projections from the two central tubules (Fig. 11.11).

The integrity of the ring of doublets is maintained by a series of circumferential ties between neighbouring doublets. According to studies of sectioned axonemes, these inter-doublet links run from a B-subfibre to the inner dynein arm or the A-subfibre of the next doublet (Warner, 1974). However, when a procedure for separating A- and B-subfibres is used to dismember whole axonemes, the A-subfibres alone are recovered, linked in a ladder-like arrangement by fine rungs at every 100 nm. The material of the links contributes a single protein band to the electrophoretic separation of axonemal proteins on urea-acrylamide gels, and has been called nexin by Stephens (1970a).

Links, whether circumferential or radial, between doublets and other axonemal components, must be expected to resist sliding. If, in the course of

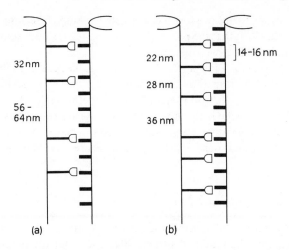

Fig. 11.11 Arrangement of radial links along subfibre A, and their relationship with central tubule projections in: (a) *Chlamydomonas* (Hopkins, 1970) and (b) *Tetrahymena* (Chasey, 1972).

bending, doublets slide past one another, these links will either stretch (elastic resistance) or may be obliged to make and break attachments according to some local cycle (viscous resistance).

Paraxial rods

Running alongside the axonemes of some flagellates (euglenids, dinoflagellates, trypanosomes) is the paraxial or paraflagellar rod, a dense structure which may double the thickness of the flagellum. Paraxial rods are often described as paracrystalline with lattice substructure, for instance in *Trypanosoma*, *Peranema* and *Euglena gracilis*, though in other cases the rod appears to be structureless. The trailing flagellum of *Entosiphon* has a large paracrystalline rod, while in the leading flagellum of the same cell the rod is smaller and of the amorphous kind (Leedale, 1967). Occasionally, links between the rod and axoneme are seen in cross-sections. Forked links from the B-subfibres of doublets 5 and 7 insert into the rod of trypanosomes. In the ventral flagella of *Giardia* (Plate 11.2a) the connections are between an amorphous rod and doublets 3, 4 and 5. The reason for the presence of paraxial rods in some flagella is obscure. The greater width given to the flagellar shaft may simply increase hydrodynamic performance, but other speculations have been made. Bovee and Jahn (1972) suggest that in response to an electro-motive force generated in the photoreceptor, the crystalline rod of *Euglena* deforms and changes the orientation of the flagellum. Conversely, mechanical bending of the crystal brings about piezoelectric charge separation, which may control the form of axoneme bending. The fact that digitonin-treated flagella retain their ATPase

299

activity after dynein arms are lost has suggested to some (Piccinni, Albergoni and Coppellotti, 1975) that the paraxial rod, where the residual activity may be localized, is independently motile.

Whatever the role(s) of the paraxial rod, if the visible bridges to the axoneme represent a real anchorage then presumably these will resist sliding between linked doublets, unless the rod material is freely deformable.

Attached or unattached, the unknown visco-elastic properties of the rod must contribute to bending resistances in certain planes and influence the shape of flagellar waves, particularly where these are three-dimensional.

11.2.4 The sliding tubule model

There is considerable confidence in the theory that muscle sarcomeres shorten by progressive interdigitation of protein filaments, that the filaments are of fixed length, and that cyclical attaching and detaching of crossbridges coupled to ATP hydrolysis causes the movement. It was perhaps inevitable that at some point the older intuitive ideas that cilia bend when peripheral fibres contract or pull in sequence would give way to explanations of bending that draw on the familiar sliding theme (Satir, 1968; Sleigh, 1968). The idea is an attractive one. ATP-splitting dynein arms, arranged in rows along the length of the axoneme are attached to and also face parallel fibres (tubules) known to be linear polymers of protein subunits with some parallels in composition and properties to muscle actin (Stephens, 1970b, 1974). Adding tubulin to preparations of muscle myosin elevates the rate of ATP hydrolysis by the enzymic protein (Alicea and Renaud, 1975).

Currently, the evidence that axoneme doublets slide over one another to cause bending is powerful and direct. Doublets are seen to slide from naked axonemes in ATP solutions, and are seen to have slid when bend patterns preserved by rapid fixation are examined in the electron microscope. Less directly, if the force causing bending is assumed to arise from doublet sliding, a description of likely mechanical properties can be written which is a sufficient, though not necessarily correct, protocol to explain initiation, propagation and the shape of flagellar waves (see next section).

The case for the sliding tubule mechanism of axoneme working has been strengthened by three important studies:

(1) Consider a cilium in which axoneme doublets are firmly rooted in the basal body. As the cilium bends, the doublet closest to the inside of the bend is more tightly curved, with a shorter arc length than its opposite number across the axoneme. If doublets cannot contract they must slide past one another by amounts specified by differences in arc length. Each doublet slides over its neighbours to some extent, except the bridged doublets 5 and 6 which behave as one unit in this respect. The slip between any two doublets appears at the tip of the cilium as a relative displacement of their free ends, and, for

every ten degrees through which the cilium is bent, adds up beyond the bend to about 10 nm (Sleigh, 1968).

Working from serial sections across a field of mussel gill cilia, Satir (1968) reconstructed the tip of an 'average' cilium bent through angles read from the stroke positions of a metachronal wave. The movements of the ends of the doublets during beating was found to correspond to the slip profile predicted from the distance in the beat plane between each doublet in the ring of nine and doublet 1 – the doublet which marks the outside of the bend in the power stroke and the inside of the recovery bend.

(2) Sea urchin sperm demembraned in the detergent Triton X100 can be caused to swim in a lifelike way in solutions of ATP. Reactivation of naked axonemes is 100 % successful under the conditions that are optimal for the ATPase reactivity of dynein *in vitro* (Gibbons and Gibbons, 1972). It seems that the chemical input to the machinery causing bending is utilized at the dynein arms. In micrographs of axonemes prepared conventionally, a gap separates dynein arms of one doublet from the B-subfibre of the next doublet. However, if ATP is rapidly washed from extracted sperm the flagella 'freeze' in a rigor-like state, and axonemes fixed at these times show complete bridging of A and B subfibres of adjacent doublets by dynein arms (Gibbons and Gibbons, 1974). The argument that swimming of sperm models depends on an ATP-dependent make-and-break cycle of dynein attachments to B-subfibres, and that this cycle causes one doublet to walk along the next, is made more compelling when sheared axonemes are first digested very briefly with trypsin. Adding ATP then brings about a spectacular disintegration of the axoneme. Individual doublets and groups of doublets slide out from the axonemal bundle, increasing its length manyfold (Summers and Gibbons, 1971). This experiment is both an elegant demonstration of active sliding, and an indication that in the intact axoneme shear forces are resisted and translated into bending forces by structures sensitive to trypsin. The structures in question are intertubule (nexin) links and radial links, which disappear from the cross-section of the axoneme at the rate at which trypsin digestion releases free-sliding doublets. Four-fifths of the dynein arms are still in place when most other links have been removed (Summers and Gibbons, 1973).

(3) In cilia sectioned along their lengths, radial spoke intervals are a convenient ruler of distance along A-subfibres. Since the spacings are the same in straight and bent axonemes, the possibility that doublets contract locally on bending can finally be discounted. By counting radial spokes on opposite sides of mussel gill axonemes bent in the section plane, Warner and Satir (1974) measured directly the accumulation at a bend of the amount of slip between doublets foreseen by the sliding hypothesis.

The behaviour of radial spokes is particularly intriguing. At the bends, spoke heads attach to the central projections, and their origin from the A-subfibre is more acutely angled as tubule slip increases around the bend. Also the spokes are stretched progressively with further bending. In the straight axoneme, spokes are mostly perpendicular to peripheral doublets, implying that here they have sprung away from their central attachment. The factors controlling spoke attachment and detachment are unknown, but attached spokes presumably resist interdoublet sliding and promote bending. The accumulation of strain in these structures also measures the extent of local bending, and is potentially a constraint regulating wave amplitude by mechanical means.

11.2.5 Mechanical properties of flagella

The direct methods of measuring mechanical properties of biological fibres that have become the stock-in-trade of muscle physiologists are largely precluded from use with cilia and flagella because of their small size, although the rigidity of cilia, and the forces developed against a glass microbalance by large compound cilia of lamellibranch gills have been directly measured (Baba, 1972; Yoneda, 1960). Instead, students of flagellar mechanics have taken the harmonic motion of a flagellum or cilium as a clue to the unknown active and passive mechanical properties and have constructed mathematical analogues of bending behaviour from assumptions about the relative sizes of internal forces and the relationship between active forces and bending.

From an early study it was appreciated, for instance, that a flagellum does not behave as an elastic beam passively waved by a rhythmic 'motor centre' at its base, since under these circumstances travelling waves are damped out by the viscous resistance of the medium (Machin, 1958). It is necessary to assume that the flagellum is active along its length in order that waves of sustained amplitude are propagated in the face of viscous resistance.

Improving the guesswork gives a closer fit of model waveforms to real waveforms. Exercises along these lines have simulated with some realism the waveforms of invertebrate spermatozoa, which have the mathematical attraction of being symmetrical, near uniform and two-dimensional.

A test of any particular model is how it behaves under the one loading condition that is readily manipulated, increasing or decreasing the viscosity of the external medium.

It may be that there are appreciable differences in mechanical properties between Protozoan flagella and sperm tails. When sperm tails are interrupted by laser beams, the fragment of flagellum beyond the lesion cannot initiate further bending waves (Goldstein, 1969). This is not true of the flagellum of *Crithidia oncopelti* which continues to initiate and propagate waves on either side of the irradiated section (Goldstein, Holwill and Silvester, 1970). One explanation of the difference in behaviour is that *Crithidia* flagella have throughout their length a sufficiently high resistance to longitudinal shear between axoneme fibres for active sliding forces to generate bends locally, whereas this resistance is thought to be low in sperm tails except at the axoneme base. For this reason, it would be premature to generalize from models of sperm tail swimming, but an outline of sperm mechanics will identify the kind of active force that produces bends, which is likely to be a general property of axonemes, and some of the internal resistances to bending which account for a significant proportion of the work done by the axoneme.

Mechanical models of flagellar bending

A notion common to a number of models is that active elements distributed along the length of a flagellum are activated in sequence by some locally-seated mechanism that responds mechanically to passing waves of bending (Machin, 1958, 1963; Brokaw, 1966a; Lubliner and Blum, 1970).

In its most recent and successful form (Brokaw, 1971), local activation of mechanochemical elements is combined with a sliding tubule explanation of bending. The local elements may be cross-bridges (dynein side-arms) generating active shear between tubules that bend if there is shear resistance at some point in the structure, for example, if they are tied together at one or both ends. The extent of sliding and bending follows from the local degree of cross-bridge activity.

Now if activation itself depends on the local curvature, bend propagation can be built into the mechanism obviating the need for a higher level control, providing the bending phase lags the rise in bending moment at each point (Machin, 1963). Real flagella operate over a range of beat frequencies, so the phase lag of the control loop must be independent of frequency and real time.

According to Brokaw, these properties are intrinsic to a mechanism producing bending moment from sliding of axonemal fibres, where the length-(rather than time-) dependent active bending moment (M_a), depends at any point (s) along the flagellum on the moment added from local shear forces (m) in a unit length (Fig. 11.12). This may be written:

$$\frac{\mathrm{d}M_a}{\mathrm{d}s} = -m(\kappa), \tag{11.14}$$

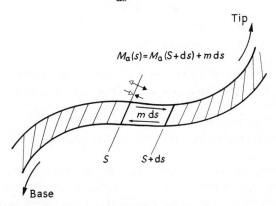

Fig. 11.12 Bending moment, M_a, generated by active sliding between two filaments in a model axoneme where free sliding is resisted by fixed cross-links. The diagram shows the balance of moments at the cross-section, S, of a two-filament system in which the active shear developed in unit length has a magnitude m.

303

specifying that local shear is itself a function of κ, the local curvature (Fig. 11.13 shows how κ is defined). Since this relation automatically solves the control problem by introducing a quarter-cycle phase shift between curvature and bending moment, the dependence of the shear moment on local curvature can be a simple one, with no further phase lag. For example, an adequate 'activation equation' in which negative curvature activates positive shear, might be:

$$m(\kappa) = -m_0 \kappa, \qquad (11.15)$$

where m_0 is a characteristic constant.

Quite opposed to the idea of local activation is Rikmenspoel's (1971) analysis of small amplitude bending waves propagated by sperm flagella. Under this rather artificial constraint, it is not necessary to assume that chemical activity rises and falls along the flagellum and that the variation is synchronized to bending. Activation may be simultaneous along the length of the flagellum and there is no requirement for a time or phase lag in bending moments. Mehanical properties are readily explained in terms of a sliding tubule mechanism. Cross-bridges spaced along one fibre pull together to buckle the elastic flagellar beam, but the different fibres either operate in two groups on opposite sides of the axoneme (in planar beating for example), or 'fire' sequentially in a 1, 2...9 cycle. Unlike Brokaw's model, the structure is not an autonomous oscillator, the rhythm of activation is set by an extrinsic pacemaker that signals the firing pattern, perhaps by a chemical signal or by membrane depolarization (Kinosita and Murakami, 1967).

In both of these cases, at any instant, the active bending moment has at each point on the flagellum an achieved magnitude (M_a) the sum of local bending forces in the shaft beyond that point. Summation of force increments is a general and important property of a sliding filament mechanism where force-producing structures are arranged in parallel, whether myosin in muscle or dynein in flagella (Rikmenspoel, 1971; Rikmenspoel and Rudd, 1973). In a locally-activated flagellum, the variation in $M_a(s)$ will be a travelling wave on the flagellum (Fig. 11.13c) since positive and negative curvatures in one wavelength generate active moments of opposite sign. In Rikmenspoel's model, because all active centres on one doublet are simultaneously active, $M_a(s)$ increases linearly from the tip towards the base where the greatest bending moment will be felt. The outstanding difficulty of the model is that though the 'forcing' role of M_a oscillating sinusoidally in time can be perceived in starting a running wave from rest, this crucial aspect of wave behaviour has not been analysed.

However activated, flagellar movements are governed by the one general equation of motion balancing the active bending moment against bending resistances from the viscosity of the medium (M_v) and the elasticity of tubules and parallel structures within the flagellum (M_e):

$$M_a(s) + M_e(s) + M_v(s, t) = 0, \qquad 0 \leqslant s \leqslant 1 \qquad (11.16)$$

The viscous moment is the quantity in the analogue that describes the biologically useful work being done by the flagellum on the external medium. Ignoring elastic properties of the flagellum results in a simplified mechanical model:

$$-M_a = M_v, \qquad (11.17)$$

or:

$$-m(\kappa) + dM_v/ds = 0, \qquad (11.18)$$

which it had been predicted (Brokaw, 1971) would propagate bending waves and overcome the difficulties in this respect encountered by earlier contracting-filament models (Machin, 1958; Brokaw, 1966a). The correctness of the model can be assessed by simulating flagellar waves by the function $\kappa(s,t)$ which is a complete description of the distribution of bends on the flagellum in space and time.

In the simulation, an arbitrary distribution of bends is specified initially, but the rate of bending at any point $(\dot{\kappa}(s) = d\kappa(s)/dt)$ is found from the moments equation (11.16 or 11.17) where it appears as a term in the viscous moment. The values of $\dot{\kappa}(s)$ are then used to find the new flagellar shape after a short time interval, and successive iterations forward in time give $\kappa(s,t)$ the movement of the model function over a period. For simplicity the rate of bending is computed only at a finite number of joints between short, straight segments along the backbone of the model flagellum. Using this routine, Brokaw (1972a) found that a headless flagellum could not swim in any stable way by the simple mechanics of Equation 11.17. It was necessary to expand the mechanical description to produce a free-swimming flagellum that would initiate and propagate bends and simulate the stable wave forms of real flagella with some verisimilitude.

If the active process is allowed an interval in which to respond to changing curvature, the model approaches stability more rapidly, without an effect on the form of the stable wave. Such a time delay causes active moment to lag the viscous moment, enabling a component of the active moment to balance the variation in elastic bending resistance (Fig. 11.13d). Wave amplitude is stabilized by non-linear elastic resistances, and this condition is satisfied by a combination of elastic parameters cubically dependent on curvature (elastic bending resistance) or amount of longitudinal shear (elastic shearing resistance).

Further resistances arise from the viscous behaviour of internal elements, either in bending or shearing. The comparison between internal and external viscosity determines the wavelength at which the movement stabilizes. Without internal viscous resistances, the ends of the model oscillate excessively because the external viscous moment near the end falls off more rapidly than the active bending moment. For realistic movements of the model function, most

305

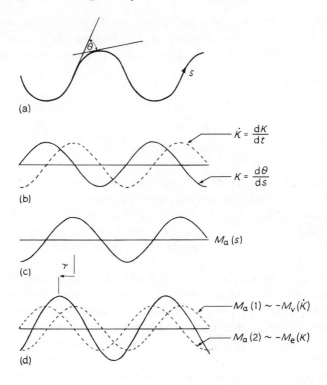

Fig. 11.13 Phase relations of mechanical properties of flagellar bends, specified by Brokaw's (1971) model. (a) Waveform of a flagellum when curvature, κ, is a sine function of distance, s, measured along the flagellum. (b) The generating function and its time derivative, $\dot{\kappa}$, when the wave travels from left to right. (c) Compared to the variation in curvature, the distribution of active bending moment, M_a, is phase shifted along the flagellum by a $\frac{1}{4}$ cycle when Equation 11.14 holds. (d) A time delay, τ, introduced into the moment-generating function allows the active moment to be resolved as two components: (1) balances viscous bending resistances, M_v ($\dot{\kappa}$), and (2) balances elastic resistances, M_e (κ).

of the internal viscous resistance is bending resistance, and the values of this parameter included in models imply that a great part of the work done by the flagellum is dissipated against internal resistance. Only about 30% is available as external work. The relative inefficiency of the model is matched by the insensitivity of real flagella to high viscosity media. In one experiment (Brokaw, 1966b) the wave velocity of sperm flagella decreased by 32–40% following a threefold rise in external viscosity. Over this range, the model predicts that wave motion would be slowed by 37%.

The presence of viscous shear resistance in the model is of interest, since it is equivalent to an active process which generates less moment when actively sliding. In this respect the sliding mechanism resembles the force-velocity behaviour of muscle (Hill's equation).

306

Observed flagellar movements at high viscosity

Raising the viscosity of the medium in which flagellated cells are swimming is an experiment that has been carried out on only a few species, and the manner in which wave motion adjusts to the increased load shows some variation between organisms.

Invertebrate spermatozoa of the genera *Lytechinus* and *Ciona* swimming normally in sea water, and glycerol-extracted *Lytechinus* in 2 mM ATP, slow wave motion in viscous media, and at the same time reduce the amplitude of waves on their flagella. By contrast, *Chaetopterus* sperm and glycerinated *Lytechinus* reactivated in low ATP solutions maintain wave amplitude by reducing the radius of curvature of bends at the slower speeds. At the time these effects were recorded, Brokaw (1966b) determined empirically that wave velocity varies inversely with the square root of external viscosity (or $V_w = C_N^{-0.5}$ where the drag coefficient is used as a measure of viscosity). There was no obvious explanation for this relationship.

Recently, it has become clear that bull sperm behave quite differently and regulate wave shape over a wide range of loading (Rikmenspoel, Jacklet, Orris and Lindemann, 1973), but this control may relate to the active or mechanical role of additional fibres in the mid-piece and around the axoneme.

Most mechanical models of the axoneme anticipate that flagellar wavelength (λ) will alter with a change in the viscosity of the medium. In the 'elastic flagellum', where it is assumed that wavelength and amplitude are fixed by the interplay of elastic and external viscous moments (Machin, 1958, 1963; Rikmenspoel, 1971; Holwill, 1966a), the relationship that is expected:

$$\lambda \propto f C_N^{-0.25} \tag{11.19}$$

includes the beat frequency. This is true of a passively vibrating beam, and following a suggestion of Machin (1958), it has been accepted that an active motion would lock in to the wave mode preferred by the passive oscillation to draw on the stored elastic energy. Similarly, a novel mechanochemical model in which reaction rates at active sites are driven by tension in the reactants is also governed by elastic processes and obeys the same rule (Miles and Holwill, 1971).

However, Brokaw (1972b) has explained that the grounds are general for predicting some kind of quartic dependence of wavelength on the reciprocal of viscosity, and can extend to the case where wavelength is set by internal viscosity. Where the moments balanced by active moments are solely or predominantly viscous, the determining rule for wavelength becomes independent of beat frequency. The 'viscous flagellum' conforms, then, to the rule:

$$\lambda \propto C_N^{-0.25} \tag{11.20}$$

but only when interfilament shear resistance (C_S) is low (Brokaw, 1972b), at which time the organelle is working at its most efficient. Under these conditions, the mechanics also predict that beat frequency will be determined by the similar relationship:

$$f \propto C_N^{-0.25}. \tag{11.21}$$

We can now see that wave velocity ($V_w = \lambda f$) will be proportional to $C_N^{-0.5}$, the originally obscure behaviour of spermatozoa swimming at high viscosity.

In Fig. 11.14, the wave speeds of three sperm genera are compared on a log.log plot to the slopes from: $V_w \propto C_N^{-0.5}$ ($C_s = 0$, efficiency of the flagellum

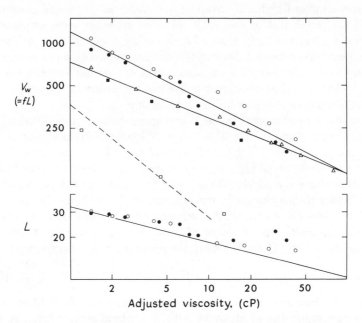

Fig. 11.14 Effect of raised viscosity on the wave speed, V_w, and arc-wavelength, L, of some flagellated micro-organisms. Data from swimming of marine sperm: o *Ciona*; △ *Chaetopterus*; ● *Lytechinus* (in sea water); ■ glycerinated *Lytechinus* in 2 mM ATP (from Brokaw, 1966b); and from one Protozoan □ *Crithidia* (from Holwill, 1965), plotted on a log–log scale. Slopes of unbroken lines are: −0.5 and −0.39 (upper figure), and −0.25 (lower figure). Note the clear step change in wavelength of *Ciona* flagella at 5–10 cP.

is 25%) and $V_w \propto C_N^{-0.39}$ ($C_s > 0$, efficiency is 10%). The swimming of *Chaetopterus* sperm, which is rather different from the other organisms, may be influenced by high interfilament shear, or more likely, by the change in bend curvature at high viscosity.

Information from Protozoan species is scant. Holwill (1965) listed parameters of the proximally-travelling waves on the flagellum of *Crithidia oncopelti* at different viscosities. The waves speeds of this organism plot anomalously, as if governed by the relationship: $V_w \propto C_N^{-0.8}$, a result which cannot satisfactorily be explained.

Rather than follow the continuous transition implied by the proportionality rule, when the computer-simulated flagellum is challenged with an increasingly viscous environment, changes in the wavelength tend to a step function. The bending pattern settles into one of a number of preferred wavelengths, and responds only at particular viscosities by adding one full wave to the flagellum (Brokaw, 1972c). There are indications from the original experiments (Fig. 11.14) that real flagella also behave in this way.

Mechanical properties of cilia

The motion in two dimensions of a cilium as it appears on film (Sleigh, 1968) has been reconstructed by a programme incorporating the assumptions that movement results from sliding of filaments, and that the balance of moments described by Equation 11.16 applies (Rikmenspoel and Rudd, 1973). The total active moment at each point on the cilium will fluctuate periodically with the beat angle. When the cilium is fairly straight, as during the effective stroke, the contribution of the elastic moment is minimal and the active moment develops to oppose viscous resistances along a time course that is similar to the change in viscous moment. Figure 11.15 shows the variation in viscous moment over one beat cycle of the compound cilium from the gill of *Sabellaria*, calculated by Rikmenspoel and Sleigh (1970) from Gray and Hancock resistive theory. At the beginning of the effective stroke, within 3–5 ms, the axoneme is activated fully at all points along the cilium. The rapidity of the

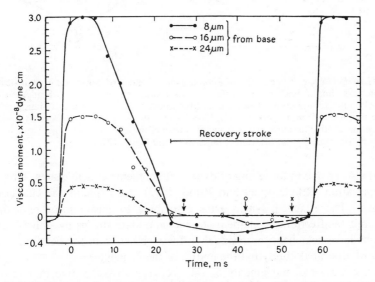

Fig. 11.15 Variation in viscous moment over the beat cycle of a cilium. Moment due to drag is calculated at three points along the compound cilium of the marine worm *Sabellaria*. Arrows indicate the instant of maximum bending at each locus (from Rikmenspoel and Sleigh, 1970).

309

signalling argues against the precept that activation of cross-bridges is by mechanical means; rather a fast travelling event such as an action potential is needed.

The simulated cilium moves too quickly at the tip and is too straight unless the active moment is allowed to fall off towards the end of the effective stroke, and the elastic moment is able to decline in step with the active moment. Intuitively, one might argue that the decay in intensity of the active moment (given a time constant, τ_{eff}) follows the kinetics of an extrinsic signal, but it

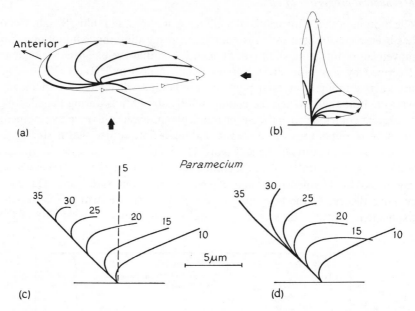

Fig. 11.16 Beat cycle of one cilium of *Paramecium*. Envelope of movement in plan view (a) and along the line of synchrony (b). Effective stroke indicated by open arrowheads, recovery stroke by solid arrows (drawn from data of Machemer, 1972). (c) Positions at 5 ms intervals of the cilium seen from the side (Sleigh, 1968). (d) Recovery bending simulated by a sliding filament model that assumes that, during this phase of the cycle, cross bridges are activated locally by bending ((c) and (d) from Rikmenspoel and Rudd, 1973).

could also be due to loosening of dynein attachments at active sites. This last interpretation (Rikmenspoel and Rudd, 1973) also explains the changes in stiffness. The attached dynein arms contribute to elastic shear resistances and limit the freedom of filaments to accommodate strains by sliding. Once dyneins have detached, the residual stiffness comes from the bending resistances of separated fibres, and the matrix of the flagellum.

In the wave-like movements of the recovery stroke, active moments of opposite sign generate a reversed bend propagating from the base of the cilium. Moment applied beyond the bend causes, in the computer model, excessive power stroke-like motion of the distal cilium. In real cilia, the travelling bend trails the tip limply in the direction of movement. This divergence is overcome when active force develops only at the bend, which then excites the active mechanism ahead of it as it travels down the cilium. Once the bend has passed, the active moment decays in each section of the axoneme with a second time constant measured by τ_{rec}. The model cilium then, is biphasically activated, by extrinsic and local mechanisms in appropriate phases of the beat cycle. For a satisfactory subjective fit of modelled beat patterns to movements of cilia (Fig. 11.16), different values were used for the ciliary constants τ_{eff} (5 ms) and τ_{rec} (20 ms). If the assumptions behind the model are reasonable, this requirement suggests that the chemistry of the active process is differently modulated over the two phases of the cycle and follows different kinetics, or that separate active sites exist for power stroke and recovery stroke. It is known that dynein *in vitro* has a dual reactivity to added ATP, attributable to high and low affinity enzymic sites (Gibbons and Fronk, 1972).

11.3 Mechanics of amoeboid movement

11.3.1 Phenomenological observations

Streaming and pseudopod formation

A whole number of phenomenologically disparate crawling movements of cells have been offered as examples of amoeboid movement (Fig. 11.17). In its most restricted sense, as in the descriptions of the familiar larger amoebae, amoeboid movement is associated with two properties: streaming of cytoplasm and the putting out and retracting of many pseudopods (polypodial organisms), finger-like extensions of the cell soma. The lobose carnivores of the *Amoeba–Chaos* group are the organisms which, by dint of their size, have become the preferred material of experiment and measurement, almost to the exclusion of other species.

Forward streaming of granules (mitochondria, crystals, and small organelles) carried in a stream of fluid-like endoplasm into a forming pseudopod, is the immediately apparent feature of movement of *Amoeba* or *Chaos* seen through the light microscope. Streaming is an autonomous and active capability of the motile cell as a whole, and as we shall see later also of isolated cytoplasm, and is the outcome of the natural motive force, wherever it resides in the organism. (Passive or induced flow, when an amoeba is drawn into a suction tube for example, is not streaming.) The endoplasm moves through a tube of more solid cytoplasm, the outer ectoplasm, towards the pseudopodial tip where it everts and becomes ectoplasm. In the tail of the organism, or in retracting pseudopods, the reverse process takes place, ectoplasmic structure breaks down and becomes endoplasm, maintaining the rough quantitative balance between the two states of amoeba cytoplasm (Fig. 11.18). Since we do not wish fortuitously to attribute particular colligitive properties to amoeboplasm we shall avoid

311

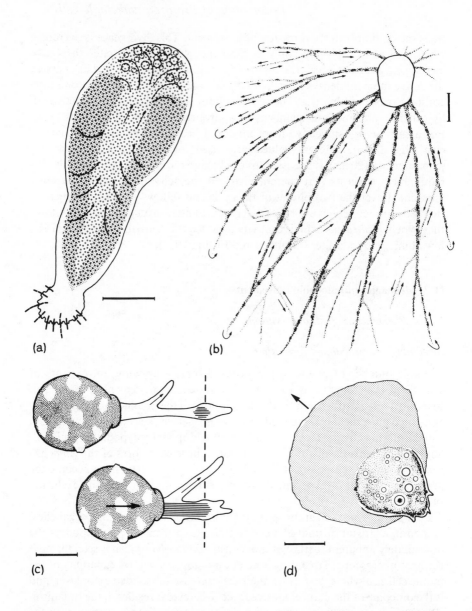

Fig. 11.17 Movements of some amoeboid organisms. (a) The herbivorous monopodial amoeba *Pelomyxa palustris* ingests filamentous algae at the tail, where there is a prominent hyaline zone – the urosphere. Anterior fountain streaming is pronounced – anterior ecto-plasm moves back toward a stationary 'girdle' region of attachment. Endoplasm is recruited from the posterior two-thirds of the organism (Griffin, 1964). (b) Countercurrent streaming in the reticulopodial net of the foraminiferan *Allogromia*. (c) The shelled amoeba *Difflugia* assembles birefringent fibrils at a point of attachment. Birefringence spreads along the pseudo-pod into the cell soma, then decays as the pseudopod retracts dragging the heavy test for-ward (Wohlman and Allen, 1968). (d) Gliding movement of *Hyalodiscus* is not accompanied by streaming. Scales: (a), (b) and (c) 100 μm; (d) 10 μm.

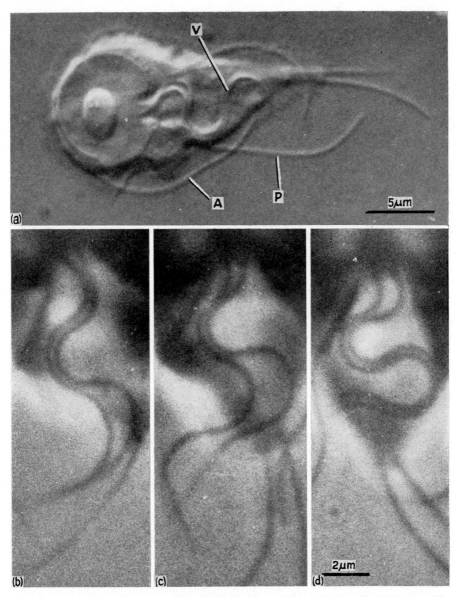

Plate 11.1 The zooflagellate *Giardia duodenalis.* (*a*) Ventral flagella (V) beat with large amplitude bending waves, while anterior (A) and posteriolateral (P) flagella show tonsate strokes or small amplitude travelling waves. Interference contrast. Bending of the two ventral flagella is not uniform and not sinusoidal, but may correspond to sine-generated waves or arc-line shapes (*b*) and (*c*), or resemble meanders (*d*). Phase contrast.

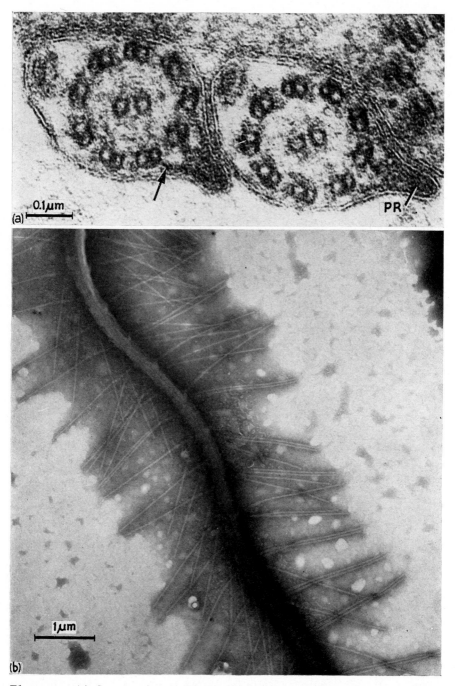

Plate 11.2 (*a*) Cross-section of the ventral flagella of *Giardia muris* as seen from their bases. The granular paraxial rod (PR) in each profile sends fine links (arrow) to the nearest doublets of the axoneme. (*b*) Pantoneme arrangement of mastigonemes on the flagellum of *Ochromonas*. Negatively stained.

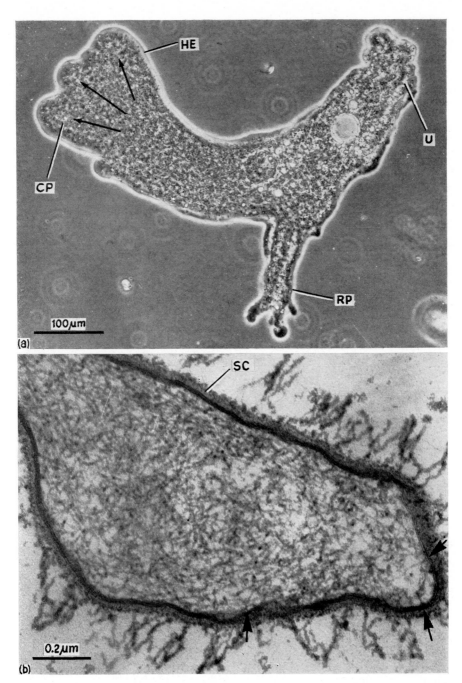

Plate 11.3 (*a*) Slightly flattened *Amoeba discoides* streaming into a compound pseudopod (CP). Endoplasm is recruited from the uroid (U) and a retracting pseudopod (RP). HE = hyaline ectoplasm. (*b*) Microfilaments in an ectoplasmic protruberance are unoriented but insert into the plasma membrane (arrows). SC = surface coat.

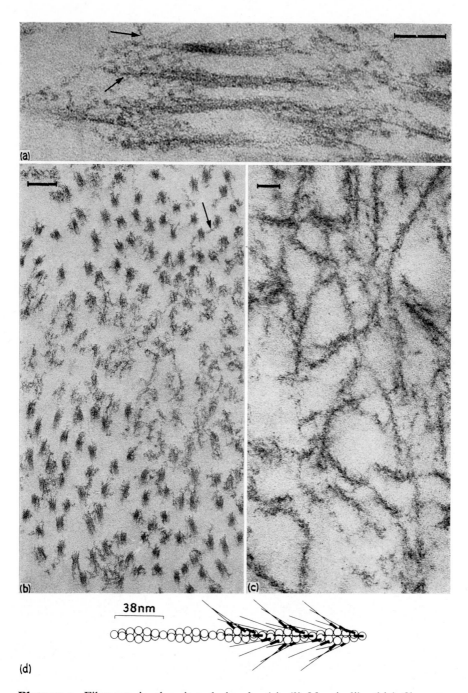

Plate 11.4 Filaments in glycerinated *Amoeba*. (*a*), (*b*) Myosin-like thick filaments assembled in 0.5mм ATP. Note bare central shaft and terminal projections (arrows). (*c*) Actin-like filaments decorated with muscle heavy meromyosin. Scale = 0.1 μm. (*d*) Interpretation of the 'arrowheads' evident in (*c*) when one HMM fragment binds to each actin monomer.

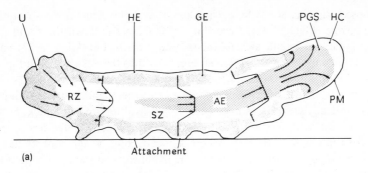

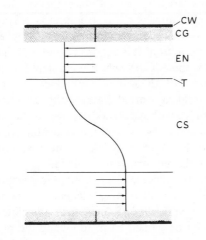

Fig. 11.18 (a) Streaming in a monopodial amoeba pictured as velocity profiles across the organism at three loci. The endoplasmic profile is parabolic in the recruitment zone (RZ), but further forward, structured axial endoplasm (AE) shows plug flow within a lubricating shear zone (SZ) (after Allen, 1974). (b) Velocity profile across a rhizoid cell of *Nitella* indicates that the motivating force of cyclosis is active shear at the ectoplasm/endoplasm interface (after Kamiya, 1959). HC = hyaline cap; PGS = plasmagel sheet; PM = plasma membrane; GE = granular ectoplasm; HE = hyaline ectoplasm; U = uroid; CW = cell wall; CG = cortical gel ectoplasm; EN = endoplasm; T = tonoplast; CS = cell sap.

the terms 'sol' and 'gel' used in Mast's (1926) description of cyclical change of consistency of cytoplasm. (For a spirited exchange of views on the usage of suitable rheological terms see: Jahn and Votta, 1972; Allen, 1972a). Understanding the exact nature of the two states of cytoplasm and the meaning of the transformation of one into the other is central to identifying the source of motive force and its point(s) of application. The tip of an advancing pseudopod, or the leading edge of a monopodial amoeba, is smooth-contoured with an appearance that suggests turgidity. By contrast, the retreating tail or uroid is

313

collapsed, the surface is highly puckered, and the structure is substantially ectoplasmic (Plate 11.3a). According to Goldacre (1952a), the uroid is to a large extent persistent, at least in *Amoeba proteus*, but Allen (1973) has observed that endoplasm will often stream back into the retracted pseudopods of *Chaos*. Ectoplasm of the uroid, and in peripheral regions of the tube is generally more free from inclusions than ectoplasm elsewhere, and than the endoplasm. These optically clear areas have been called hyaloplasm (Komnick, Stockem and Wohlfarth-Bottermann, 1973) although it is more satisfactory to restrict this description to cytoplasm where a locally high water content has been measured by interferometry, such as the hyaline cap at the tip of a pseudopod (Allen, Cowden and Hall, 1962; Allen and Francis, 1965). In an active *Amoeba* or *Chaos* that has not been compressed by a coverslide, the ectoplasmic tube may be sculptured into a series of longitudinal ridges or fins on the upper and lateral surfaces. The significance of these structures is unknown.

The streaming cycle describes internal motion of cytoplasm, but adhesion is causal for forward locomotion. It is against regions of adhesion that motive force must act. Seen from the side, *Amoeba proteus* steps forward on new pseudopods that make contact with the ground and adhere, the tail region and retracting pseudopods are always lifted clear of the surface (Haberey, 1971). The same organism seen from above shows an ectoplasmic tube that is stationary. Ahead of an adhesion it grows by accretion from the everting endoplasmic stream, and behind an adhesion it shrinks as new endoplasm is recruited from dissolution of ectoplasmic structure in the tail.

Despite earlier doubts (Goldacre, 1952b, 1961; Goldacre and Lorch, 1950), it is probable that the surface of a large amoeba is a relatively permanent structure (Wolpert and O'Neill, 1962). Some impression of its behaviour during amoeboid movement can be gained from following the motion of particles adhering to the surface; for example Haberey and his collaborators (1969) have recently filmed the local displacements of carbon particles attached to *Amoeba proteus*. The surface (the plasma membrane and its mucus coat probably move as one) move forward more or less at pseudopodial speeds, flowing around immobilized regions at the attachment points. In polypodial amoebae the surface for new pseudopod formation is unfolded from the sink of convoluted surface in retracted pseudopods and in the tail. In each pseudopod the surface rolls forward over the stationary ectoplasmic tube on a lubricating layer of hyaloplasm dispersed from the hyaline cap.

When pseudopods fail to attach to the substrate during normal movement (Kanno, 1965), or when the adhesiveness is reduced by mild trypsinization, cytoplasmic motion is uncoupled from cell locomotion. Under these circumstances granule markers in the ectoplasm move backward relative to the observer as endoplasmic granules stream forward. The 'fountain streaming' pattern demonstrated in this way is also seen when amoebae drawn into tight glass capillaries are prevented from moving forward, and in freshly demembraned preparations (Allen, Cooledge and Hall, 1960).

314

Gliding movements

The exchange of weak membrane-substrate adhesions for stronger adhesions (haptotaxis) has been offered as one explanation for the migration of tissue cells explanted into culture. Theories of locomotion based on the physical properties of the membrane, or on the turnover of membrane (Abercrombie, Heaysman and Pegram, 1972), are attractive to students of these cells, which move very slowly and do not stream.

Some small amoebae like *Hyalodiscus simplex* also glide slowly over the substrate without streaming, and this fact taken with ultrastructural observations from amoebae and tissue cells of a similar fibrillar cytoplasm, of which one component is an actin-like 5–7 nm filament (Ishikawa, Bischoff and Holtzer, 1969; Perdue, 1973; Comly, 1973), has encouraged the belief that the whole gamut of crawling and gliding movements of animal cells are closely related at a mechanistic level (Komnick *et al.*, 1973).

Hyalodiscus (Fig. 11.17) is one of a number of small amoebae with a lamellipodium, a broad fan-like leading edge that is entirely ectoplasmic (see Bovee, 1964; Page, 1968 for similar genera). The endoplasm is confined to a hump trailed by the lamellipodium, and it is probable that no cyclical conversion of endoplasm to ectoplasm is involved in the forward movement of the organism (Komnick *et al.*, 1973). Any broad hypothesis of amoeboid movement that depends on streaming or the eruption of endoplasm through an ectoplasmic tube will not readily explain the movements of this amoeba.

11.3.2 *Motive force*

The source of motive force

The simple notion that primitive cytoplasm can contract and develop motive force (Ecker, 1849; Schulze, 1875) has outlasted a variety of other speculations: that amoeboid cells spread forward by surface tension (Butschli, 1894), or by active extension of membrane (Bell, 1961); that a motive force derives from the jet stream of pumping reticula (Kavanau, 1963), or from a posterior-anterior gradient of transmembrane potential (Bingley and Thompson, 1962); and now gains considerable support from the identification of muscle-like proteins in the plasms of amoebas and slime moulds (Pollard, 1973).

Our understanding of sarcomere shortening, and the induced sliding of microtubules that can be observed from demembraned flagella in ATP, suggests the likelihood that 'contraction' of cytoplasm will eventually be resolved at a fine structural level as some active form of sliding motion between filaments. This possibility has engendered suggestions that at grosser level of interaction, streaming movements may result from the bulk transport of one cytoplasmic layer across the surface of another.

Subirana (1970) has formalized a model for the locomotion of large *proteus-*

type amoebae in which active shear generated by structures at the interface drives the endoplasm forward over the more rigid ectoplasm. The model is unacceptable because it predicts, quite contrary to well-documented observations (Allen and Roslansky, 1959; and see Fig. 11.18), that endoplasm will stream faster at the interface, where the force is developed, than along the centreline of the amoeba.

Active shear is a sufficient explanation for the rotational streaming (cyclosis) seen in cells of algae belonging to the family Characeae, and accounts for the jump in the velocity profile at the boundary between moving endoplasm and a stationary layer of cortical gel lining the cell wall (Fig. 11.18; Kamiya, 1959). The shear force has been measured directly by applying enough hydrostatic pressure across the cut ends of a *Nitella* cell to just halt streaming in one direction (Tazawa, 1968), and by extrapolating from the perturbing effects of a range of high pressures on the velocity field in the endoplasm and cell sap (Donaldson, 1972; and see Table 11.4). The second study localized the motive force to a shallow layer about 1 μm thick at the cortical plasm-endoplasm interface, coincident with a structural zone of 5 nm microfilaments packed side-by-side in groups of 50–100 (Nagai and Rebhun, 1966).

Two-way, or countercurrent, streaming is a feature of cytoplasm movement through the very fine (about 1 μm) pseudopodial strands that branch and fuse within the net-like reticulopodia of Foraminifera, and might well be caused by active shear (Jahn and Rinaldi, 1959). Since microtubules have been found in these strands (McGee-Russell and Allen, 1971), the mechanism may have more in common with the mechano-chemistry of flagella than with motive force generation in other cytoplasms. Wohlfarth-Bottermann (1964) has pointed out that strands in which bi-directional streaming is seen in the light microscope are found under the electron microscope to be made up from finer, separately membrane-bound, cores, with the possibility that streaming is uni-directional in any one element.

Following Pantin (1923), force generation in the larger amoebae has been attributed repeatedly to contraction of the ectoplasm. That amoeba cytoplasm can contract is established by experiments in which amoebae spherulated by high pressures are decompressed. Invariably, as the solated cytoplasm reconstitutes a 'gel' layer there is a rapid centripetal contraction of the granular cytoplasm which weeps hyaline fluid (syneresis) into the space beneath the membrane (Landau, Zimmerman and Marsland, 1954; Marsland, 1964). At about the same time it was demonstrated (Hoffman-Berling, 1954) that glycerolated amoebae, like muscle, could be caused to contract in nucleoside triphosphates, particularly ATP.

The site of contractile forces

Circumstantial observations have been made of particles or granules in the ectoplasm moving closer together, particularly in the tail region (Goldacre,

316

1961; Yagi, 1961; Haberey, 1972) but such movements should not be adduced as evidence of localized active contraction. A similar result may be expected from passive shortening, or compression of cytoplasm, or migration of material from between the reference points in response to forces generated more remotely. It has become a matter of controversy where motive force is specifically localized in organisms of the *Amoeba–Chaos* group, and the means by which forces are harnessed to produce forward movement.

The tail contraction, or contraction-hydraulic, theory holds that active contractions in the tail squeeze fluid and smaller particles from the interstices of the ectoplasm into the endoplasmic stream and forward toward the hyaline cap. Motility is coupled to active contraction by an internal hydraulic force pressurizing the endoplasm. Contact between the membrane and granular ectoplasm may regulate the supply of ATP by some enzyme-substrate interaction, and limit contraction to the tail and retracting pseudopods (Goldacre, 1961, 1964). This theory has been updated and reviewed recently by Jahn (1964), Jahn and Bovee (1969), and Komnick *et al.* (1973).

The opposing frontal (fountain-zone) contraction theory of Allen (1961) followed a reappraisal of the non-Newtonian properties of endoplasm (see below). As it approaches each pseudopod tip, the structured, viscoelastic endoplasm contracts and develops tension, everts and in the process acquires the rheological condition of ectoplasm. The endoplasm is responsible for its own displacement, a hypothesis that grew originally from the need to explain the continued streaming of naked cytoplasm released from its membranous casing. Critics of this experiment (Marsland, 1964b) believe that the movements of demembraned cytoplasm are not a fair representation of streaming patterns *in situ*, and merely confirm the native contractility of cytoplasm.

Marsland's suggestion that contraction occurs at any point on the ectoplasmic tube is in keeping with the shuttle streaming seen in veins of the acellular slime mould *Physarum polycephalum*. The wall of these thick protoplasmic strands is a stationary ectoplasm that contracts periodically and at intervals along the strand, driving the fluid internal plasm back and forth, with overall progress depending on net volume flow.

One test of the pressure gradient hypothesis is to set up a competing gradient by sucking through a capillary on one end of an amoeba. When this is done (Allen, Francis and Zeh, 1971), although cytoplasm in the capillary flows rapidly toward the negative pressure source, streaming elsewhere continues in directions away from the tube. The criticism levelled at this simple experiment amounts to a lack of faith in the original observation that the capillary draws in both endoplasm and ectoplasm, and not ectoplasm alone (Jahn and Votta, 1972c), though the experiment has since been repeated (Kirby, Rinaldi and Cameron, 1972). In defence of the hydraulic theory, one might hypothesize that the region of the amoeba just outside the capillary entrance is induced to form a tail squeezing endoplasm in both directions. In this case the two poles should have a lower internal pressure than the middle, and the gradient in either direction should be of the order normally present in an amoeba (Kirby *et al.*, 1972). For as long as the pseudopod in the suction tube remains intact, it is possible that a large part of the suction force applied to the amoeba (up to 35 cm of water) is dissipated in stretching and bending the membrane and ectoplasm (diaphragm stresses) and is not felt by the endoplasm. However, in the experiment reported, the superficial layers were eventually ruptured in the capillary and endoplasm flowed freely from the cell without disturbing streaming against the pressure gradient in pseudopods outside the tube.

The movement of particles in the cytoplasm

If a moving particle is seen to accelerate we can assume that it has been acted on by a force the direction of which we can detect, even if we are unsure of the origin of the force. If a series of particles, more or less on the same streamline, accelerate in turn, we are no better informed since it is a fallacious interpretation that the particle accelerating first is necessarily closest to the origin of the force.

The advance of a pseudopod is usually unsteady in the larger amoebae. Granules in the endoplasm jerk forward in a series of sporadic movements as waves of acceleration pass backward from the tip. Mast (1926) accounts for the fact that movements often start at the front, by supposing that a thin plasmagel sheet across the tip of a forming pseudopod gives way under the internal pressure, allowing endoplasmic particles to erupt through the breach. The first sign of movement is the accumulation, at the tip, of hyaline fluid squeezed under pressure through the plasmagel sheet. In broad, flat pseudopods the endoplasm stream may divide into streamlets advancing independently, according to the rhythm set by hyaline eruption at the front of each channel.

The front end contraction theory readily explains particles accelerating first at the tip, and reasonably identifies the hyaline cap with syneretic fluid expelled from the hydrated endoplasmic proteins as they contract and re-orient in the fountain zone. The refractive interface between compressed gel at the tip, and hyaline fluid, appears under simple optics as the sheet-like structure described by Mast.

When, in *Chaos*, a pseudopod halts and begins to retract, evacuation of endoplasm is noticed as a wave of reversed streaming, the reversal appearing first at the base of the pseudopod and spreading back to the tip (Fig. 11.19). For a few seconds during reversal, particles at the tip and base may be moving in opposite directions, as if under the influence of separate motive centres (Allen, 1973). The observation draws attention to the local character of motive force in keeping with the front-end contraction thesis, but although it discounts the idea of a simple and single pressure gradient in the pseudopod, the test is not definitive for the reasons stated above.

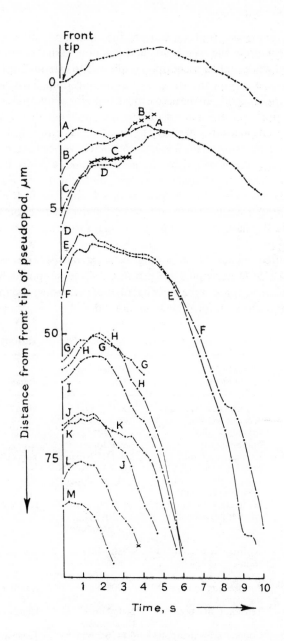

Fig. 11.19 A reversal in the direction of streaming endoplasm as a pseudopod of *Chaos* begins to retract is monitored by plotting the changing positions of endoplasmic particles. Endoplasm closest to the base (particle M for instance) is reversed seconds before the tip changes direction. After 3 s endoplasm is streaming in opposite directions in different regions of the pseudopod (B − D versus G − M) (from Allen, 1973).

Measurement of motive force

Contraction of amoeba cytoplasm can produce a motion, streaming, that is entirely confined within the membrane and has no external component (and does no external work) until a pseudopod is allowed to extend. The form of the coupling between extension and streaming is still uncertain. To a large extent, then, the forces produced from the active mechanism have internal components during locomotion and are inaccessible to measurement in ways that give clear insight into the mechanochemistry of streaming. Against the background of the debate over mechanism, the few attempts that have been made to measure a motive force are compromised by the need to make *a priori* and partisan assumptions of where and what active force is, and how it acts. For example, Goldacre (1961) has found that the prongs of a small glass fork pushed into the tail of *A. proteus* are squeezed together by an 'active contraction' equivalent to a 3 μg weight. By the same technique he could measure no forces in the fountain zone.

Using measured values of streaming velocity in the Hagen–Poiseuille equation, Yagi (1961) attempted an estimate of the supposed hydrostatic pressure gradient in a *proteus*-type amoeba, on the (erroneous) assumption that endoplasm behaves as a uniform Newtonian fluid.

Table 11.4 Motive force of cytoplasmic streaming measured from various organisms

Organism	Method of measurement	Measured motive force	Author
Amoeba proteus	Movement against an acceleration field	5×10^{-8} N (whole amoeba)	Allen (1960)
	Implanted glass microbalance	2.9×10^{-7} N (local force)	Goldacre (1961)
	Estimated internal pressure to account for streaming velocity of viscous cytoplasm	2.9–11.5 Nm^{-2} or 0.6–2.3 $\times$ 10^{-8} N (per pseudopod)*	Yagi (1961)
	Hydrostatic pressure to counter movement	~100 Nm^{-2} or 2×10^{-7} N (per pseudopod)*	Kamiya (1964)
Chaos chaos	Hydrostatic pressure to counter movement	~150 Nm^{-2} or 3×10^{-7} N (per pseudopod)*	Kamiya (1964)
Physarum	Air pressure to balance movement	~2000 Nm^{-2}	Kamiya (1968)
	Tension transducer	5000 Nm^{-2} (maximum)	Kamiya *et al.* (1972)
Nitella	Counter pressure applied to cell sap	0.14–0.2 Nm^{-2}	Tazawa (1968)
	Counter pressure effect on velocity profile	0.36 Nm^{-2} (at the shear interface)	Donaldson (1972)

* A pseudopod 50 μm in diameter is assumed.

The motive force of *A. proteus* has been calculated by Allen (1960) from the speed with which whole cells move in the acceleration field of a centrifuge microscope (Table 11.4).

In the elegant 'double-chamber' experiments of Kamiya (1964), an amoeba is forced into a short agar or glass capillary and extends pseudopods from both ends into closed pressure chambers. The hydrostatic pressure difference across the two chambers is manipulated to counter streaming in either direction through the capillary. The method relies on the null principle, the pressure difference at balance is monitored and plotted as a continuous record (a dynamoplasmogram) of the magnitude and direction of motive force. The technique assumes that internal hydrostatic pressure is motivating and can be opposed by an externally applied pressure, and is challengable on that count.

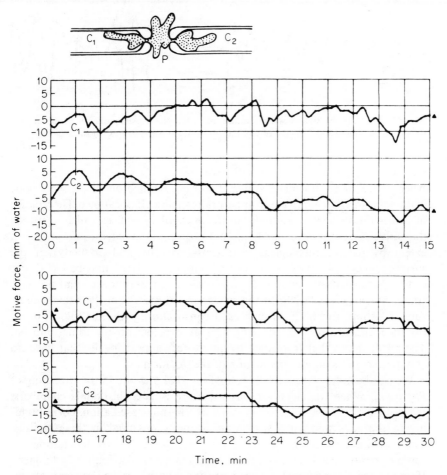

Fig. 11.20 Simultaneous dynamoplasmograms recording the balance pressures, C_1 and C_2, on two pseudopods of a single *Chaos* isolated in a triple chamber (from Kamiya, 1964).

321

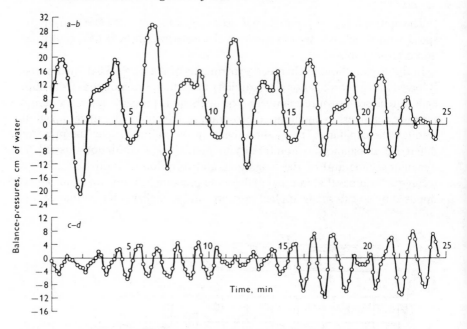

Fig. 11.21 Oscillating dynamoplasmograms recorded from plasmodia of *Physarum*, enriched (upper trace) and depleted of endoplasm (lower trace) (from Kamiya, 1968).

Allen (discussion following Kamiya, 1964) has pointed out that pseudopods in the pressure chambers have an unnaturally spherical appearance. The balance of forces on motionless endoplasm at the capillary orifices will be between the active streaming force, and external pressure attenuated to an unknown extent by passive rheological properties (viscosity and elasticity) of the surface and cytoplasm of the swollen pseudopods. A large amoeba like *Chaos* can be persuaded into two capillaries connecting three independent pressure compartments (Fig. 11.20). In this way, parallel records of the balance pressures resisting streaming through each constriction give some indication of the independence of force development at two loci on the same amoeba. The result shows that from time to time forces appear locally that have no counterpart in the more remote limb of the organism.

We can be more certain of the role of internal pressure in the streaming of *Physarum*. When fragments of plasmodia are isolated in double chambers, the motive force dynamoplasmogram shows rhythmic oscillations corresponding to the back and forth streaming normally observed in these strands. The magnitude of the motive force variation is matched to the endoplasm/ectoplasm ratio in a way that attempts to regulate by redistributing endoplasm. If a polarized plasmodium is bisected into endoplasm-enriched and endoplasm-depleted fragments, force oscillations of the predominantly endoplasmic

322

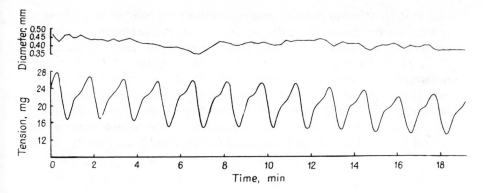

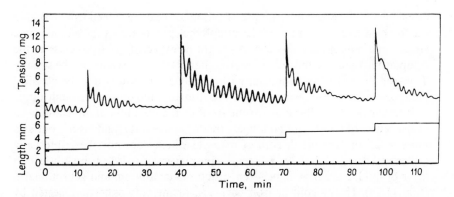

Fig. 11.22 Isometric tension in strands of *Physarum* oscillates rhythmically. The rising tension is biphasic (upper figure). The amplitude of the oscillation is augmented each time the strand is stretched (lower figure) (from Kamiya *et al.*, 1972).

preparation are of much greater amplitude than those from the endoplasm-poor preparation (Fig. 11.21; Kamiya, 1968).

Capitalizing on the large size of the slime moulds, some recent experiments (Kamiya, Allen and Zeh, 1972) have directly measured tension along the axes of excised strands of *Physarum* suspended from a tension transducer. At constant length (isometric contraction) tension changes follow a uniform oscillation with the phase of rising tension significantly longer than the fall in tension. A 'shoulder' on the rising tension trace suggests that the active mechanism may resolve into two separately timed tension-producing stages. When, at intervals, the preparation is stretched to new lengths, the rise in mean tension at each step is accompanied by an increase in the amplitude of the oscillation, even though the strand is thinner than before (Fig. 11.22). Further-

323

more, strands free to contract (isotonic contraction) oscillate in length to a greater extent when more heavily loaded. The contractile mechanism appears to sense the increased load and responds by augmenting the contractile force.

There are other indications that the motile cytoplasm of *Physarum* self-organizes under tension. Young's modulus for whole strands is about 50% higher during tension development than during relaxation. Newly assembled, or redeployed, microfilaments have been resolved in the electron microscope in contracting strands along the axes of greatest stress (Wohlfarth-Bottermann, 1964). Stretched strands show enhanced birefringence (Kamiya *et al.*, 1972).

11.3.3 Physical properties of amoeba cytoplasm

Rheology of cytoplasm in situ

Some idea of the complex structure of endoplasm can be gained from tracking the forward movement of granules at different points across the endoplasmic stream in a monopodial amoeba. As might be expected, particles close to the ectoplasmic tube wall move more slowly than those at the centre of the stream. Towards the tail of the amoeba, particle movement traces out the classical parabolic velocity profile of a simple (Newtonian) fluid flowing through a tube as if under pressure. But close to the front of the pseudopodium, endoplasm at the centre of the stream no longer streams as separate fluid layers, but as a more solid, or structured, core of material which does not shear internally (Allen and Roslansky, 1959). The central plug is carried forward on a lubricating layer of fluid, a shear zone between structured endoplasm and gel ectoplasm (Fig. 11.18). Heavy gold or iron particles, or mercury spheres, ingested by *Amoeba* or *Chaos* do not fall through, but rather around, the axial structure of the endoplasm (Allen, 1961). A pioneering observation made with the centrifuge microscope by Harvey and Marsland (1932) can now be interpreted in the same way (Allen, 1960). These authors noted that in a steady acceleration field, granules faltered in their movement through the cytoplasm, as if obstructed by invisible components.

Endoplasm is structurally heterogeneous and certainly not a uniform, low viscosity 'sol'. Its behaviour recalls a pseudoplastic material, like paint.

Optical properties of cytoplasm in situ

On the whole, electron microscopy has been unsuccessful in uncovering patterns of organization of fibres or filaments in whole amoebae that suggest an arrangement for producing motive force or movement (Plate 11.3b). On the other hand, ordered packing of thin (5–7 nm) or thick (ca. 14 nm) filaments have been seen in strands of cytoplasm pooled and gelled in ATP (Wolpert, Thompson and O'Neill, 1964; Pollard and Ito, 1970) or glycerinated amoebae caused to contract in ATP (Plate 11.4; Holberton and Preston, 1970; Schäfer-Daneel, 1967). Probably, these structures interact transiently during cyclical streaming *in situ*, and such organization as is present is labile to fixation. Under these circumstances, dynamic order might be detected more readily by optical methods, as birefringence patterns under the polarizing microscope. It would be particularly interesting if variations in birefringence mapped at different regions of the cytoplasm, indicating changes in the degree of structural anisotropy, or if during movement photoelasticity (strain birefringence) appeared locally, signalling dynamic realignment of structures by local tensions.

Amoebae are weakly birefringent, even when not moving. The sign of this intrinsic or form birefringence is positive on the long axis of the cell, indicating a preferential alignment in that direction of those linear elements that contribute to the barely anisotropic structure. As cytoplasm streams, the positive axial birefringence is enhanced. The additional optical retardation extends to the tail endoplasm, but is strongest in the fountain zone. Ectoplasm retains diffuse positive birefringence except for areas of reversed sign bordering the endoplasm shear zone (Allen, 1972). The earliest reliable study with refined optics (Allen, Francis and Nakajima, 1965) uncovered a cycle of photoelasticity at the tip of a sporadically erupting pseudopod in *Chaos*. Cytoplasm approaching the fountain zone first became more intensely positively birefringent, and then, as it everted, decreased in birefringence or

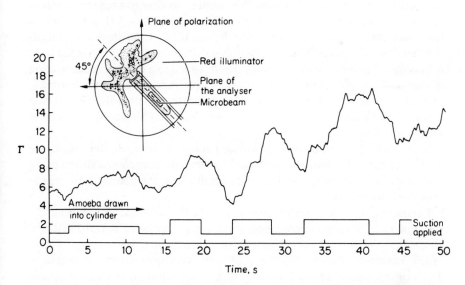

Fig. 11.23 By sucking on a pseudopod in a capillary tube, endoplasmic birefringence (retardation, Γ, measured in Angstrom units) is increased. The enhanced birefringence is not proportional to the magnitude of the suction force but to the length of time it is applied (from Francis and Allen, 1971).

became negatively birefringent in patches on the inside of the 'bend'. The impression was given of cytoplasm being stretched in the endoplasmic stream and then compressed as it was added to the tip of the ectoplasmic tube.

To rule out the possibility that the increase in birefringence was due to the passively orienting influence of flow, Francis and Allen (1971) measured quantitatively the birefringence changes in endoplasm induced to stream sporadically by sucking on a pseudopod in a capillary. In keeping with the viscoelastic interpretation of cytoplasm, the absolute magnitude of birefringence increased as long as suction was applied. Released from suction, the endoplasm recoils, and though streaming may stop immediately, the birefringence persists for a short while (Fig. 11.23). Flow birefringence is characteristic of a solution of macromolecules in a shear field, but the birefringence detected by Francis and Allen arose in the plug zone of the endoplasm where there is no shear, and in all probability was strain birefringence of a material sufficiently structured to transmit tensile forces.

Isolated cytoplasm

The cytoplasm of *Chaos* shelled out of its membrane into a 'stabilizing' salt solution containing minimal calcium, stiffens, and assumes life-like viscoelastic properties. Strain birefringence appears when it is stretched with a micropipette, and decays as strands recoil elastically when the tension is interrupted. During this experiment, the increase in optical retardation is not completely lost, indicating that stress has induced some permanent microscopic deformation or 'set' (Taylor, Condeelis, Moore and Allen, 1973). Under tension, this preparation develops visible fibrils, 0.1–0.4 μm thickness. Fibrils shorten dramatically, and pull on glass fragments, when the cytoplasmic mass is transferred to a new solution with calcium added to a free ion concentration in excess of 7×10^{-7} gram ions l^{-1}. In the electron microscope, the fibrils are seen to be assembled from side-to-side aggregates of actin-like (7 nm) thin filaments, with adhering thick myosin-like (15–25 nm) filaments, and smaller 4.5 nm filaments.

If, instead, stabilized cytoplasm is placed in low calcium solution with added ATP, it 'relaxes'. The fibrils no longer cohere when the preparation is drawn out, and strain birefringence is no longer detectable as if the filaments are free to slide over one another.

Furthermore, the preparation reproduces features of amoeboid movement – streaming and pseudopod extension – that encourage belief that the essential structures have been preserved, at least temporarily, in a native state. Streams of granuloplasm erupt in fountain and loop formations from the surface of drops of cytoplasm freshly isolated into a 'flare' solution, in which a critical balance of calcium and ATP has been achieved. Added magnesium prolongs the motility of flare cytoplasm. Convincingly natural behaviour is seen at the tip of these naked pseudopods, suggesting that the mechanochemistry of intact pseudopods is operating in the model. Cytoplasm streams towards the bend of the flare and, on turning, contracts, decelerates and increases in refractive index before migrating back into the central mass. If the slowly returning limb adheres temporarily, the pseudopod will locomote over several millimetres.

Acknowledgements

This review was written during tenure of a Science Research Council Grant for a study of flagellate behaviour. I would like to thank Mr. T. M. Preston whose electron micrograph appears in Plate 4c, and Mr. J. Marshall for preparing material used in Plate 2.

References

Abercrombie, M., Heaysman, J.E.M. and Pegram, S.M. (1972) Locomotion of fibroblasts in culture. V. Surface marking with concanavalin A. *Exp. Cell Res.* **73**, 536–539.

Alicea, H.A. and Renaud, F.L. (1975) Actin-tubulin homology revisited. *Nature* **257**, 601–602.

Allen, R.D. (1960) The consistency of amoeba cytoplasm and its bearing on the mechanism of amoeboid movement. II. The effects of centrifugal acceleration observed in the centrifuge microscope. *J. Biophys. Biochem. Cytol.* **8**, 379–397.

Allen, R.D. (1961) A new theory of amoeboid movement and protoplasmic streaming. *Exp. Cell Res.*, Supplement 8, 17–31.

Allen, R.D. (1972a) Reply to Jahn and Votta. *J. Mechanochem. Cell Motility* **1**, 247–249.

Allen, R.D. (1972b) Pattern of birefringence in the giant amoeba, *Chaos carolinensis*. *Exp. Cell Res.* **72**, 34–45.

Allen, R.D. (1973) In: *The Biology of Amoeba*, Jeon, K.W., Ed., Ch. 7. pp. 202–249.

Allen, R.D., Cooledge, J.W. and Hall, P.J. (1960) Streaming in cytoplasm dissociated from the giant amoeba, *Chaos chaos*. *Nature* **187**, 896–899.

Allen, R.D., Cowden, R. and Hall, P.J. (1962) Syneresis in amoeboid movement: its localization by interference microscopy and its significance. *J. Cell Biol.* **12**, 185–189.

Allen, R.D. and Francis, D.W. (1965) Cytoplasmic contraction and the distribution of water in the amoeba. *Symp. Soc. Exp. Biol.* **19**, 259–271.

Allen, R.D., Francis, D.W. and Nakajima, H. (1965) Cyclic birefringence changes in pseudopods of *Chaos carolinensis* revealing the localization of the motive force in pseudopod extension. *Proc. natn. Acad. Sci. U.S.A.* **54**, 1153–1161.

Allen, R.D., Francis, D. and Zeh, R. (1971) Direct test of the positive pressure gradient theory of pseudopod extension and retraction in amoebae. *Science* **174**, 1237–1240.

Allen, R.D. and Roslansky, J.D. (1959) The consistency of amoeba cytoplasm and its bearing on the mechanism of amoeboid movement. I. An analysis of endoplasmic velocity profiles of *Chaos chaos*. *J. Biophys. Biochem. Cytol.* **6**, 437–446.

Amos, W.B. (1971) Reversible mechanochemical cycle in the contraction of *Vorticella*. *Nature* **229**, 127–128.

Amos, W.B., Routledge, L.M. and Yew, F.F. (1975) Calcium-binding proteins in a vorticellid contractile organelle. *J. Cell Sci.* **19**, 203–213.

Baba, S.A. (1972) Flexural rigidity and elastic constant of cilia. *J. exp. Biol.* **56**, 459–467.

Bardele, C.F. (1972) A microtubule model for ingestion and transport in the suctorian tentacle. *Z. Zellforsch. mikrosk. anat.* **126**, 116–134.

Bell, L.G.E. (1961) Surface extension as the mechanism of cellular movement and cell division. *J. Theor. Biol.* **1**, 104–106.

Bingley, M.S. and Thompson, C.M. (1962) Bioelectric potentials in relation to movement in Amoebae. *J. Theor. Biol.* **2**, 16–32.

327

Mechanics and energetics of animal locomotion

Blake, J.R. (1971a) A spherical envelope approach to ciliary propulsion. *J. Fluid Mech.* **46,** 199–208.

Blake, J.R. (1971b) Infinite models for ciliary propulsion. *J. Fluid Mech.* **49,** 209–222.

Blake, J.R. (1971c) Self propulsion due to oscillations on the surface of a cylinder at low Reynolds number. *Bull. Aust. math. Soc.* **5,** 255–264.

Blake, J.R. (1972) A model for the micro-structure in ciliated organisms. *J. Fluid Mech.* **55,** 1–23.

Blake, J.R. (1973) A finite model for ciliated micro-organisms. *J. Biomech.* **6,** 133–140.

Blake, J.R. (1974) Hydrodynamic calculations on the movements of cilia and flagella. I. *Paramecium. J. Theor. Biol.* **45,** 183–203.

Blake, J.R. and Sleigh, M.A. (1974) Mechanics of ciliary locomotion *Biol. Rev.* **49,** 85–125.

Bouck, G.B. (1972) Architecture and Assembly of Mastigonemes. *Adv. Cell Molec. Biol.* **2,** 237–272.

Bovee, E.C. (1964) In: *Primitive Motile Systems in Cell Biology*, Allen, R.D. and Kamiya, N., Eds., pp. 189–220. New York: Academic Press.

Bovee, E.C. and Jahn, T.L. (1972) A theory of piezoelectric activity and ion movements in the relation of flagellar structures and their movements to the phototaxis of *Euglena. J. Theor. Biol.* **35,** 259–276.

Brokaw, C.J. (1965) Non-sinusoidal bending waves of sperm flagella. *J. exp. Biol.* **43,** 155–169.

Brokaw, C.J. (1966a) Bend propagation along flagella. *Nature* **209,** 161–163.

Brokaw, C.J. (1966b) Effects of increased viscosity on the movements of some invertebrate spermatozoa. *J. exp. Biol.* **45,** 113–139.

Brokaw, C.J. (1971) Bend propagation by a sliding filament model for flagella. *J. exp. Biol.* **55,** 289–304.

Brokaw, C.J. (1972a) Computer simulation of flagellar movement. I. Demonstration of stable bend propagation and bend initiation by the sliding filament model. *Biophys. J.* **12,** 564–586.

Brokaw, C.J. (1972b) Viscous resistances in flagella: analysis of small amplitude motion. *J. Mechanochem. Cell Motility* **1,** 151–155.

Brokaw, C.J. (1972c) Computer simulation of flagellar movement. II. Influence of external viscosity on movement of the sliding filament model. *J. Mechanochem. Cell Motility,* **1,** 203–211.

Brokaw, C.J. and Wright, L. (1963) Bending waves on the posterior flagellum of *Ceratium. Science* **142,** 1169–1170.

Brooker, B.E. (1965) Mastigonemes in a bodonid flagellate. *Exp. Cell Res.* **37,** 300–305.

Butschli, O. (1894) *Investigations on microscopic foams and on protoplasm.* A. & C. Black, London.

Chasey, D. (1972) Further observations on the ultrastructure of cilia from *Tetrahymena pyriformis. Exp. Cell Res.* **74,** 471–479.

Chwang, A.T. and Wu, T.Y. (1971) A note on the helical movement of micro-organisms. *Proc. R. Soc. Ser. B* **178,** 327–346.

Chwang, A.T., Winet, H. and Wu, Y. (1974) A theoretical mechanism of spirochetal locomotion. *J. Mechanochem. Cell Motility* **3,** 69–76.

Cleveland, L.R. and Cleveland, B.T. (1966) The locomotory waves of *Koruga, Deltotrichonympha* and *Mixotricha. Arch. Protistenkunde* **109,** 39–63.

Coakley, C.J. and Holwill, M.E.J. (1972) Propulsion of micro-organisms by three-dimensional flagellar waves. *J. Theor. Biol.* **35,** 525–542.

Comly, L.T. (1973) Microfilaments in *Chaos carolinensis*. Membrane association, distribution and heavy meromyosin binding in the glycerinated cell. *J. Cell Biol.* **58,** 230–237.

Cox, R.G. (1970) The motion of long slender bodies in a viscous fluid. 1. General theory. *J. Fluid Mech.* **43,** 641–660.

Donaldson, I.G. (1972) The estimation of the motive force for protoplasmic streaming in *Nitella. Protoplasma* **74,** 329–344.

Ecker, A. (1849) Zur Lehre vom Bau und Leben der contractilen substanz der niedersten thiere. *Z. wiss. Zool.* 218–245.

Francis, D.W. and Allen, R.D. (1971) Induced birefringence as evidence of endoplasmic viscoelasticity in *Chaos carolinensis. J. Mechanochem. Cell Motility* **1,** 1–6.

Frey-Wyssling, A. (1965) In: *Ultrastructure in Plant cytology*, Frey-Wyssling, A. and Möhlthaler, K., Eds., pp. 329–41. Amsterdam: Elsevier.

Gibbons, I.R. and Fronk, E. (1972) Some properties of bound and soluble dynein from sea urchin sperm flagella. *J. Cell Biol.* **54**, 365–381.

Gibbons, B.H. and Gibbons, I.R. (1972) Flagellar movement and adenosine triphosphatase activity in sea urchin sperm extracted with Triton X-100. *J. Cell Biol.* **54**, 75–97.

Gibbons, I.R. and Gibbons, B.H. (1974) The fine structure of rigor wave axonemes from sea urchin sperm flagella. *J. Cell Biol.* **63**, 110a.

Goldacre, R.J. (1952a) The action of general anaesthetics on amoebae and the mechanism of the response to touch. *Symp. Soc. Exp. Biol.* **6**, 128–143.

Goldacre, R.J. (1952b) The folding and unfolding of protein molecules as a basis of osmotic work. *Int. Rev. Cytol.* **1**, 135–164.

Goldacre, R.J. (1961) The role of the cell membrane in the locomotion of amoebae and the source of the motive force and its control. *Exp. Cell Res.* Supplement 8, 1–16.

Goldacre, R.J. (1964) In: *Primitive Motile Systems in Cell Biology*, Allen, R.D. and Kamiya, N., Eds., pp. 237–256. New York: Academic Press.

Goldacre, R.J. and Lorch, I.J. (1950) Folding and unfolding of protein molecules in relation to cytoplasmic streaming, amoeboid movement and osmotic work. *Nature* **166**, 497–499.

Goldstein, S.F. (1969) Irradiation of sperm tails by laser microbeam. *J. exp. Biol.* **51**, 431–441.

Goldstein, S.F., Holwill, M.E.J. and Silvester, N.R. (1970) The effects of laser microbeam irradiation on the flagellum of *Crithidia* (*Strigomonas*) *oncopelti*. *J. exp. Biol.* **53**, 401–409.

Gray, J. and Hancock, G.J. (1955) The propulsion of sea-urchin spermatozoa. *J. exp. Biol.* **32**, 802–814.

Griffin, J.L. (1964) In: *Primitive Motile Systems in Cell Biology*, Allen, R.D. and Kamiya, N., Eds., pp. 303–321. New York: Academic Press.

Grimstone, A.V. (1959) Cytology, homology and phylogeny – a note on 'organic design'. *Am. Nat.* **93**, 273–282.

Haberey, M. (1971) Bewegungsverhalten und untergrundkontakt von *Amoeba proteus*. *Mikroskopie* **27**, 226–234.

Haberey, M. (1972) Cinematography of cell membrane behaviour and flow phenomena in *Amoeba proteus*. *Acta Protozoologica* **11**, 95–102.

Haberey, M., Wohlfarth-Bottermann, K.E. and Stockem, W. (1969) Pinocytose und Bewegung von Amöben. VI. Kinematographische untersuchungen über das Bewegungsverhalten der Zelloberfläche von *Amoeba proteus*. *Cytobiologie* **1**, 70–84.

Hancock, G.J. (1953) The self-propulsion of microscopic organisms through liquids. *Proc. R. Soc. Ser. A* **217**, 96–121.

Harvey, E.N. and Marsland, D.A. (1932) The tension at the surface of *Amoeba dubia* with direct observations on the movement of cytoplasmic particles at high centrifugal speeds. *J. Cell. Comp. Physiol.* **2**, 75–98.

Hawkes, R.B. and Holberton, D.V. (1975) Myonemal contraction of *Spirostomum*. II. Some mechanical properties of the contractile apparatus. *J. Cell. Physiol.* **85**, 595–602.

Hoffman-Berling, H. (1954) Adenosintriphosphat als Beitriebstoff von Zellbewegung. *Biochim. Biophys. Acta* **14**, 182–194.

Holberton, D.V. (1973) Fine structure of the ventral disk apparatus and the mechanism of attachment in the flagellate *Giardia muris*. *J. Cell Sci.* **13**, 11–41.

Holberton, D.V. and Ogle, W.S. (1975) Conditions for contraction and extension of *Spirostomum* cell models. *J. Protozool.* **22**, 39A.

Holberton, D.V. and Preston, T.M. (1970) Arrays of thick filaments in ATP-activated *Amoeba* model cells. *Exp. Cell Res.* **62**, 473–477.

Holwill, M.E.J. (1965) The motion of *Strigomonas oncopelti*. *J. exp. Biol.* **42**, 125–137.

Holwill, M.E.J. (1966a) Physical aspects of flagellar movement. *Physiol. Rev.* **46**, 696–785.

Holwill, M.E.J. (1966b) The motion of *Euglena viridis*: the role of flagella. *J. exp. Biol.* **44**, 579–588.

Holwill, M.E.J. (1974) In: *Cilia and Flagella*. Sleigh, M.A., Ed., Ch. 8, pp. 143–176. London: Academic Press.

Holwill, M.E.J. and Burge, R.E. (1963) A hydrodynamic study of the motility of flagellated bacteria. *Arch. Biochem. Biophys.* **101**, 249–260.

Holwill, M.E.J. and Miles, C.A. (1971) Hydrodynamic analysis of non-uniform flagellar undulations. *J. Theor. Biol.* **31**, 25–42.

Holwill, M.E.J. and Sleigh, M.A. (1967) Propulsion by hispid flagella. *J. exp. Biol.* **47**, 267–276.

Hopkins, J.M. (1970) Subsidiary components of the flagella of *Chlamydomonas reinhardii. J. Cell Sci.* **7**, 823–839.

Ishikawa, H., Bischoff, R. and Holtzer, H. (1969) Formation of arrowhead complexes with heavy neromyosin in a variety of cell types. *J. Cell Biol.* **43**, 312–328.

Jahn, T.L. (1964) In: *Primitive Motile Systems in Cell Biology*, Allen, R.D. and Kamiya, N., Eds., pp. 279–302. New York: Academic Press.

Jahn, T.L. and Bovee, E.C. (1969) Protoplasmic movements within cells. *Physiol. Rev.* **49**, 793–862.

Jahn, T.L., Landman, M.D. and Fonseca, J.R. (1964) The mechanism of locomotion in flagellates. II. Function of the mastigonemes of *Ochromonas. J. Protozool.* **11**, 291–296.

Jahn, T.L. and Rinaldi, R.A. (1959) Protoplasmic movement in the foraminiferan, *Allogromia lamicollaris*; and a theory of its mechanism. *Biol. Bull.* **117**, 100–118.

Jahn, T.L. and Votta, J.J. (1972a) Locomotion of Protozoa. *Ann. Rev. Fluid Mech.* **4**, 93–116.

Jahn, T.L. and Votta, J.J. (1972b) Birefringence as an index of amoeboid movement. *J. Mechanochem. Cell Motility* **1**, 245–249.

Jahn, T.L. and Votta, J.J. (1972c) Capillary suction test of the pressure gradient theory of amoeboid motion. *Science* **177**, 636–637.

Kamiya, N. (1959) Protoplasmic streaming. *Protoplasmatologia* **8**, 3a. 1–199.

Kamiya, N. (1964) In: *Primitive Motile Systems in Cell Biology*, Allen, R.D. and Kamiya, N., Eds., pp. 257–278. New York: Academic Press.

Kamiya, N. (1968) The mechanism of cytoplasmic movement in a myxomycete plasmodium. *Symp. Soc. exp. Biol.* **22**, 199–214.

Kamiya, N., Allen, R.D. and Zeh, R. (1972) Contractile properties of the slime mould strand. *Acta protozoologica* **11**, 113–124.

Kanno, F. (1965) An analysis of amoeboid movement. IV. Cinematographic analysis of movement of granules with special reference to the theory of amoeboid movement. *Annotationes Zoologicae Japonenses* **38**, 45–63.

Kavanau, J.L. (1963) A new theory of amoeboid locomotion. *J. Theor. Biol.* **4**, 124–141.

Kinosita, H. and Murakami, A. (1967) Control of ciliary motion. *Physiol. Rev.* **47**, 52–63.

Kirby, G.S., Rinaldi, R.A. and Cameron, I.L. (1972) Capillary suction test of the pressure gradient theory of amoeboid motion. *Science* **177**, 637–638.

Knight-Jones, E.W. (1954) Relations between metachronism and the direction of ciliary beat in Metazoa. *Q. J. micros. Sci.* **95**, 503–521.

Komnick, H., Stockem, W. and Wohlfarth-Bottermann, K.E. (1973) Cell Motility: Mechanisms in Protoplasmic streaming and amoeboid movement. *Int. Rev. Cytol.* **34**, 169–252.

Landau, J.V., Zimmerman, A.M. and Marsland, D. (1954) Temperature pressure experiments on *Amoeba proteus*; plasmagel structure in relation to form and movement. *J. Cell. Comp. Physiol.* **44**, 211–232.

Leedale, G.F. (1967) *Euglenoid Flagellates.* New Jersey: Prentice-Hall.

Lighthill, M.J. (1969) Hydromechanics of aquatic animal propulsion. *Ann. Rev. Fluid Mech.* **1**, 413–446.

Lowndes, A.G. (1941). On flagellar movement in unicellular organisms. *Proc. Zool. Soc. Lond.* **111**, 111–134.

Lowndes, A.G. (1944) The swimming of unicellular flagellate organisms. *Proc. Zool. Soc. Lond.* **113**, 99–107.

Lubliner, J. and Blum, J.J. (1970) Model for bend propagation in flagella. *J. Theor. Biol.* **31**, 1–24.

Machemer, H. (1972) Ciliary activity and the origin of metachrony in *Paramecium*: effects of increased viscosity. *J. exp. Biol.* **57**, 239–259.

Machemer, H. (1974) In: *Cilia and Flagella*, Sleigh, M.A., Ed., Ch. 10, pp. 199–286. London: Academic Press.

Machin, K.E. (1958) Wave propagation along flagella. *J. exp. Biol.* **35**, 796–806.

Machin, K.E. (1963) The control and synchronization of flagellar movement. *Proc. R. Soc. Ser. B* **158**, 88–104.

Marsland, D. (1964a) In: *Primitive Motile Systems in Cell Biology*, Allen, R.D. and Kamiya, N., Eds., pp. 173–188, New York: Academic Press.

Marsland, D. (1964b) In: *Primitive Motile Systems in Cell Biology*, Allen, R.D. and Kamiya, N., Eds., Free Discussion, pp. 323–327, New York: Academic Press.

Mast, S.O. (1926) Structure, movement, locomotion and stimulation in *Amoeba. J. Morphol. Physiol.* **41**, 347–425.

McGee-Russel, S.M. and Allen, R.D. (1971) Reversible stabilization of labile microtubules in the reticulopodial network of *Allogromia. Adv. Cell Molec. Biol.* **1**, 153–184.

McIntosh, J.R. (1973) The axostyle of *Saccinobaculus*. II. Motion of the microtubule bundle and a structural comparison of straight and bent axostyles. *J. Cell Biol.* **56**, 324–339.

Miles, C.A. and Holwill, M.E.J. (1971) Experimental evaluation of a mechanochemical model of flagellar activity. *J. Mechanochem. Cell Motility* **1**, 23–32.

Nagai, R. and Rebhun, L.I. (1966) Cytoplasmic microfilaments in streaming *Nitella* cells. *J. Ultrastruct. Res.* **14**, 571–589.

Page, F.C. (1968) Generic criteria for *Flabellula, Rugipes* and *Hyalodiscus*, with descriptions of species. *J. Protozool.* **15**, 9–26.

Pantin, C.F.A. (1923) On the physiology of amoeboid movement. I. *J. Mar. Biol. Ass. U.K.* **13**, 24–69.

Parducz, B. (1967) Ciliary movements and co-ordination in ciliates. *Int. Rev. Cytol.* **21**, 91–128.

Perdue, J.F. (1973) The distribution, ultrastructure, and chemistry of microfilaments in cultured chick embryo fibroblasts. *J. Cell Biol.* **58**, 265–283.

Piccinni, E., Albergoni, V. and Coppellotti, O. (1975) ATPase activity in flagella from *Euglena gracilis*. Localization of the enzyme and effects of detergents. *J. Protozool.* **22**, 331–335.

Pítelka, D.R. and Schooley, C.N. (1955) Comparative morphology of some protistan flagella. *Univ. Calif. (Berkeley) Publ. in Zool.* **61**, 79–128.

Pollard, T.D. (1973) In: *The Biology of Amoeba*, Jeon, K.W., Ed., Ch. 9, pp. 291–318. New York: Academic Press.

Pollard, T.D. and Ito, S. (1970) Cytoplasmic filaments of *Amoeba proteus*. I. The role of filaments in consistency changes and movement. *J. Cell Biol.* **46**, 267–289.

Preston, J.T., Jahn, T.L. and Fonseca, J.R.C. (1970) Helical form of ciliary beat in *Tetrahymena pyriformis. J. Cell Biol.* **47**, A161.

Rikmenspoel, R. (1971) Contractile mechanisms in flagella. *Biophys. J.* **11**, 446–463.

Rikmenspoel, R., Jacklet, A.C., Orris, S.E. and Lindemann, C.B. (1973) Control of bull sperm motility. Effects of viscosity, KCN and thiourea. *J. Mechanochem. Cell Motility* **2**, 7–24.

Rikmenspoel, R. and Rudd, W.G. (1973) The contractile mechanism in cilia. *Biophys. J.* **13**, 955–993.

Rikmenspoel, R. and Sleigh, M.A. (1970) Bending movements and elastic constants in cilia. *J. Theor. Biol.* **28**, 81–100.

Satir, P. (1968) Studies on Cilia. III. Further studies on the cilium tip and a 'sliding filament' model of ciliary motility. *J. Cell Biol.* **39**, 77–94.

Schäfer-Daneel, S. (1967) Strukturelle und funktionelle Voraussetzungen für die Bewegung von *Amoeba proteus. Z. Zellforsch. mikosk. Anat.* **78**, 441–462.

Schreiner, K.E. (1971) The helix as a propeller of micro-organisms. *J. Biomechan.* **4**, 73–83.

Schulze, F.E. (1875) Rhizopodien Studien IV. *Arch. mikrosk. Anat. EntwMech.* **11**, 329–353.

Silvester, N.R. and Holwill, M.E.J. (1972) An analysis of hypothetical flagellar waveforms. *J. Theor. Biol.* **35**, 505–523.

Sleigh, M.A. (1962) *The Biology of Cilia and Flagella*, Pergamon Press, Oxford.

Sleigh, M.A. (1964) Flagellar movement of the sessile flagellates *Actinomonas, Codonosiga, Monas* and *Poteriodendron. Q. J. Microsc. Sci.* **105**, 405–414.

Sleigh, M.A. (1968) Patterns of ciliary beating. *Symp. Soc. exp. Biol.* **22**, 131–150.

Sleigh, M.A. (1972) A classification of mechanisms of motility found in Protozoa. *J. Protozool.* **19** (Supplement) Abstract 69, p. 29.

Sleigh, M.A. (1974) In: *Cilia and Flagella*, Sleigh, M.A., Ed., Ch. 1, pp. 1–7. London: Academic Press.

Stebbings, H., Boe, G.S. and Garlick, P.R. (1974). Microtubules and movement in the Archigregarine *Selenidium fallax*. *Cell and Tissue Res.* **148**, 331–345.

Stephens, R.E. (1970a) Isolation of nexin – the linkage protein responsible for maintenance of the nine-fold configuration of flagellar axonemes. *Biol. Bull.* **139**, 438.

Stephens, R.E. (1970b) On the apparent homology of actin and tubulin. *Science*, **168**, 845–847.

Stephens, R.E. (1974) In: *Cilia and Flagella*, Sleigh, M.A., Ed., Ch. 3, pp. 39–76. London: Academic Press.

Subirana, J.A. (1970) Hydrodynamic model of amoeboid movement. *J. Theor. Biol.* **28**, 111–120.

Summers, K.E. and Gibbons, I.R. (1971) Adenosine triphosphate-induced sliding of tubules in trypsin-treated flagella of sea-urchin sperm. *Proc. natn. Acad. Sci., U.S.A.* **68**, 3092–3096.

Summers, K.E. and Gibbons, I.R. (1973) Effects of trypsin digestion on flagellar structures and their relationship to motility. *J. Cell Biol.* **58**, 618–629.

Taylor, G.I. (1952) The action of waving cylindrical tails in propelling microscopic organisms. *Proc. R. Soc. Ser. A* **211**, 225–239.

Taylor, D.L., Condeelis, J.S., Moore, P.L. and Allen, R.D. (1973) The contractile basis of amoeboid movement. I. The chemical control of motility in isolated cytoplasm. *J. Cell Biol.* **50**, 378–394.

Tazawa, M. (1968) Motive force of the cytoplasmic streaming in *Nitella*. *Protoplasma* **65**, 207–222.

Tucker, J.B. (1972) Microtubule arms and propulsion of food particles inside a large feeding organelle in the ciliate *Phascolodon vorticella*. *J. Cell Sci.* **10**, 883–903.

Warner, F.D. (1974) In: *Cilia and Flagella*, Sleigh, M.A., Ed., Ch. 2, pp. 11–38. London: Academic Press.

Warner, F.D. and Satir, P. (1974) The structural basis of ciliary bend formation. Radial positional changes accompanying microtubule sliding. *J. Cell Biol.* **63**, 35–63.

Weinberger, H.F. (1972) Variational properties of steady fall in Stokes flow. *J. Fluid Mech.* **52**, 321–344.

Weis-Fogh, T. and Amos, W.B. (1972) Evidence for a new mechanism of cell motility. *Nature* **236**, 301–304.

Wessenberg, H. and Antiba, G. (1970) Capture and ingestion of *Paramecium* by *Didinium nasatum*. *J. Protozool.* **17**, 250–270.

White, C.M. (1946) The drag of cylinders in fluids at slow speeds. *Proc. R. Soc. Ser. A* **186**, 472–479.

Witman, G.B., Carlson, K. and Rosenbaum, J.L. (1972) *Chlamydomonas* Flagella. II. The distribution of tubulins 1 and 2 in the outer doublet microtubules. *J. Cell Biol.* **54**, 540–555.

Wohlfarth-Bottermann, K.E. (1964) In: *Primitive Motile Systems in Cell Biology*, Allen, R.D. and Kamiya, N., Eds., pp. 79–109. New York: Academic Press.

Wohlman, A. and Allen, R.D. (1968) Structural organization associated with pseudopod extension and contraction during cell locomotion in *Difflugia*. *J. Cell Sci.* **3**, 105–114.

Wolpert, L. (1965) Cytoplasmic streaming and amoeboid movement. *Symp. Soc. Gen. Microbiol.* **15**, 270–293.

Wolpert, L. and O'Neill, C.H. (1962) Dynamics of the membrane of *Amoeba proteus* studied with labelled specific antibody. *Nature* **196**, 1261–1266.

Wolpert, L., Thompson, C.M. and O'Neill, C.H. (1964) In: *Primitive Motile Systems in Cell Biology*, Allen, R.D. and Kamiya, N., Eds., pp. 143–172. New York: Academic Press.

Yagi, K. (1961) The mechanical and colloidal properties of *Amoeba* protoplasm and their relations to the mechanism of amoeboid movement. *Comp. Biochem. Physiol.* **3**, 73–91.

Yoneda, M. (1960) Force exerted by a single cilium of *Mytilus edulis*. I. *J. Exp. Biol.* **37**, 461–468.

Index

Index

334

Index

Index

341